6

GEOMORPHOLOGY TEXTS

Tectonics and landforms

General Editor: K. M. Clayton, University of East Anglia

Other titles in this series

6

GEOMORPHOLOGY TEXTS

Tectonics and landforms

Cliff Ollier

Professor of Physical Geography
University of New England
Armidale, Australia

55/
4
049

Longman
London and New York

Edited by
K. M. Clayton

Longman Group Limited
Longman House
Burnt Mill, Harlow, Essex, UK

Published in the United States of America
by Longman Inc., New York

© Longman Group Limited 1981

First published 1981

British Library Cataloguing in Publication Data

Ollier, Cliff
    Tectonics and landforms - (Geomorphology texts; 6).
    1. Landforms
    2. Plate tectonics
    I. Title   II. Series
    551.4'01   GB401.5   79-41722

    ISBN 0-582-30032-0
    ISBN 0-582-30033-9   Pbk

Printed in Great Britain by
William Clowes (Beccles) Ltd.
Beccles and London

Contents

Acknowledgements

This book was written while I was a Research Fellow in the Department of Biogeography and Geomorphology, Research School of Pacific Studies, at the Australian National University, and I am very grateful for that opportunity.

For comments on early drafts of some chapters I thank John Barrie, Joe Jennings, Brian Ruxton and Bob Wasson. For typing and all the office side of book production I thank Marlene Arney and Mary Horgan.

Most of the illustrations were drawn by Rudi Boskovic, with assistance from Neil Kennan and Mike Roach, and I am grateful for their elegant style.

To Janeta, Katy and Chris, thanks for putting up with my absences while preparing the book, and my presence when writing it.

Cliff Ollier

We are grateful to the following for permission to reproduce copyright material:

A.B.P. (U.K.) Ltd. for our fig. 9.5 from *Outlines of Structural Geology* by E. S. Hills, published by Methuen & Co. Ltd.; American Association for the Advancement of Science and the authors for our fig. 18.3 by Meservey from *Science* Vol. 166 pp. 609–611, 31 October 1969 and our fig. 2.3 based on figs. 2 and 3 by F. J. Vine from *Science* Vol. 154 pp. 1407–1409, Copyright 1966, 1969 by the American Association for the Advancement of Science; American Association of Petroleum Geologists for figures from the *Bulletin of the American Association of Petroleum Geologists*: our fig. 5.11 by Evamy *et al.*, Vol. 62 (1979), our fig. 10.4 by Trusheim, Vol. 44 (1960), our figs. 14.20 to 14.24 by Krebs, Vol. 59 (1975); American Journal of Science and the authors for our figs. 6.2 and 14.16 by Gastil and Miyashiro from *American Journal of Science*, 1960; Artemis Press Ltd. for our fig. 8.4 after J. B. Wright from *Understanding The Earth*, published by Artemis Press Ltd.; Australian and New Zealand Association for the Advancement of Science Incorporated and the author for an extract and our figs. 1.5 and 1.6 by Carey from the *Australian Journal of Science*, Vol. 25 (1963); Australian Geographic Studies for our table 17.1 and fig. 17.1 by Williams and our fig. 21.1 by Ollier from *Australian Geographical Studies*, 1967 and 1973; Australia National University Press for our figs. 4.4, 8.1, 8.8, 8.10 and 8.11a from *Volcanoes* by C. D. Ollier and figs. 8.3, 12.4 and 12.10 by C. D. Ollier from *Landform Evolution In Australia* edited by Davies and Williams; Australian Petroleum

Exploration Association Ltd. for our fig. 9.11 by Findlay from the *Australian Petroleum Exploration Association Journal* (1974); A. A. Balkema (Pty.) Ltd. for our figs. 6.5 and 7.1 by Brock from *Global Approach*, published by A. A. Balkema Ltd.; Blackwell Scientific Publications Ltd. for our fig. 14.3 by Finlayson *et al.* from *Geophysical Journal of the Royal Astronomical Society*, Vol. 44 (1976); Cambridge University Press for our fig. 21.2 by Kennedy from *Geological Magazine*, Vol. 99; Department of Geology, University of Tasmania for our figs. 13.2 and 13.3 by Heezen and Drake from *Syntaphral Tectonics and Diagenesis* and the author, W. S. Carey for our figs. 6.12 and 6.14 from *Continental Drift, A Symposium*; Department of Scientific and Industrial Research for our fig. 5.16 by Kingma from p. 271 of the *New Zealand Journal of Geology and Geophysics*, Vol. 1 (1958); Dover Publications Inc. for our fig. 2.1 by Wegener from *The Origin of Continents and Oceans*, 1929; Dowden, Hutchinson & Ross, Inc. for our tables 15.1 and 15.2 from *The Encyclopedia of Geomorphology* edited by Rhodes W. Fairbridge, Copyright © 1968 by Dowden Hutchinson & Ross, Inc.; Gebrüder Borntraeger Verlagsbuchhandlung for our fig. 17.2 by Slaymaker and McPherson from the *Zeitschrift für Geomorphologie*; The Geological Society of America and the authors for our fig. 16.3 by Bond from *Tectonophysics* Vol. 61 (1979) and figures from *Geology*: our figs. 16.4, 16.5 and 16.6 by Bond, Vols. 4 & 6 (1976 & 1978), our fig. 4.7 by Farhoudi and Karig, Vol. 5 (1977), our fig. 20.6 by Molnar and Tapponnier, Vol. 5 (1977), our fig. 4.5 by Noble and McKee, Vol. 5 (1977), our fig. 4.9 by Roeder, Vol. 5 (1977), our fig. 16.1 by Winker and Howard, Vol. 5 (1977), our fig. 6.7 by Woodward, Vol. 5 (1977), and for figures from the *Bulletin of the Geological Society of America*: our fig. 15.2 by Chappell, Vol. 85 (1974), our fig. 9.12 by Deiss, Vol. 54 (1943), our fig. 9.10 by Korn and Martin, Vol. 70 (1959), our fig. 4.11 by Mammerickx, Vol. 89 (1978); The Geological Society of Australia for our figs. 6.15 and 6.16 by E. S. Hills from the *Journal of the Geological Society of Australia* Vol. 24 (1977); The Geological Society of London and the authors for figures from the *Journal of the Geological Society of London*: our fig. 6.6 by Baker and Mitchell (1976), our fig. 17.3 by Boulter, Walsh *et al.*, Vol. 128 (1972), our fig. 9.3 by Dingle, Vol. 134 (1977), our fig. 20.7 by Gansser, Vol. 129 (1973), our fig. 9.4 by Hollingsworth, Taylor and Kellaway, Vol. 100 (1944) our fig. 6.10 by Kent, Vol. 134 (1977), our figs. 8.11 b,c,d and 11.7 by LeBas, Sutherland and King, Vol. 128 (1972), our fig. 19.4 by Moseley, Vol. 128 (1972), our fig. 4.12 by Phillips, Stillman and Murphy, Vol. 132 (1976), our fig. 14.1 by Ringwood, Vol. 130 (1974), our fig. 5.15 by Scholz, Vol. 133 (1977), and our fig. 2.6 by Thompson, Vol. 133 (1977); Geological Society of South Africa for our fig. 10.6 by MacGregor from the *Proceedings of the Geological Society of South Africa*, Vol. 54 (1951); Harper & Row, Publishers, Inc. for an extract from p. 410 and our fig. 17.4 from p. 379 of *Geology: A Survey of Earth Science* (T. Y. Crowell), Copyright © 1965 by Harper & Row, Publishers Inc.; the author, M. J. LeBas for our fig. 6.4 from 'Per-alkaline Volcanism, Crustal Swelling, and Rifting' from *Nature Physical Science*, Vol. 230, No. 12, March 1971; Liverpool Geological Society for our fig. 8.12 by Rast from *Mechanism of Igneous Intrusion* edited by Newall and Rast, published by Seel House Press Ltd.; Longman Cheshire Pty. Ltd. for our fig. 1.1 from *Earth History In Maps And Diagrams* by C. D. Ollier; Macmillan Journals Ltd. and the authors for our table 1.1 by Dietz and Sproll from *Nature* 212 (1976), our fig. 6.9 by Reeves from *Nature* 273 (1978) and our fig. 20.8 by Myers from *Nature* 254 (1975); Thomas Nelson & Sons Ltd. for our figs. 1.4, 5.5 (simplified) and 20.1 from *Principles of Physical Geology* by Arthur Holmes (1965); W. W. Norton & Co. Inc. for our table 17.2 and our fig. 19.1 from *Evolution of Sedimentary Rocks* by Robert M. Garrels and Fred T. Mackenzie. Reproduced by permission of W. W. Norton & Co. Inc, Copyright © 1971 by W. W. Norton & Company, Inc.; Pergamon Press Ltd. for our figs. 18.1 and 19.2 from *The Expanding Earth* by Jordan; Princeton University Press for our figs. 3.8 and 3.9 from 'Tectonics of Quaternary Time In Middle North America' by Philip B. King in *The Quaternary of the United States* edited by H. E. Wright, Jr. and David G. Frey (Copyright © 1965 by Princeton

University Press): Figs. 4 and 14, pp. 836 and 850. Reprinted by permission of Princeton University Press; A. H. & A. W. Reed Ltd. for our fig. 5.8 from *Rugged Landscape* by Graeme Stevens, published by Reed Publishers Ltd.; D. Reidel Publishing Company for our fig. 20.9 (simplified) by Elston from *Petrology and Geochemistry of Continental Rifts* edited by Newmann and Ramberg; Mrs. Judith Rose for our fig. 11.5 from *Structure, Surface And Drainage In South-Eastern England* by S. W. Wooldridge and D. L. Linton; The Royal Society of London and the authors for figures from the *Philosophical Transactions of the Royal Society*: our fig. 2.2 by Bullard, Everett and Smith, Vol. 258 (1965), our fig. 4.6 by Gass and Masson-Smith, Vol. 255 (1963) and our fig. 4.8 by Barker and Griffiths, Vol. 271 (1972); The Royal Society of New Zealand for our fig. 8.9 by Joyce from *Bulletin 13 of the Royal Society of New Zealand*; The Royal Society of Victoria for our fig. 5.17 from 'The Tawonga Fault' by F. C. Beavis from the *Proceedings of the Royal Society of Victoria* Vol. 72 (1960); Rutgers University Press for our fig. 14.26 by Maxwell and our fig. 20.10 by Gilluly from *Megatectonics of Continents and Oceans* edited by Helgi Johnson and Bennett L. Smith, Copyright © 1970 by Rutgers University, the State University of New Jersey. Reprinted by permission of Rutgers University Press; W. H. Freeman and Co. for an extract and our fig. 13.1 by D. P. Cox and H. R. Cox from *Geology: Principles and Concepts*, 1974 and for our fig. 2.9 from 'Sea-Floor Spreading' by J. R. Heirtzler from *Scientific American*, Copyright © 1968 by Scientific American Inc., our fig. 2.15 from 'The Breakup of Pangaea' by R. S. Dietz and J. C. Holden from *Scientific American* Copyright © 1970 by Scientific American Inc. and our fig. 4.13 from 'Geosynclines, Mountains and Continent-Building' by R. S. Dietz, Copyright © 1972 by Scientific American Inc.; Springer-Verlag and the authors for our fig. 6.3 by Coleman and Irwin (U.S. Geological Survey), our fig. 13.5 by Curray and Moore, our figs. 14.4 and 14.11 by Hatherton, our fig. 20.3 by Stoneley, our fig. 6.11 by Veevers, our fig. 16.8 by Wise, all from *The Geology of Continental Margins* edited by Burke and Drake; Sydney University Press for our fig. 12.16 by C. D. Ollier from *Australia: A Geography* edited by Jeans; The University of Chicago Press for tables and figures from the *Journal of Geology*: our fig. 5.10 by Fuller and Waters Vol. 37, our fig. 13.4 by Dietz Vol. 71, our figs. 15.4 and 15.5 and tables 15.3 and 15.4 by Inman and Nordstrom Vol. 79, our figs. 13.6 by Van Houton & Brown and 2.14 by Shields Vol. 85, our fig. 12.18 by Potter Vol. 86, Copyright 1929, 1963, 1971, 1977 and 1978 by The University of Chicago; the author's widow, Mrs I. Valentin for our fig. 15.3 by Dr Hartmut Valentin; Whitcombe and Tombs for our fig. 12.13 from *Physiography of Victoria* by E. Sherbon Hills; John Wiley & Sons Ltd. for our fig. 20.2 from *The Evolving Continents* (1977) by Windley and for an extract and our fig. 12.2 from *Principles of Geomorphology* by Thornbury, our figs. 5.9 and 19.5 from *An Outline of Structural Geology* by Hobbs, Means and Williams, and our fig. 15.6 from *Principles of Geochemistry* by Mason.

Whilst every effort has been made to trace the owners of copyright material, in some cases this has proved impossible and we take this opportunity to apologise to any copyright holders whose rights we may unwittingly have infringed.

1 Introduction

The world we live in is constantly changing. It is being worn away by erosion and its debris deposited in the sea. If this process were to go on long enough could the continents be worn away completely? From what we know of erosion rates and the age of the earth there has been ample time for the continents to be worn away many times over. Yet they are still there. There seems to be a 'law of conservation of continents', some mechanism that keeps the land in existence despite the constant wear. Furthermore the land is not flat all over, but has mountains and plateaus, frequently arranged in distinct chains. How are mountain chains formed, and why do they occur in some places and not others?

In a great many parts of the earth it is easy to see where rocks that were originally flat have been crumpled into folds, and the idea seems natural that the pressure that folded the rocks could also have heaved up the fold belts of the mountains, just as a shrinking apple develops fold wrinkles on its skin. We shall see that there are no fold mountains, and that the earth is not shrinking – some other mechanism must be found for mountain building.

There is now overwhelming evidence that the continents drift around the earth's surface, and that some ocean basins are gradually spreading from submarine ridges. Continents move a few centimetres each year, and over a few hundred million years at this rate the face of the earth could experience great changes: this period covers the time span of evolution of life on land (about 400 m.y.), and continental drift had a considerable effect on the course of evolution. Much of the movement of continents and creation of mountain belts relates to a smaller period of only the last one or two hundred million years.

By historical accident the majority of early studies of landscape evolution took place in Europe and North America where the Ice Ages had a great effect on concepts. Landscapes were thought of in terms of time spans appropriate for the last Ice Age, and time periods of less than one million years – frequently only a few thousand years – were thought appropriate for the development of landforms. However, it is now known that the history of landscapes goes back over a hundred million years in some places – places that have had a continuous history as dry land since the Mesozoic or earlier.

This means that the time scale for landscape evolution is now seen to be the same geological time scale that is appropriate for discussion of mountain building, continental drift, and the evolution of terrestrial life.

Geomorphology is thus in a new setting. We are not dealing with a static world of the ancients, or with a Davisian world of periodic uplift. We are dealing with a mobile world of spreading sea floors and continental fragments which, while drifting, retain some features of pre-drift physiography. They may also reveal in their landforms a story of the manner of drift with its associated buckling, cracking and vulcanicity interspersed with periods of erosion.

On this modern scene some of the successful techniques used to elucidate geomorphic history turn out to be very old-fashioned. The study of erosion surfaces and drainage patterns has been very out of favour for the past quarter of a century, but now seems likely to provide insights for our deciphering the broad features of landform history.

Some definitions

Geomorphology is concerned with the description, the form and the origin of landforms, from continent – ocean relationships to minor landforms such as terracettes and waterfalls.

Tectonics is concerned with the form, pattern and evolution of the globe's major features such as mountain ranges, plateaus, fold belts and island arcs.

Structural geology concerns smaller structures such as anticlines, faults and joints.

Orogenesis originally meant the formation of mountains.

In the belief that mountains were made by compressive folding, the term became extended to mean a major folding of rock.

Nowadays orogeny means a period of folding, and an orogenic belt is a belt of folded rock.

If we really want to talk about the formation of mountains we have to use a phrase such as 'mountain building', because orogeny has quite lost this meaning.

Tectogenesis means the study of deformation.

Geomorphic history and geological time

Geomorphology is concerned with landforms, their description, history, and genesis. Most landforms are produced by erosion, and of these eroded landforms most are produced by fluvial erosion – that erosion related to the work of rivers.

The material eroded from the land is carried to the sea by rivers and deposited in the sea as sediment, ususally in layers. Sedimentology is concerned with the sediments themselves: stratigraphy is concerned with the layers or strata.

The basic principle of stratigraphy is superposition, which simply means that with a few unusual exceptions, a younger layer of sediment will always be deposited on top of pre-existing sediments or rocks. In any succession of rocks the oldest will be at the bottom, the youngest at the top.

The ideas are shown at their simplest in Fig. 1.1.

With the principle of superposition we can get the relative date of rocks in a succession at any one place.

A great advance in geology was made by William Smith when he discovered that some strata could be identified by the assemblage of

Fig. 1.1 Stratigraphic principles. (*a*) Shows deposition of sand on the sea floor. The sand, derived from the land by erosion, is younger than the bedrock beneath. (*b*) Shows a later stage when sand deposition has been replaced by clay deposition. Since it is above the sand, the clay is younger. (*c*) Shows strata in a quarry. The original sediments were pebbles (turned into conglomerate), sand (sandstone), and shelly deposits (limestone). Using the law of superposition we know that the conglomerate is the oldest rock represented and the limestone is the youngest. The area has been uplifted above sea level since the strata were deposited, but we cannot see from the picture whether there has been any natural erosion. (*d*) Although not geologically very complicated, indicates a minimum geological history as follows: deposition of pebbles, sand and clay which become conglomerate, sandstone and clay. These strata were folded, uplifted and eroded to a plain. Since then there has been renewed subsidence, and deposition of limestone above an unconformity that separates it from the folded strata beneath. (Figures from Ollier, 1973).

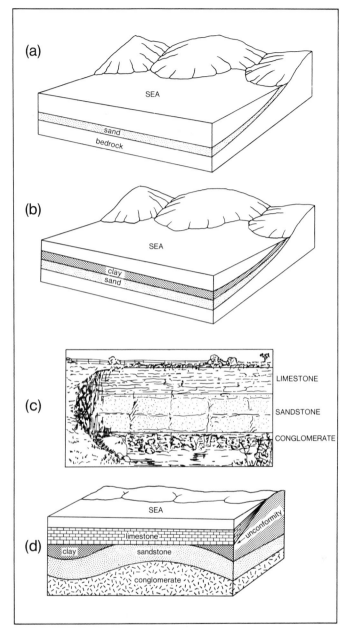

fossils they contain. A limestone, for example, at one end of the country could be matched with a limestone at the other end if they both contained the same assemblage of fossils. This process is called correlation. Long and repeated success with the method enabled stratigraphers to construct a stratigraphic column which can be used for world–wide correlation, for all the rocks that have fossiliferous remains. The column is shown in Fig. 1.2.

The stratigraphic column is divided into units of various types. A division into three separated the Palaeozoic (old life), the Mesozoic (middle life) and the Cenozoic (also known as the Cainozoic). These three main groups are divided into smaller groups called Periods. The Cenozoic is frequently divided into the Quaternary – the uppermost

Fig. 1.2 The stratigraphic time scale.

ERA	PERIOD	EPOCH	M. Yrs Ago
CENOZOIC	QUATERNARY	RECENT - about 10,000 years / PLEISTOCENE	
CENOZOIC			1.8
CENOZOIC	TERTIARY	PLIOCENE	5.5
CENOZOIC		MIOCENE	22.5
CENOZOIC		OLIGOCENE	38
CENOZOIC		EOCENE	54
CENOZOIC		PALAEOCENE	65
MESOZOIC	CRETACEOUS		
MESOZOIC			135
MESOZOIC	JURASSIC		
MESOZOIC			190
MESOZOIC	TRIASSIC		225
PALAEOZOIC	PERMIAN		280
PALAEOZOIC	CARBONIFEROUS		
PALAEOZOIC			345
PALAEOZOIC	DEVONIAN		395
PALAEOZOIC	SILURIAN		440
PALAEOZOIC	ORDOVICIAN		500
PALAEOZOIC	CAMBRIAN		
			570
PRECAMBRIAN			

Period – and the Tertiary. It is also sometimes found convenient to split the Cenozoic into the Palaeogene, which includes the Palaeocene, Eocene and Oligocene, and the Neogene which includes the Miocene, Pliocene and Pleistocene. The Quaternary is split into the Pleistocene, and the Recent. All the rocks down to the Cambrian are known together as the Phanerozoic rocks and the normal principles of correlation by fossils can be used throughout this period.

Older rocks are known as Precambrian, and because they are normally without fossils quite different methods of correlation are used. The Precambrian has been divided into many units but the main ones are the Archaen (the very oldest known rocks, pre 2 500 m.y. old), and the Proterozoic – the younger Precambrian rocks.

Volcanic activity complicates the geological column locally. Field geologists can recognize volcanic rock, and although it contains no fossils it can be dated relative to underlying and overlying fossiliferous rock.

Volcanic rock reaches the ground surface through dykes – vertical or steep sheets of volcanic rock that filled cracks. The dykes clearly are younger than the sedimentary rocks they cut across.

If one dyke cuts across another one, it must be younger than it. Volcanic rock can also be injected into layers of sedimentary rock as a sill. In this situation the volcanic rock is younger than the sedimentary rock above the sill, so the law of superposition does not apply.

Fortunately there are plenty of tests that enable the field geologist to distinguish sills from flows (for field methods of geology see Compton, 1962).

Deep in the crust the formation and emplacement of granite presents a further complication. Granite will clearly be younger than the rocks it replaces or pushes aside, or which it alters by heat.

Sedimentary rocks are frequently found to be folded, and if (as is usually the case) the strata were originally deposited near horizontal, the folding must be younger than the deposition of the sediments.

Erosion usually takes place above sea level, so if folded sediments are eroded, a period of uplift above sea level is indicated. It is often worth noting if erosion creates an irregular surface – perhaps valleys and hills, bays and headlands – or if erosion is to a plain. If erosion is followed by subsidence, the sea spreads on the land – a process called transgression – and new sediments may be deposited on the erosion surface. The break between the two sets of rocks separated by the erosion surface is called an unconformity.

An elementary account of the way geologic and geomorphic histories are derived from sections and maps is provided by Ollier (1973).

A stratigraphic history is preserved in the strata of seas bordering a landmass. On the land a related history may be determined from the history of the landscape, with periods of erosion to a plain, periods of uplift and renewed erosion, and so on. The study is best called geomorphic history. An older term was denudation chronology, but this does not take sufficient account of the depositional features of the landscape (lake deposits, volcanic rocks, river and glacial deposition) or of other changes (earth movement, climatic change, or changes in the style of weathering), so the term geomorphic history is preferred. Of course the history found on land should be closely related to the history found in nearby sedimentary sequences – periods of rapid erosion should relate to periods of rapid deposition, times when the land was worn down to a plain should match periods of little or no deposition in the sea, possibly with unconformities.

The relative dating of early stratigraphy has been aided in modern times by techniques of absolute dating. Most of these methods are based on radioactive decay, and the most useful today are carbon dating (effective to a maximum of about 50 000 years), potassium–argon dating (from the oldest rocks to about 50 000 years), uranium series dating (about 250 000 to 50 000), and uranium-lead dating for older rocks. Various text books deal with the technical side of these methods, including Harper (1973).

Other methods such as palaeomagnetism (Ch. 2) may also be used to provide a more accurate time scale to certain events. Advances in the use of these methods has led to absolute dating of geological events, and the stratigraphic time scale can now be printed with dates in years for the major events.

A geological time scale is shown in Fig. 1.2. In general we know more of recent events in earth history than of remote times, in our landscapes recent events have more impact than remote ones, and the more recent tectonics are visible at the surface while ancient tectonic features may be preserved in rocks but not in landscapes. We thus have a perspective of time, with a clearer view and larger scale for those features near us in time, and a more limited view of events and features in the distant, remote past.

Scale of geomorphic features

At the very highest scale is the whole earth. Carey (1963a) has provided a summary of views and says:

The shape of the earth has been compared with a disc (Homer), a ball (Pythagoras), an egg (Cassini), an orange (Newton), a peach (Carey), an eccentric sphere (Love), a pentagonal dodecahedron (Elie de Beaumont), a pear (Sir James Jeans), a tetrahedral peg-top (J. W. Gregory) and a potato (Sir George Darwin). Sir John Herschel said it was earth-shaped and it is now called a geoid (Listing), which merely begs the question.

Geomorphologists leave this question to geodesists, and start an order lower down.

Salisbury (1919) introduced the concept of orders of scale into geomorphology, with three divisions as follows:

First-order features are the major divisions of the earth's surface, continents and oceans. In the modern terminology of plate tectonics, the main individual plates (which include both continental and oceanic components) would be first-order features.

Second-order features are the major features of individual continents or oceans, such as mountain ranges, plateaus, plains, or island arcs, deep-sea trenches, or mid-ocean ridges.

Third-order features are hill and valleys, escarpments, volcanoes and similar landforms.

This classification is crude and difficult to apply, but it has long proved useful in the vocabulary of geomorphology.

So far as this book is concerned, most of the tectonic features discussed will be first- and second-order features, while most of the geomorphology will be concerned with second- and third-order features. Nevertheless earth surface features of different orders are commonly related genetically. For example a diverted river (third order) may result from uplift along the rim of a rift valley (second order), and the rifting may lead to the splitting of a continent into two smaller continents (first order).

Von Engeln (1942) compared the first-order features to theatres, permanent structures in which geological dramas are enacted. The second-order features were compared with stages that might be rebuilt from time to time, and the third-order features to the portable stage scenery that is relatively mobile and temporary. This useful analogy is no longer true at the first-order level, for the 'fixist' ideas of 1942 are no longer tenable. The continents and oceans are *not* permanent, but grow, move and split. It is essentially the size, and not the permanence, that determines the order of a given feature.

Carey (1963a) has a rather different approach to the ordering

Fig. 1.3   The first-order polygons of Carey (1976): 1. Pacific; 2. North America; 3. South America; 4. Africa; 5. Europe; 6. Siberia; 7. India; 8. Antarctica; 9. Australia.

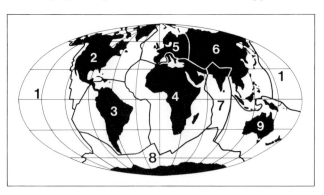

system. He uses first-order polygons for the plates isolated by mid-oceanic ridges (Fig. 1.3). If some of these were split at subduction sites (Carey does not believe in subduction) the first-order polygons would correspond to the plates of plate tectonics (Ch. 4). Second-order polygons are such features as cratonic basins (Fig. 1.4), often bounded by raised rims and major faults, and commonly about 1 000 km across. First-order continental polygons involve the whole

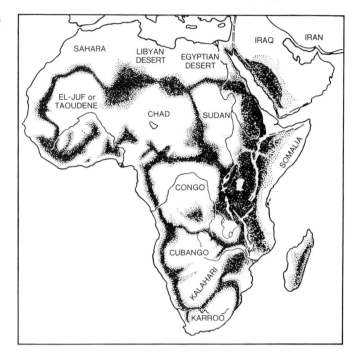

Fig. 1.4 The basins and swells of Africa – second-order landscape features (after Holmes, 1965).

mantle. Second-order basin and swell polygons are relieved at the asthenosphere. Third-order polygons are blocks about 10 km across, such as the fault blocks of Japan, which move independently (Fig. 1.5). Master joints create fourth-order polygons about 10 m apart, and ordinary joint blocks are fifth-order polygons usually less than a metre across.

Another proposal for developing scale concepts is the *G* scale,

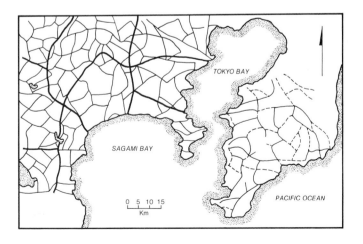

Fig. 1.5 Tilt blocks in central Honshu defining third-order polygons (after Carey, 1963*a*).

based on the earth's surface, *Ga*, with a scale of measurements derived from successive subdivisions of this standard by the power of ten (Haggett, Chorley and Stoddart, 1965). The value can be determined from the formula

$$G = \log (Ga/Ra)$$

where *Ga* is the earth's surface area, and *Ra* is the area under investigation. The larger the value of *G* the smaller the feature, as indicated by the following values: USSR 1.35; USA 1.82; Texas 2.87; UK 3.31; Yorkshire 4.51; Rhode Island 5.27; one square kilometre 8.71; ten hectares 10.1.

If people were sufficiently familiar with it this scale might prove useful, but it does not seem to have been put to any use so far.

One of the most surprising features of first-order landforms is the antipodal relationship of land and sea. Antarctica is antipodal to the Arctic Ocean, Australia antipodal to the North Atlantic, North America to the Indian Ocean, Europe and Africa to the Pacific, and the South Atlantic Ocean is not far from antipodal to eastern Asia. The equator is a plane of antithesis, with ocean maxima in one hemisphere corresponding remarkably well with land maxima in the other (Fig 1.6). Since the oceans have approximately twice the area of the land masses some antipodal relationship might arise by chance, but 95 per cent of land is antipodal to oceans, which is very much better than chance.

Fig. 1.6 Antipodal relationship of land and sea. Percentage of continental crust plotted against latitude (*from* Carey, 1963*a*). Seas shallower than 2 000 m are plotted as continental. AA': South polar continent, north polar ocean; BB': Northern land girdle, southern ocean girdle; CC': Lands taper south, oceans taper north.

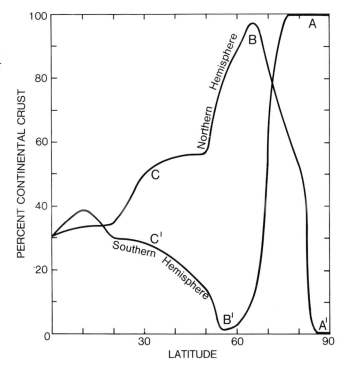

At a scale between continental plates and the whole earth there has been a repeated suggestion that before continental drift the continents were arranged in two supercontinents, Laurasia in the north and Gondwanaland in the south. Dietz and Sproll (1966) showed that the areas of these two supercontinents are remarkably similar (Table 1.1), inviting speculation about some remarkable features of earth history.

Table 1.1 Areas of Laurasia and Gondwana* (after Dietz and Sproll, 1966)

Laurasia†	Area	Gondwana	Area
Eurasia	70.49	Africa‡	37.80
North America	35.39	Australia and New Guinea	13.31
		New Zealand	2.59
	105.88	Antarctica	16.91
India	−5.19	South America	22.36
	100.69		92.97
		India	+5.19
		Himalayan Overlap§	+2.44
			100.60

* Areas include continental shelves and upper continental slopes to 1 000 fathom isobath. All values in million square kilometres.
† Includes Greenland. Does not include the Philippine Islands and southern Central America, an area of 1.35×10^6 square kilometres. Possible accretion to continents by orogeny after the mid-Mesozoic is not taken into account.
‡ Includes Arabia, Malagasy, and the northern portion of the Seychelles Plateau.
§ That part of the Indian subcontinent presumably underlying the Tibetan Plateau, which must be two continents thick to satisfy isostasy.

Most modern continental reassemblies make not two continents but one, Pangaea, but the date of this assembly is about Jurassic. Is it possible that the Jurassic Pangaea was created by the temporary collision of Gondwanaland and Laurasia which had earlier been separated by an equatorial belt of sea? If so when did the two earlier continents form, and in what position? If we can conceive of an expanding earth (Ch. 18) it is possible that a primary earth-covering continent covered the whole earth, and was broken into two equal halves by the creation of new seas along an equatorial shear. This would create Laurasia and Gondwanaland, which collided later to form the Pangaea continent of the Jurassic (although continents had been in that position probably throughout the Palaeozoic, at least around the Atlantic).

It is also appropriate to consider the time scale involved in earth history and landform evolution. Some features may be formed almost instantly, such as scarplets associated with earthquake activity. Some may be created in days or weeks, like lava flows. Yet others may take hundreds or thousands of years to form, such as river terraces, and many features take millions of years or hundreds of millions of years.

The earth's surface contains many relics of former geomorphic processes – landforms that were created long ago, and remain at the earth's surface. So in thinking of the time scale we are concerned not only with the formation, but also the preservation of landforms. There are places where actual landforms, such as river valley systems, have been preserved for hundreds of millions of years.

Eventually we come to rates of geomorphic processes, which vary from the very fast (perhaps avalanches, at speeds exceeding 50 m/s; or *nuées ardentes* with speeds up to 200 km/h; intrusion of kimberlite at the speed of sound) to the slow rates of movement of continents, erosion of plains and deposition in deep sea floor (less than 1 mm/1 000 yr). The rates of earth processes are so important that a separate chapter (17) will be devoted to the topic. For ease of comparison many rates will be converted to Bubnoff units, symbolized B. One Bubnoff unit is equivalent to 1 mm in 1 000 years, or 1 m in a million years, which will generally be abbreviated to m.y.

2

Continental drift and sea floor spreading

Perhaps the greatest geological discovery of this century is that the ocean floors are spreading. More than any other fact this gave a positive need for new ideas and the geology after this discovery is so different from what went before that it is sometimes called the 'New Global Geology'.

As soon as the voyages of discovery had revealed the similarity in shape of the opposite sides of the Atlantic, the idea that opposing continents had drifted apart was bound to arise. Francis Bacon appears to have been the first to note the 'conformable instances' in 1620. Later writers including Lilienthal (1756), Snider (1858) and Taylor (1910) wrote briefly on the 'fit' of opposing continents and speculated on its meaning (see Rubke, 1970), but it was Wegener (1929) who brought scientific knowledge and scholarship together in a serious and concentrated exposition of continental drift (Fig. 2.1).

In Wegener's day the concept of drift could be supported by evidence from the distribution of plants and animals, the distribution of fossils, ancient deserts and glaciated rocks, and by matching other geological evidence. Wegener also thought that the mountains along the Pacific coast of the Americas were pushed up at the front edge of drifting continents where the continental slab buckled against the Pacific.

Continental drift was not accepted readily by the geologists of the northern hemisphere, although geologists familiar with the southern continents were much more impressed by the incredible details of some of the matching evidence on opposite sides of the South Atlantic. At the time of a famous conference on continental drift in 1926 it would be true to say that geological authority was marshalled against the concept, sometimes with great vigour and bitterness, for few concepts have led to such emotion in geology. Only a few geologists were prepared to concede the possibility and 'drifters' continued to compile evidence of matching data on opposite sides of the Atlantic and elsewhere.

Fig. 2.1 Wegener's reconstructions of the map of the world for three periods (after Wegener, 1929).

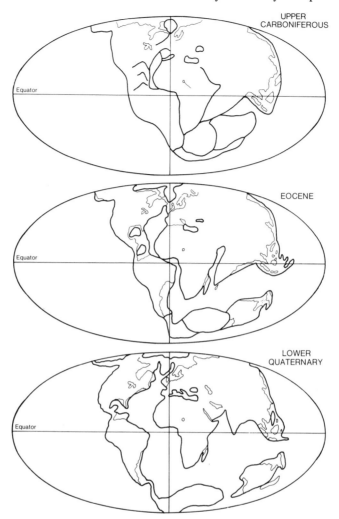

The Atlantic

By 1958 Carey had shown that the 'fit' of the South Atlantic was remarkably good, within half a degree over 45 degrees of latitude, which could hardly be accidental, and by 1965 Bullard and others had a computer fit that showed the entire Atlantic could be closed to give an equally good fit (Fig. 2.2).

To achieve these good fits it is necessary to match not the present coastlines, which result from local accidents of overlap of sea onto the continental shelves, but to use the continental slope, a line that bounds the continents in a geological sense. This is not a new idea, and although the misconception that Wegener used coastlines in his continental reassemblies has been perpetuated to the present day, he made it quite clear that he used the margin of the continental slope in the deep sea (Drake, E. T., 1976).

However, the evidence that was to change our views on drift came not from the continents but from the study of the ocean floor. Ocean surveys in the nineteenth century showed that the Atlantic was shallowest along the centre; there is a mid-Atlantic ridge running the full length of the Atlantic and remaining very close to the middle, emerging above sea level in Iceland. Of course this ridge could be 'explained' either with drift or without.

Fig. 2.2 The fit of the continents with closure of the Atlantic. Black areas are overlaps (after Bullard, Everett and Smith, 1965).

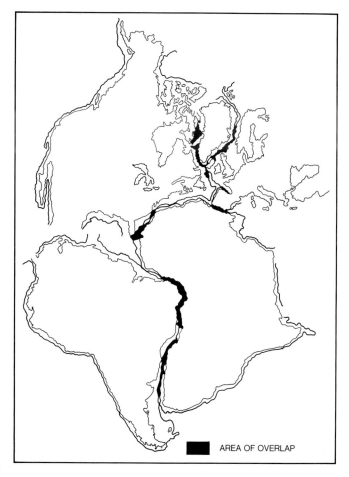

AREA OF OVERLAP

More detailed surveys showed that the ridge is split by a great crack or rift, so there is a huge valley-like feature running along the centre of the Atlantic, bounded by the high shoulders of the mid-Atlantic Ridge.

Palaeomagnetism

Geophysics provided the next evidence. It was found that if a magnetometer were towed across the ocean it recorded areas of higher and lower magnetic intensity and these, when plotted, formed stripes parallel to the mid-Atlantic Ridge. The stripes were first found in the Pacific where they appeared to be parallel to the coast, and various ideas of pressure zones parallel to the coast and possibly perpendicular to the stresses of the mountain belts were invoked. The key to understanding the sea-floor magnetism came with the discovery that the magnetic stripes are distributed symmetrically about a central ridge, with broad bands, narrow bands, and sequences of thick and thin bands being fairly easy to detect on opposite sides (Fig. 2.3).

At the same time techniques for determining the age of rocks had improved, and dating of ocean floor and of volcanoes rising from the floor suggested that the sea floor became progressively younger towards the ridge, progressively older away from it.

Meanwhile work on the magnetism of old rocks showed a remarkable feature. Some rocks were magnetized in a direction completely reversed from the normal. A compass at the time these

Fig. 2.3 The symmetry of the magnetic
pattern across the Reykjanes Ridge south
of Iceland.

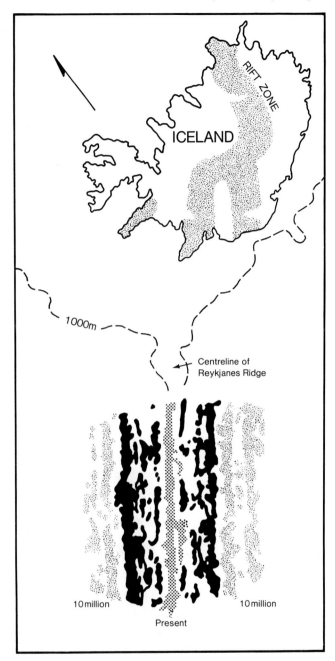

rocks were deposited would have pointed to the south instead of north.
Many special-case explanations were tried to account for this reversed
magnetism, but as more results were collected it became clear that the
reversals were world-wide events, synchronous in time, and that a
magnetic history of the last few million years of earth history could be
built up. It was found, for instance, that the present 'normal' magnetic
direction has been around for 700 000 years, and before that the
earth's magnetism was reversed from about 2 400 000 years ago.
These are known as the Brunhes (normal) epoch and the Matuyama
(reversed) epoch, and each is now known to contain a number of

short-lived 'events' when the magnetic field of the earth was reversed for very short periods. The upper part of the table is shown in Fig. 2.4 and a more complete table is shown in Fig 2.5.

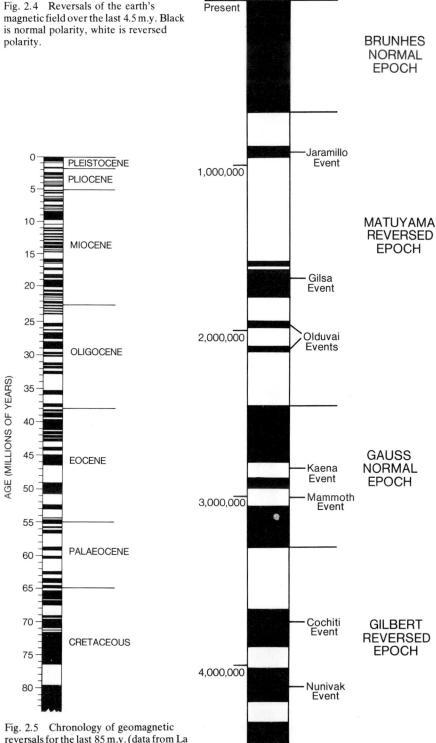

Fig. 2.4 Reversals of the earth's magnetic field over the last 4.5 m.y. Black is normal polarity, white is reversed polarity.

Fig. 2.5 Chronology of geomagnetic reversals for the last 85 m.y. (data from La Breque, Kent and Cande, 1977).

Palaeomagnetism will eventually be used as a common tool in the Quaternary problems which occupy the bulk of geomorphologists at present. A magnetic stratigraphy can be built onto a series of lake deposits, providing a rough guide to age and enabling the deposits to be related in some way to the surrounding geomorphology. In Lake George, New South Wales, for instance, the Matuyama–Gauss reversal of 2.4 m.y. is found at about 31 m. Since the lake sediments are 71 m thick, it seems reasonable to suppose that the lake originated in Miocene time. Fig. 2.6 shows the application of the palaeomagnetic

Fig. 2.6   Polarity time scale and stratigraphic subdivisions of the Quaternary in the Netherlands and East Anglia (from Thompson, 1977).

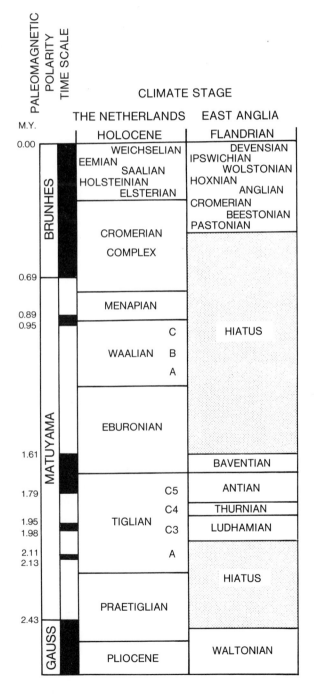

polarity time scale to the climatic stages of Europe, and Thompson (1977) discusses the application of this technique in some detail.

Magnetic reversals can also be detected in the basalts of the sea floor, and the same sort of sequence that is determined in a stratigraphic sequence can be found with increasing distance from the mid-Atlantic Ridge. It is as if the ocean floor were a huge tape recorder, carrying the signals of the earth's changing magnetic direction with time on two tapes that emerge from the mid-Atlantic rift and move slowly sideways in both directions towards the continents. The sea floor is spreading (Fig 2.7).

Fig. 2.7   The tape recorder model, and symmetry of magnetic reversals across a spreading ridge.

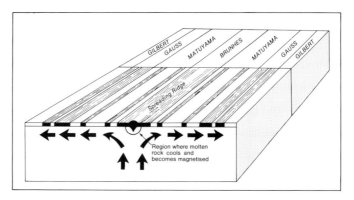

Knowing the age of different parts of the ocean floor and the distance from the ridge, it is easy to work out the rate of sea-floor spreading, and it turns out to be about 2 cm/y; that is the opposed continents are drifting apart at about 4 cm/y.

Ideas of sea-floor spreading were supported by evidence from the palaeomagnetism of continental rocks. Recent rocks of Europe have a magnetic direction pointing to the present magnetic pole. Older rocks, it was found, point somewhat away from the present-day pole, and so either the pole or the continent has moved. When the apparent position of the magnetic pole had been determined from rocks of many ages it was found that the direction was a long way off in, say, Triassic times (200 m.y. ago) and had moved along a curve to finish up at the position of the present pole. A similar curve could be determined from North America, where the curve was found to finish at the same position as the European curve of polar wander, but became further away with increasing age (Fig. 2.8). At any earlier time the positions of the apparent poles could be reconciled if the continents were placed closer together, and indeed the curves fit best if the Atlantic is opened, with Europe starting adjacent to North America and then drifting apart, just as determined from sea-floor data.

Above a certain temperature (the Curie point) rocks are not magnetized. As hot rock cools it acquires the prevailing magnetic direction. It seems that hot magma is injected into the mid-Atlantic Ridge and as this cools to basalt it acquires the magnetic direction of the earth's field. Later this basalt splits, moves apart, and new volcanic material is injected in the middle which becomes magnetized when it cools, and so on. New sea floor is created at the mid-Atlantic Ridge, and the ocean floor gets progressively older towards the continents.

When points move apart on a sphere they move along 'latitudes' relative to poles of rotation, known as Eulerian poles after the

Fig. 2.8   Palaeomagnetic polar-wander
curves for Europe and North America.
PC = Precambrian; Ca = Cambrian;
S = Silurian; D = Devonian;
C = Carboniferous; P = Permian;
T = Triassic; K = Cretaceous;
M = Miocene.

mathematician who described the geometry of motion on a sphere. As
shown in Fig. 2.9, as a gap widens there will be greater motion near the
equator of motion than near the poles. So far as the opening of the
Atlantic is concerned, this means there will be faster motion in some
places than others, so the sea floor is split into bands that spread at
slightly different rates.

Fig. 2.9   Formation of spreading ridges
with offsets along transform faults with
reference to a pole of spreading. In each
ocean the fracture zones are
perpendicular to the spreading axis, and
the rate of spreading indicated by the
length of the arrows is directly
proportional to distance from the pole.

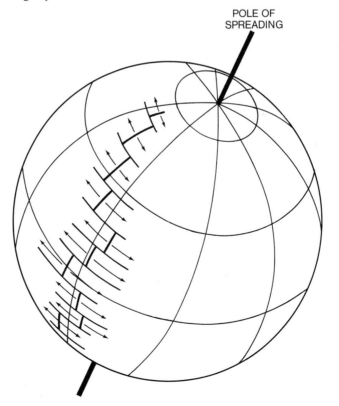

Furthermore the mid-Atlantic Ridge is not a continuous line, but is offset along faults (Fig. 2.10). Along simple faults with horizontal movement, the material on opposite sides of the fault moves in opposite directions. These are called transverse faults. As shown in Fig 2.11, the faults that offset the mid-Atlantic Ridge are more

Fig. 2.10 The mid-Atlantic Ridge, its transform faults, and the age of ocean floor (data from Heirtzler, 1968).

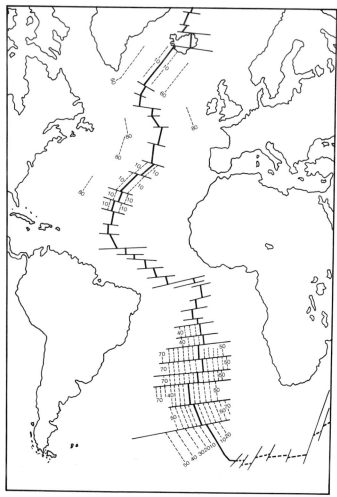

Fig. 2.11 A fracture zone offsets spreading axes. Between B and C the plates on opposite sides move in opposite directions, generating earthquakes symbolized by asterisks. Between A and B and between C and D plates on opposite sides of the fault move in the same direction. The fault along such a fracture is a transform fault.

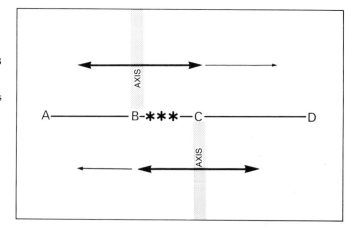

complicated. Only in the fairly short section of the fault between the ridges are fault movements opposed, and beyond them the material on opposite sides of the fault is moving in the same direction. Earthquakes are only associated with that part of the fault where movements are opposed. Such faults are called transform faults in distinction from the more familiar transverse faults found on continents, where all the rocks move in opposed directions on opposite sides of the fault. Great fracture zones follow the transform faults, marked by huge submarine escarpments.

There is very little sediment on the Atlantic Ridge, but sediment thickens with distance from the rift, as would be expected from a spreading ocean where the outer parts are older and so have had more time to accumulate sediment. No pre-Jurassic sediments are known, and it seems that the Atlantic Ocean only started to open in the Triassic, about 200 m.y. ago. Maps showing the position of the Atlantic at various times since then can be prepared with confidence today.

In summary there is overwhelming evidence that the Atlantic Ocean has been formed by the drifting apart of the continents on opposite sides. The sea floor grows at the centre – the mid-Atlantic rift – and spreads sideways, apparently gently. The ocean is split into numerous roughly latitudinal strips that move at slightly different rates, inevitable with spreading on a sphere, and the strips are separated by great fracture zones along transform faults.

Sea floor spreading involves the sequence:
1. heating of spreading ridge
2. formation of a crack
3. injection of new rock, with spreading
4. associated movement on transform faults.

The strips nearer to the Euler Pole move slower than those far from the pole. If the spreading resulted from forceful injection along the length of a crack this could not happen, and it seems that the ridges are response features, not the driving force.

Mountains are not being pushed up along the shores of the Atlantic. Indeed when geomorphologists speak of an Atlantic type of coastline, they mean one where the grain of the mountain ranges runs at a high angle to the direction of the coast. The few mountains that are roughly parallel to the Atlantic coast, such as the Appalachians, are much older than the age of the opening of the Atlantic (see p. 50).

Since there is no sign of collision between the spreading ocean and the American continent it seems that the American continent must be drifting along with the spreading ocean; continent and ocean are part of the same 'plate' and move as one, though the fracture zones of the ocean do not extend onto the continents.

Looking at the cordillera or mountain ranges running parallel to the west coast of the Americas – the Coastal Ranges, Rockies, Central American mountains and the Andes – we might suppose (as Wegener did) that continental drift was creating mountains where the leading edges of the drifting plates collided with the Pacific (though we might be suspicious because there are no similar mountains on the eastern side of Africa to match the Andes of South America). Before considering this further, we shall look at the evidence from the Pacific Ocean.

The Pacific

If the earth remains the same size and the Atlantic Ocean is growing by sea-floor spreading, then the Pacific Ocean must be shrinking by an amount equal to the increase in area of the Atlantic. But there is plenty of evidence that the Pacific is also spreading.

A spreading site ridge crosses the Pacific, but since it runs ashore against North America the term mid-ocean ridge is hardly appropriate. As shown on Fig. 2.12 this ridge leaves a circum-Antarctic ridge south of New Zealand and heads for California where it disappears under the continental mass. The southern part is known as the Pacific–Antarctic Ridge and the rest is known as the East Pacific Rise.

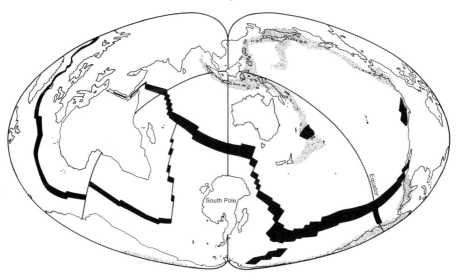

Fig. 2.12   New crust created in the last 10 m.y. This equal-area projection shows that most new crust formation has been along the East Pacific Rise, the Pacific–Antarctic Ridge and the Indian Ocean Ridge (simplified from Menard, 1974).

Submarine topography, magnetic evidence, the age of the sea floor and sedimentary evidence combine, as they did in the Atlantic to confirm that the Pacific is spreading. Magnetic bands indicate a spreading rate of up to 6 cm/y. Essentially the north Pacific is moving north west and is roughly perpendicular to the coasts of the Asian and Pacific countries that it abuts. The southern plate moves south. Off South America a Y-shaped junction of spreading ridges is formed, a so-called triple junction, and new plate, the Nazca plate is created off the northern part of South America. The Californian coast is complicated by many transform faults. In the southwest Pacific the direction of spreading is roughly parallel to the Fiji–New Guinea line, so great shearing might be expected.

Supporting evidence for spreading comes from the composition of volcanic rocks erupted in the Pacific. In the central part the volcanoes consist of basalt, typical of true oceanic areas, but around the edges andesitic volcanoes prevail. A distinct andesite line (Fig. 8.1, p. 105) separates the inner areas with no andesites, and this corresponds well with the supposed area of new sea floor created by sea-floor spreading.

The great fracture zones along transform faults were well known as topographic features long before the ideas of sea-floor spreading were proposed. It was by trying to match the magnetic intensity patterns on opposite sides of the Mendocino fracture zone (Fig. 2.13) that the first ideas of the amount of lateral movement on these fractures became clear – it turned out to be 1 100 km.

The Pacific sediments are rather complex, but in brief it may be said

Fig. 2.13 Fracture zones of the
northeast Pacific.

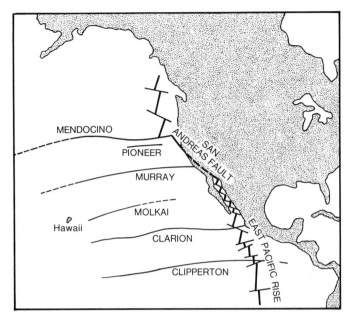

that sediments are absent or thin at the spreading sites, restricted to
oceanic sediments for most of the sea floor, and that as in the Atlantic
terriginous sediments derived from the erosion of continents are
restricted to within about 1 000 km of the continental edge. Unlike the
Atlantic, the Pacific is rimmed by many active volcanoes, and a distinct
wedge of volcanic ash can also be determined in the sediments around
the edge of the ocean.

Volcanoes in the mid-Pacific area also shed light on the spreading of
the oceans. Pacific islands are commonly grouped in lines or island
chains, of which Hawaii is perhaps the best known. The most active
volcanicity is found on the island of Hawaii itself, the most easterly of
the islands. Further west the islands bear evidence of being extinct
volcanoes that finished eruption, became eroded, grew coral reefs, and
slowly subsided to form atolls. The islands are oldest to the west,
youngest to the east. This might mean that the centre of volcanicity is
moving east, or that the earth's crust is moving west over a deep-set
'hot spot' from which the volcanoes are erupted. The apparent rate of
migration of the volcanicity is the same as the rate of sea-floor
spreading, which appears to be too much for coincidence.

The locus of volcanism in the Hawaiian chain has migrated towards
the southeast at a rate of about 11 cm/y. Duncan and McDougall
(1976) have found the same direction and rate for the Marquesas,
Society and Austral Islands.

These results are consistent with the idea of the motion of a large
coherent slab of sea floor over several hot spots that are fixed relative
to each other.

As mentioned earlier, if the Atlantic is growing, the Pacific should
be shrinking. But the Pacific is itself growing by sea-floor spreading, at
an even faster rate than the Atlantic. As in the Atlantic there are no
areas of pre-Jurassic sea floor (older than 200 m.y.), so both the Pacific
and the Atlantic have apparently been formed within the last 200 m.y.
This new knowledge explodes much of older traditional geology (that
before ideas of sea-floor spreading, roughly pre-1960) when the
permanence of continents and oceans was generally assumed.

Sea-floor spreading presents a space problem. If the Atlantic is spreading and the Pacific is spreading, and millions of square kilometres of new crust are being formed (about 200 million km² in 200 m.y.), where it is all going? One possibility is that the earth is actually expanding, an idea carried furthest by Carey (1976). An idea more generally favoured at present is that the sea floor disappears down the trenches that bound the Pacific (and a few other places) and is somehow recycled. These sites are called subduction sites. It is worth noting that the mountains around the Pacific do not fall into one simple category, any more than do those of the Atlantic, though it is true that most mountain ranges are roughly parallel to the Pacific coast. But the mountains appear to be much too complicated and varied to explain them all by simple collision of continent with the Pacific.

Other oceans

Growth of the Indian Ocean began in the Jurassic, but details of the pre-drift arrangement of continents are still debatable. One configuration places India against Australia (Fig 2.14), an argument with much geological support from matching geological features.

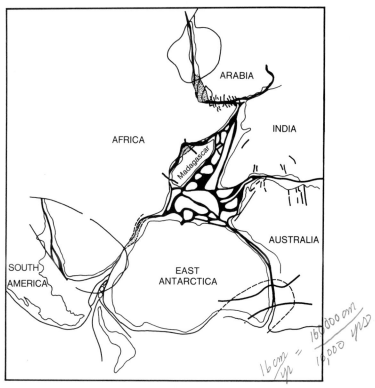

Fig 2.14   A diagrammatic reconstruction of Gondwanaland for the Late Jurassic (after Shields, 1977).

Another configuration (Fig. 2.15) has India against Antarctica, filling a gap between Australia and Africa. This gives the best computer fit of the continents.

Whatever the start, it is clear that India, once isolated, sped north at high speed (16 cm/y or 160 000 B) (Johnson, Powell and Veevers, 1976). The eastern Indian Ocean is evidently split into several minor plates (Fig. 2.16). The Carlsberg spreading site, continuous with the Red Sea, marks one area, bounded to the east by the Seychelles Ridge,

Fig. 2.15   A reconstruction of Pangaea about 200 m.y. ago (after Dietz and Holden, 1970).

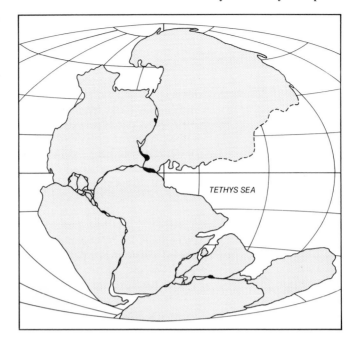

Fig. 2.16   Features of the Indian Ocean.

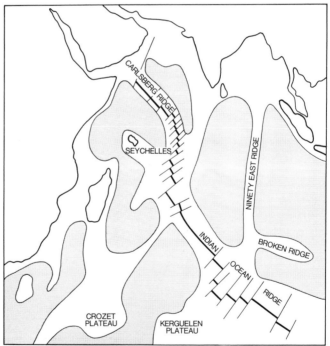

an isolated strip of sial. Another spreading site lies between this and Madagascar, which is itself separated from mainland Africa by further spreading sites. The 90° Ridge has been described as all sorts of features, but the latest evidence is that it has been produced by the passage of ocean crust over a hot spot presently located at the island of St Paul (Luyendyk and Rennick, 1977). This is rather similar to the Hawaiian and other island chains of the Pacific, but instead of forming a chain this hot spot has traced out a continuous ridge.

The Indian Ocean is separated from the Western Pacific by the Indonesian arc. According to Carey (1976) the rapid movement of India away from Australia created a huge gap, into which the Pacific welled, creating the great arc of Indonesia in Cenozoic times.

Palaeomagnetic results may not always show drift, but nevertheless put important constraints on palaeogeography. For instance, Haile, McElhinney and McDougall (1977) showed that since the Middle Cretaceous the Malay Peninsula and West Kalimantan have behaved as a unit, have remained in much their present latitude, but have rotated anticlockwise about 50°.

Antarctica is an isolated continent, or perhaps rather two halves of a continent. It is covered by an ice cap over 4 km thick which depresses the crust below sea level. Western Antarctica has the Antarctic Peninsula stretching towards the Scotia arc. Eastern Antarctica appears to have a simple outline resulting from splitting of the earlier Godwanaland. Between Antarctica and Australia the splitting effect appears very simple. Drift between these two continents started as late as the Eocene, though swelling, rifting, and intrusion of sills may go back to the Jurassic. Incidentally Wegener's maps of continental drift showed Australia and Antarctica splitting very late, a remarkable guess considering how little was known of the geology of Antarctica at the time. The mid-ocean ridge appears to continue between Africa and Antarctica, but to the west this is replaced by the Scotia arc, an example of an island arc to be discussed later. Significantly Antarctica appears to be surrounded by spreading sites (Fig. 2.17), but there are no subduction sites against the Antarctica continent. Yet Antarctica appears to be a very high continent. To some extent the height of the edges is due to depression in the centre by the weight of ice, but there is no obvious tectonic force to keep Antarctica so high.

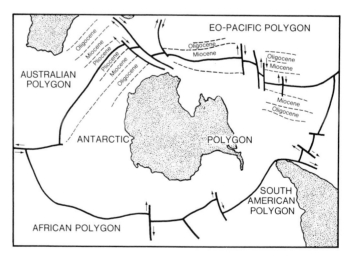

Fig. 2.17 Antarctica, surrounded on all sides by ocean derived from the circum-Antarctica spreading sites (after Carey, 1976).

The Arctic is a wedge-shaped spreading area (Fig. 2.18). Greenland and the Lomonosov Ridge separate an active spreading ridge that is continuous with the mid-Atlantic Ridge, and an older inactive spreading ridge.

No subduction zones are associated with either spreading ridge.

As data accumulated about sea-floor spreading it became clear that not only the world's major oceans are spreading, but that there are many minor spreading sites in smaller seas and basins, some of which

Fig. 2.18   Spreading ridges of the Arctic.

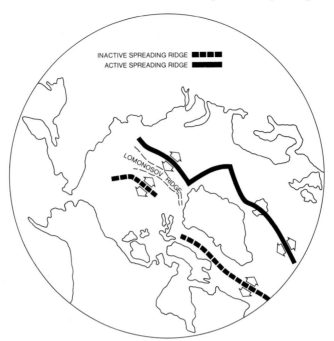

are currently active and some of which stopped spreading at various times in the past. Continuing work on sea-floor spreading, together with palaeomagnetic reconstructions of latitudes and orientations of the continents is presenting an ever more accurate picture of continental drift – a major feature of the earth that was dismissed as too improbable for serious consideration only 50 years ago.

3

The layered earth

A cross-section of the earth, like a cross-section of an onion, shows it to be made of successive layers or concentric spheres, except for a discontinuous outer layer that forms the continents. To understand what is happening at the earth's surface it is necessary to know what is going on at depth. Of course even the deepest bores go only a very short way towards the earth's centre, and our knowledge of the deeper structures must come from indirect evidence. We cannot see into the earth with light, but we can use earthquake waves in the manner of light waves to 'see' deep earth structure, and by using other geophysical phenomena such as gravity, magnetism, and heat flow we can build up a picture of the whole earth. The geochemistry and petrology of rocks adds further data for those parts of the earth that are directly available for study, and the topographic expression of major features such as mountain chains and island arcs adds further information concerning the outer layers.

Earthquakes

Suppose a rock is subjected to great pressure. For a while it may be deformed, but if the pressure reaches breaking-point the rock will fracture, and pieces of rock on opposite sides of the crack will move relative to each other. On a large scale the fracture is called a fault, that is a fracture along which there has been differential movement. The release of energy upon fracture causes an earthquake.

The place where an earthquake happens is the focus of the earthquake, and the point on the earth's surface above the focus is the epicentre of the earthquake (Fig. 3.1). All earthquakes produce vibrations or waves which travel through the earth in all directions from the focus. Most of our knowledge of the internal structure of the earth comes from seismology, the study of earthquake waves.

Earthquakes produce several kinds of waves. The P, or primary, wave is usually the first wave to be recorded on a seismogram because it travels fastest. The S, or secondary waves, usually arrive later and the L waves last of all (Fig. 3.2). Early in the twentieth century Mohorovicic noted that two distinct sets of P and S wave reflections appeared on the seismographs when an earthquake within a radius of

Fig. 3.1 Location of epicentre and focus of an earthquake. Isoseismal lines run through points on the surface with equal earthquake intensity.

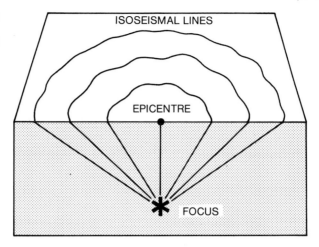

Fig. 3.2 (*a*) Diagrammatic representation of different earthquake waves travelling at different rates. (*b*) Trace of a typical earthquake record. The time differences between the arrival of different waves enables the distance of the earthquake from the recorder to be calculated.

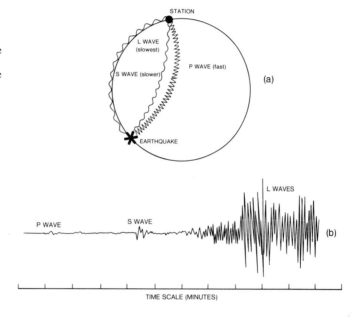

800 km and with a focus within 40 km of the surface was recorded. He concluded that this could happen only if waves travelled from the focus to the recording station by two different paths:

(*a*) direct travel path

(*b*) refracted waves

He inferred that within 100 km of the earth's surface there are two zones, an upper one where seismic waves travel slowly and a lower one where they travel faster. Later when it was found that this seismic discontinuity was present over most of the earth it was named the Mohorovicic discontinuity, or Moho for short.

The Moho is now accepted as a major dividing boundary in the structure of the earth; the part above is called the crust, and the part below is called the mantle.

The crust

Under the ocean basin the crust is thin, averaging 6–8 km in most places, whereas under the continents it averages 30–45 km, and reaches 70 km under the Andes and 80 km under the Himalayas.

The continental crust is made of a great variety of igneous, sedimentary and metamorphic rocks, but because of the dominance of the chemical elements silica (Si) and aluminium (Al) it is commonly called sial. It has the average composition of granite.

Under the oceans sial is not present, the crustal rocks are of average basalt composition, and as silica (Si) and magnesium are important constituents it is called sima. Within the continental crust there is another seismic break, the Conrad discontinuity, which is by no means as distinct or as continuous as the Moho but conveniently divides the crust into upper and lower parts.

In newly-created oceanic areas the crust mantle boundary is mostly at 6–8 km; in the Hercynian fold belt of Europe it is seismically sharp, below a thicker crust (20–30 km), under the Andes it reaches 70 km and 80 km below the Himalayas; it is gradational below Precambrian shield regions and young fold belts with crust over 35 km thick.

The mantle

The sudden change in seismic velocities at the Moho may be due to a change in chemical composition or to a phase change (change in physical properties without chemical change). The likeliest mantle rock is peridotite, consisting mainly of olivine and pyroxene but with a similar chemical composition to basalt. The Moho may mark an isochemical phase transition from gabbro to eclogite, but the preferred theory now is a chemical discontinuity from granulite to peridotite (Dawson, 1977). However, modern granulites are inaccessible to study, and inferences have to be drawn from ancient ones.

The low velocity layer or asthenosphere

As early as 1926 Gutenberg noted that seismic waves from earthquakes having foci at depths of between 50 and 250 km took slightly longer to arrive at recording stations than expected on theoretical grounds. He postulated the existence of a low-velocity layer between these depths. There is some doubt as to whether this represents a gradual diminution of wave velocities over the whole zone (50–250 km) or several very narrow layers within the zone where seismic velocities are slow.

One of the currently-favoured explanations depends on an effect of temperature, which increases with depth, giving the earth's thermal gradient. If there is water in the peridotite of the mantle (even as little as 0.1 per cent which seems very probable) then the melting point of the mantle is lowered appreciably and will intersect with the thermal gradient. This means peridotite will begin to melt in this zone. If 5 per cent of the peridotite melted to give a liquid rock this would lower the seismic wave velocities to those of the low-velocity layer. The low-velocity layer can thus be conceived as a crystal mush with a small percentage of liquid.

The low-velocity layer is a much more important and significant layer in tectonics than the Moho, which has nothing to recommend it as a surface at which movement could occur since both the crust and the mantle below must both be solid. Modern terminology for the upper layers of the earth may be summarized as follows (Fig. 3.3):

crust + upper mantle above low-velocity layer = lithosphere

low-velocity layer = asthenosphere

mantle below the low-velocity layer = mesosphere

The East African rift and the United States between the Rocky Mountains and the Sierra Nevada have virtually no lithosphere

Fig. 3.3   Layers in the crust and upper mantle.

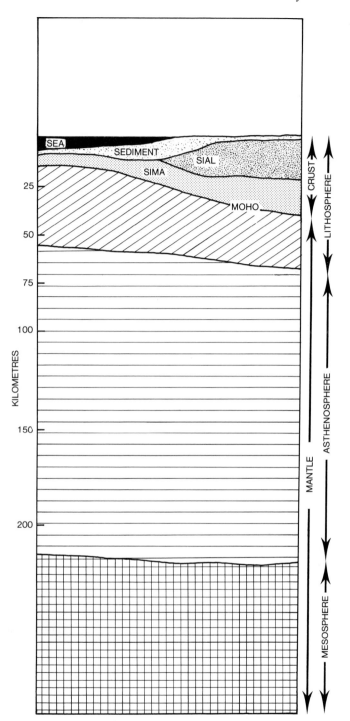

beneath the crust; the low-velocity zone seems to come almost to the Mohorovicic discontinuity (Knopoff, 1974).

Chapman and Pollack (1977) predict a lithospheric thickness of a few tens of kilometres in young oceans and continental fold belts, and more than 300 km in shield areas. This suggests that plates with shields will have more drag and move slowly. They suggest that in the future the lithosphere will continue to thicken, plate motion will diminish and eventually cease, so bringing to an end the plate tectonic phase of the earth's evolution.

Major earthquakes send waves that can be detected right around the earth, but beyond a distance of 120° from the epicentre neither P nor S waves are recorded which suggests that deeper waves are stopped or deflected by another discontinuity in the interior. The S

Fig. 3.4   Refraction of P waves and elimination of S waves by a liquid core.

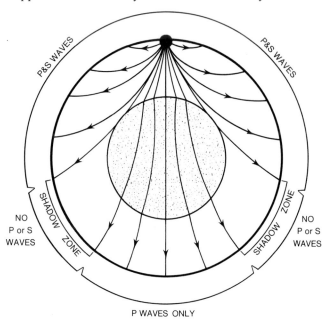

Fig. 3.5   The layered earth.

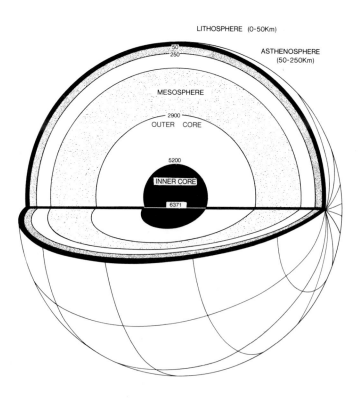

waves never reappear, but the P waves emerge strongly at a distance 143° from the epicentre. The emerged P waves have longer than normal travel time indicating that they have passed through a zone of low seismic velocity. The zone in which no P or S waves can be detected is called the shadow zone. These facts can be explained by the supposition that the earth has a liquid core that cannot transmit S-waves, but which refracts P waves (Fig. 3.4). More detailed work suggests that there is also a solid inner core, so the final picture of the layered earth is as shown in Fig. 3.5.

Gravity and isostasy

At sea level the value of gravity is dependent only on the latitude, and is less at latitudes near the equator than at latitudes near the poles. This value for gravity at a particular point on the spheroid is called the theoretical value for that point, and can be calculated simply by knowing the latitude of the point.

Subtracting the theoretical value of gravity from an observed value of gravity at a point gives a difference called the gravity anomaly. If this value is corrected to remove the effects of the elevation of the gravity instrument above the spheroid, the result is a 'free-air gravity anomaly'. The free-air anomaly is almost always positive because it does not take into account the gravitational pull of the rock that lies between the instrument and the spheroid. Such rocks between the instrument and the spheroid will exert an attraction depending on their density. The measured value of gravity, corrected for both elevation and the gravitational attraction of the rocks between the instrument and the spheroid, minus the theoretical value, equals the 'Bouguer gravity anomaly'. Bouguer gravity anomalies are generally strongly negative over mountains and plateaus, and zero or positive over oceans.

The word 'isostasy' is used to describe the principle of flotation as applied to continents and oceans.

There are two theories concerning the way in which isostasy acts to buoy up or compensate mountain masses. Pratt's theory assumes that different parts of the lithosphere have different density and float on a uniformly dense substratum (Fig. 3.6). The less dense crustal blocks float higher, forming mountains and the more dense blocks form basins and lowlands. This seems to be essentially the case for the difference between continents and oceans. Airy's theory assumes that the various parts of the lithosphere have approximately the same

Fig. 3.6 (*a*) The Pratt theory of isostasy explains topography by blocks of different density above a general level of compensation. (*b*) The Airy theory of isostasy explains topography by varying thicknesses of material of uniform density over an irregular base.

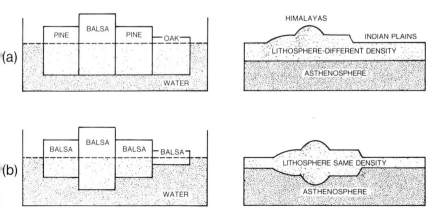

(a)

(b)

density but have different thicknesses. High mountains therefore not only project upwards, but have roots extending into the denser substratum. Thick lithosphere should form mountains, and thin lithosphere should form lowlands. Both theories assume the presence of a dense fluid or plastic layer in which the lithospheric blocks float – the layer that we now call the asthenosphere. Both theories account for the deficiency of mass under high mountains, but Airy's theory is now known to be a better explanation of mountains within continental regions.

The geophysical root of the Alps is postulated to explain the negative Bouguer anomaly over much of it. The negative anomaly corresponds to a thickening of the crust, but not of the upper, sialic layer. The granite-sedimentary layer above the Conrad discontinuity hardly deviates from normal thickness underneath the Alps. It is the 'basaltic' or 'intermediate' layer between the Conrad and Moho discontinuities that shows considerable thickening. There is no evidence for tectonic mixing of granitic and intermediate material. In northern Europe there is no trace of a similar root beneath the Hercynian uplands.

Compelling evidence for the general existence of isostasy in the southwestern United States is shown by the very good correlation of generalized topography and the raw Bouguer anomalies, as indicated by Gilluly (1970). The greatest negative anomalies coincide precisely with the topographic highs of the San Juan Mountains and the Mt Elbert mass in Colorado; topographic and anomaly gradients are concordant around the southern and western sides of the Colorado and Mogollon Plateaus; even the Colorado River and Death Valley are reflected.

Fig. 3.7 Isostatic recovery of Fennoscandia since glacial times. Contours in metres are reconstructed from raised shorelines, marine fossils and other evidence.

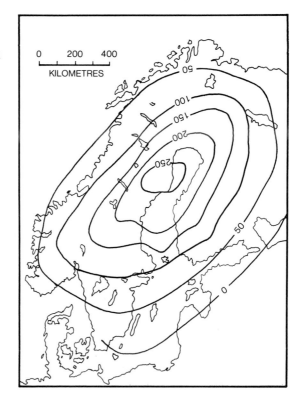

The laws of flotation act on the continents just as they would on a raft or on an iceberg. If we melt the top of an iceberg, the berg rises further out of the water. If we increase the load on a raft, it sinks further into the water. If we erode our continents they will rise anew. If we load our continents with a heap of sediments or an ice cap they will be depressed.

The adjustment time will not be the same for a continent as for a raft or iceberg. The raft sinks as soon as a sailor climbs aboard, and rises as soon as he jumps off. Loading and unloading continents is a slower process, depending on the viscosity of the earth materials.

Perhaps the best-documented information on crustal movements in response to load comes from those areas which had ice caps in the last Ice Age and are now 'rebounding' after the ice has melted. In the Baltic, for instance, contours on the amount and rate of uplift can be drawn (Fig. 3.7), and similar maps can be prepared for Canada (Fig. 3.8), and many other glaciated areas. Greenland is a

Fig. 3.8 Isostatic uplift in northeastern North America (after King, 1965).

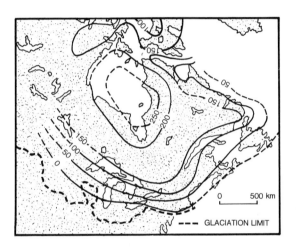

saucer-shaped country with a rim of mountains and a deep central area depressed by the Greenland ice cap. If the ice-cap melts we must not expect the saucer shape to remain, for the central area will start to rise as the load of the overlying ice is removed. Isostatic rebound can also follow removal of the weight of a large lake such as Lake Bonneville (Fig. 3.9).

The rebound of the Bonneville Basin after unloading of its water load indicates a mantle viscosity nearly an order of magnitude smaller than that deduced from the rebound of Fennoscandia (Crittenden, 1963). This might be explained by the very much smaller area involved, so a thinner zone of the mantle flowed.

For the vertical movements there have to be compensating horizontal movements of material in the underlying layers. If a boat sinks deeper upon loading, water under the boat moves out of the way. If Greenland sinks under the weight of ice, some compensating sub-crustal material under Greenland must move sideways out of the way. If an iceberg rises as it melts, more sea water moves in to occupy the space beneath the iceberg. If a mountain range rises the space under the range must be filled by the inflow of material below.

Gravity anomalies may result from two quite different causes: the rocks may not be in isostatic equilibrium, so they are tending to slowly

Fig. 3.9   Isostatic recovery in the Lake Bonneville area (after King, 1965).

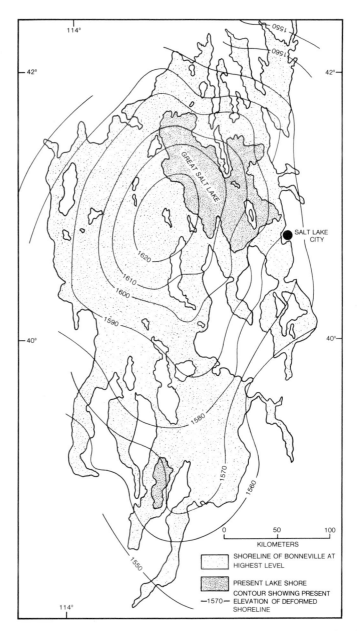

rise or sink towards a new equilibrium; or the rocks may simply be less or more dense than the usual.

Most gravity anomalies seem to tell us whether the underlying earth materials are denser or lighter than the average. Continents are generally lighter, being siallic, and ocean floors are denser, made of basalt sima. Mountain ranges and high plateaus have negative gravity anomalies (lighter) as they have a great thickness of lighter materials. The Red Sea has a positive anomaly because it is floored by dense sima. The rift valleys of Africa have negative anomalies because they are floored by thousands of metres of light sediment.

Deep sea trenches such as those off Chile and Peru, or in front of the island arcs of Indonesia and the West Pacific have marked negative anomalies. The simplest explanation for this is that they are full of light sediment, like the rift valleys, but various dynamic explanations have also been proposed.

Dynamic explanations of gravity anomalies provide quite different solutions to earth problems, so gravity data alone cannot give firm answers to tectonic problems. For instance, when the negative anomaly over rift valleys was first reported, it was thought that to produce this anomaly the fallen block in the middle of the rift must be held down against a tendency to rise and so it was proposed that the rift valleys were actually bounded by thrust faults. Abundant geological evidence is against such an interpretation and the much simpler hypothesis that the negative anomaly results from the thickness of accumulated sediment is preferred.

Deep sea trenches can be explained dynamically too. The first hypothesis was that some sort of down-fold (a so-called tectogen) was holding down sediments, and nowadays downthrusting at a subduction zone is a preferred explanation, though the simple effect of the sediment mass may in fact provide sufficient explanation (Worzel, 1976, and see p. 209).

One simple model for relating the effect of isostasy to mountain building is the idea of cymatogeny originated by Lester King. Assume a continental edge is being eroded and the resultant sediments deposited in the sea. The weight of the sediments depresses the crust, creating space for more sediments. The mountain mass, being reduced in thickness as a result of erosion, rises isostatically. Transfer of subcrustal material takes place by some sort of flow mechanism. Once initiated this process might be self sustaining, with the sedimentary belt getting thicker and thicker, and the mountainous edge of the continent repeatedly being uplifted anew. A significant point of this idea is that it might be checked by comparing the denudation chronology of the mountains, worked out by geomorphic methods, against the sedimentary history worked by stratigraphic methods. Thus periods of planation and reduced erosion should correspond with periods of reduced sedimentation in marine basins, while periods of marked uplift and erosion should correspond to increased deposition and formation of less mature sediments. A long and narrow sedimentary basin like this approximates to a geosyncline, a concept that is described further in Chapter 13.

Gravity can have other effects where materials of different density rise as plugs or 'diapirs'. The most obvious examples are salt domes. It seems that some granite plutons rise through the surrounding sediments in a similar way, pushing aside and doming up the neighbouring rock. Such domes and diapirs are discussed further in Chapter 10.

Isostasy and sediment accumulation

As sediment accumulates on a continental margin, the depositional surface subsides isostatically in response to the sediment load, and the thickness of sediment that can accumulate will exceed the original depth of the sea. For any given rise in sea level the maximum thickness that can be deposited is given by the equation

$$2.4t = h + 3.4t - 3.4h$$

where h is the rise of sea level, t is the thickness of sediments, 2.4 g/cm is the density of sediment, and 3.4 g/cm is the density of the mantle. This simplifies to

$$t = 2.4h$$

Using this formula, Bond (1976) found that for the Late Cretaceous transgression when *h* was estimated to be 310, *t* was calculated to be 700 m.

The equation assumes a purely isostatic response of the crust to the sediment load, but in reality there is both viscous and elastic resistance to the subsidence (Walcott, 1970a). A correction for these factors may reduce the calculated thickness, so by ignoring the correction an overestimate is calculated. Wallcott (1970b) has documented the correction factors.

Sedimentation causes loading, and uplift in neighbouring erosional areas. If the wavelength of the topography – that is the distance between erosional area and sedimentation area – is great, arching will result. If the wavelength is small a critical situation may develop when stress exceeds the elastic limit of the basement rocks, and faulting leads to the formation of horst and graben structures (Walcott, 1970).

Heat from the earth

It does not seem likely that the earth was formed as a hot body, and its first internal heat came from impaction and gravitational compaction of its formative particles.

The early heat energy was absorbed by a rise in temperature of the interior and much was lost by radiation to space. Because the earth is a poor conductor there is still some of this early heat left after the 5 000 m.y. of earth history, perhaps as much as a fifth. The rest of the heat comes from the decay of radioactive isotopes, especially of uranium, thorium and potassium. Granite has about four times as much uranium as basalt, about three times as much potassium, and perhaps ten times as much thorium.

Heat flow

Temperature in the earth increases with depth. Near the surface temperatures are influenced by the weather and groundwater, but beyond these effects the temperature increases at a rate of about 20 °C/km, but ranges from about 10 °C to 50 °C, largely depending on the thermal conductivity of the rock.

The relatively steep temperature gradient near the surface does not continue at depth. Diamonds found in kimberlite pipes would have been converted to graphite if the temperature was above 1 100 °C, yet pressure considerations of associated minerals suggest the rock was formed at a depth of 150 km. The formula for heat flow is:

$$Q = -Kr$$

Where Q = heat flow, K = thermal conductivity, and r is the thermal gradient. The minus sign signifies that heat flows from high temperature to a low temperature.

Since heat always flows from the hotter to the colder region it is clear that the earth's heat is being lost to space. Typical conductivities of the rocks of most interest are:

| Granite | $5–7 \times 10^{-3}$ cal/s/cm/per °C |
| Basalt | 4.0×10^{-3} cal/s/cm/per °C |

So the continents have considerably higher conductivity than the oceans. Heat flow is measured in heat-flow units. One HFU = 1μ cal/cm/s. The HFU is still a working unit. The SI unit for heat flow is mW/m, which is converted to the HFU by multiplying by 23.9. The SI unit for vertical temperature gradient is K/m which is just a thousandth of the working unit, °C/km.

Some average values for heat flow on continents are:

	(HFU)
Precambrian regions	< 1
Palaeozoic fold belts	1.5
Cenozoic fold belts	1.7
Cenozoic volcanic areas	2.1

Some average values for heat flow in oceans are:

	(HFU)
Atlantic	1.4
Indian	1.4
Pacific	1.7
Marginal seas	2.13

Episodes of erosion will affect heat flow, as described by Lee (1979). Depending on the rate and duration of erosion the transient heat flux may increase while the near-surface rocks that contain higher amounts of heat-generating elements are stripped off. When erosion stops it takes about 20 per cent of the duration of the erosional episode for the transient heat to attain its steady-state value.

Low values of heat flow are found in the Sierra Nevada, and the intermontane plateau country, but high heat flow is found in the California Coast Ranges, the Rocky Mountains, and the Basin and Range Province, Idaho and eastern Utah. Ocean-bottom heat flow has an average of about 1.4 HFU with high values of 1.9 HFU and over on ocean ridges, abnormally low values about 1.1 over trenches, but with higher flow values on either side of the trench some distance away.

Heat-flow measurements in back-arc basins may be used to distinguish the active ones that are still spreading, with high heat flow, and the extinct ones that are no longer spreading, have thicker sediments, and lower heat flow.

As the number of measurements increases the precision of observations in relation to earth features becomes more precise, but at present problems seem to be increasing rather than decreasing. For example in the eastern Pacific, Anderson and Hobart (1976) were able to use over 800 heat-flow measurements which permitted them to relate heat flow to age zones in the ocean. They found that the mid-ocean ridge crestal regions have lower heat flow than expected, there was an increase in heat flow some distance from the crest in a zone which appears to mark a transition from dominantly convective to conductive heat transfer, and still older zones have a heat flow that approximates that predicted by conductive cooling models. The same sort of distribution is found in the Galapagos spreading centre, but in the Galapagos the age of the transition zone is 5–6 m.y., whereas on the East Pacific Rise it occurs in 10–15 m.y., crust. To understand these results it is necessary to take sedimentation rates into account as well as spreading rates.

The world averages for heat-flow are:

All oceans and seas	1.65
All continents	1.49

This conclusion, that the heat flow is greater in oceans than on land, makes the source of the ocean's heat even more mysterious.

Distribution of heat and radioactivity

The minor variations in heat flow between physiographic provinces of continents and oceans are not too hard to explain, but the general equality in heat flow between continent and oceans poses real problems.

Radioactive elements, having large size, tend to be segregated into the granites, and continents are more radioactive than any other layer in the earth. If half the thickness of continents were of granite, it could account for all the observed heat flow.

But the basalts of the ocean floor are poor in radioactive elements, and could not provide the observed heat. The heat therefore must come from lower down, from the mantle. But the mantle is present under the continents too, so why is there not an additive effect in the continents, with a heat flow equal to the sum of that from the mantle and that from the granites?

If excess heat is being produced beneath the continents and cannot escape, it may be responsible for increasing the thickness of the asthenosphere, so that the asthenosphere beneath the continents is thicker than that beneath the oceans allowing more ready isostatic and other tectonic movement to take place.

According to Sass (1971), from the energy point of view heat flow is the most impressive of all terrestrial phenomena, being many orders of magnitude greater in terms of energy than all earthquakes, volcanic activity and tectonic processes put together.

It is very important to realize that the heat flow from the earth is minute compared with the energy received from the sun, about 0.5 kcal/cm/s. Solar energy provides perhaps a thousand times as much energy to the earth as geothermal heat. The solar energy absorbed by the atmosphere will also be available for geological work such as erosion. Altogether it is probable that external energy is about 5 000 times as great as that from the earth.

Furthermore, England and Richardson (1977) draw attention to the powerful effects of erosion upon temperatures within thick continental crust. Convection of heat by uplift and erosion is as potent a means of energy supply to the upper lithosphere as radiogenic heating or transfer of heat from beneath the lithosphere. Consequently rocks in eroded mountain regions will generally reflect temperatures higher than those predicted from the latter effects alone. In regions of thick crust, erosion in many instances may be the most important factor in determining the pressure and temperature which are recorded by mineral assemblages.

It thus seems likely that the external energy is more than adequate to drive the tectonic machine by eroding continents, transporting sediments, filling depositional basins, and so driving internal flow within the earth, causing isostatic uplift and subsidence, and creating the surface features of the continents.

But there seems no way in which external energy can cause the major features of the oceans. Sea-floor spreading seems to be undoubtedly driven from inside the earth, and with known rates of energy production in the earth if internal processes did nothing else it would seem to use most available energy.

4 Plate tectonics

In consideration of the first-order features of the earth, historically the first modern concept was continental drift. In this the continents were thought to drift around on a 'sea' of sub-continental material, but the ocean floors were thought to be quite passive in the process. The next idea was that of sea-floor spreading, in which new sea floor is generated at spreading sites, and destroyed beneath the marginal trenches and island arcs which are the 'sinks' in the system. If such a spreading site were to originate beneath a continental area, the continent would be rifted and drifted apart and a new ocean basin formed in the middle.

Plate tectonics replaced the older ideas of continental drift and sea-floor spreading when it was realized that typical plates include both continental and oceanic crust. Moreover it was realized that the sial–sima boundary, the Moho, was not a suitable surface for continents to move on, and that movement took place in the low-velocity layer, the asthenosphere. The lithosphere with its appreciable strength and rigidity includes both sial and sima, continent

Fig. 4.1 The major plates, ridges and subduction zones of plate tectonics.

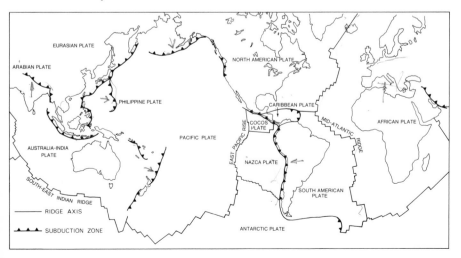

RIDGE AXIS

SUBDUCTION ZONE

and sea floor, so 'plates' which may likewise contain both components convey a better image of the process than continental drift or sea-floor spreading which emphasize one or the other. Plate tectonics is a name for the general idea that the earth's crust can be divided into several plates (Fig. 4.1) and the world's main tectonic features are related to activity at the edges of the plates. It is supposed that new crust is created at spreading sites, and that the crust is destroyed at subduction zones (Fig. 4.2). Beyond this simple exposition many complications are possible, and variations from writer to writer, and from place to place are almost numberless. For the past decade plate tectonics has been the ruling theory in geology, and has been used to account for almost everything in earth science.

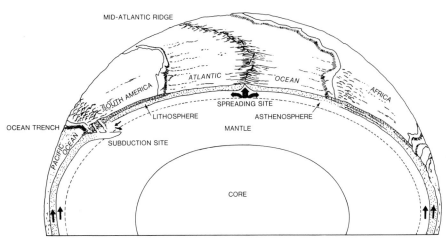

Fig. 4.2 The dynamics of plate tectonics. Material from the asthenosphere rises below the crest of an ocean ridge, erupting to form new sea floor. The lithospheric plates diverge, with South America and the western half of the South Atlantic making one plate, Africa and the eastern half of the South Atlantic Ocean making another. The growth of new lithosphere is balanced by the destruction of an equal amount at a subduction zone, where a lithospheric plate descends into the mantle. Note however that although there is subduction on the edge of South America there is no subduction bordering Africa, although the ocean east of Africa is apparently moving towards Africa from the Atlantic–Indian Ridge (Fig. 4.1). Presumably not only is Africa drifting away from the mid-Atlantic Ridge, but the Atlantic–Indian Ridge is drifting away too, at about twice the rate.

Firstly let us look at some distribution maps of major earth phenomena.

Figure 4.3 shows the distribution of earthquakes. These follow two main sets of lines. One set, all shallow earthquakes, follows the line of the sub-oceanic ridges, the spreading sites. Another set, including most of the deep earthquakes, follows the line of subduction sites around the Pacific rim and lines of deep sea trenches. A few earthquakes follow the great fracture zones, and others the rift valleys which might be incipient spreading sites. A residue of other earthquakes remains unexplained by this hypothesis, but clearly the distribution of earthquakes, by and large, divides the earth's surface into a number of areas that correspond very well with the plates of plate tectonics. Indeed plate tectonics provides a very useful simplification in accounting for earthquakes. Earthquakes occur where the action takes place – at spreading sites and subduction zones at the borders of plates.

Looking now at a map of the distribution of volcanoes (Fig. 4.4) we see a very similar pattern. Volcanoes are distributed along lines of sea-floor spreading, and on the areas bordering subduction zones. The type of volcanicity is even more diagnostic, with essentially basaltic volcanicity on spreading sites, and andesitic volcanicity at subduction sites. Alkaline volcanic activity follows some rift valleys. As with earthquake activity the plate-tectonic theory provides a satisfactory explanation for the location and type of much volcanic activity.

Other evidence, including geochemical, gravity, and heat-flow measurements all helps to confirm the same boundaries of the major

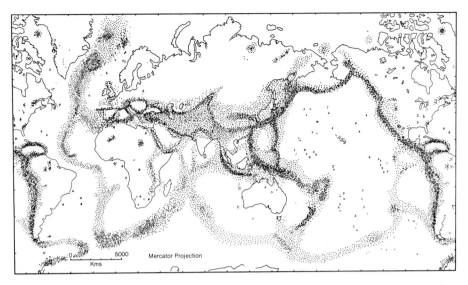

Fig. 4.3   Distribution of Earthquake
Activity.

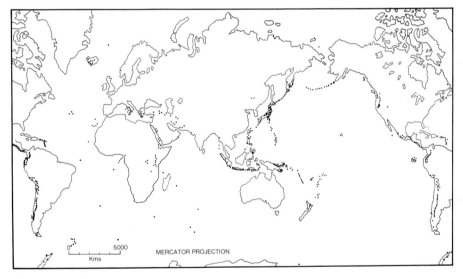

Fig. 4.4   Distribution of volcanoes.

plates of the earth's crust, and there is no doubt that the view of the earth in terms of plates has made a major advance in geological thinking, even though details may be disputed.

For many earth processes, and for many tectonic processes, the action takes place at plate edges. These fall into three types: the spreading edges, the collision edges, and transverse edges where plates slide past one another.

Spreading sites

Spreading sites are the simplest. Here sub-ocean rises are presumably related in some way to the rise of magma. On these sites there are many volcanic eruptions, vertical dykes are intruded in great numbers (parallel to the rift), and submarine lava flows and hills are formed. Earthquakes occur in the top few kilometres of the crust, and show essentially tensional and vertical movement. The geomorphology of the rift area can be seen in fair approximation in Iceland, which may be regarded as a piece of the mid-Atlantic Ridge above sea level. The rocks get older away from the rift, the rocks are dominated by fissure eruptions, and the topography, where not primarily volcanic, is marked by many faults and fissures parallel to the main mid-ocean (or mid-Iceland) rift.

It would seem that the volcanic intrusion is permissive. That is, dyke rocks come in to fill a space created by tension. We should not imagine a thin dyke forcing its way into a crack and pushing aside a plate many thousands of kilometres across. Rather there is some force pulling the plates apart and the dykes come in to fill the space created when a crack appears.

Some Atlantic islands are on or close to the mid-ocean ridges. The active Ascension Island is virtually on the ridge, St Helena 700 km from the ridge and extinct. A very rough correlation of the age of oceanic islands with distance from the ridge led Tuzo Wilson to propose that the Atlantic islands were formed on the rift and then drifted away with the spreading of the ocean. This idea still appears roughly true, but is slightly complicated by the fact that the islands remain active for several million years and for a considerable distance from the rift. Tristan da Cunha is still active (the last eruption was in 1961), although it is 350 km from the mid-Atlantic rift.

Collision sites

Continental–ocean collision with subduction

Collision zones provide more varied tectonic settings than spreading sites, including the following types:

This is the commonest sort of collision envisaged in the plate tectonic theory, but can itself be divided into two:

Simple subduction (Andes type)   This is typified by the plate junction on the west of the southern half of South America. The advancing continent overrides the sea floor, which is thrust down under the continental plate. The downgoing slab melts to produce andesitic magma which is intruded and erupted as andesitic volcanoes. The course of the slab's progress is recorded by a series of earthquakes that show a fairly consistent increase in depth with distance from the plate edge making an inclined earthquake zone known as a Benioff zone after its discoverer. The continental edge is uplifted to form mountains (the Andes) and the collision zone is marked by a deep ocean trench (Dewey and Bird, 1970). These ocean trenches are the deepest parts of the ocean, often over 10 000 m.

Another possibility is that a subducted plate may flatten, causing perhaps a cessation of volcanism and renewed uplift. This has been proposed for the subduction of the Nazca Plate beneath Peru by Barazangi and Isacks (1976) and elaborated by Noble and McKee (1977) as shown in Fig. 4.5.

Island-arc subduction   Around the western edge of the Pacific the map shows many islands arranged in festoons of arcs, either single lines or double rows of islands, making a quite clear pattern of curves. These lie some distance off the continent, and in front they have a deep

Fig. 4.5 Stages of igneous and tectonic evolution of central and northern Peru during Late Cenozoic time and their relation to possible changes in subduction geometry. The stipple indicates lithosphere (after Noble and McKee, 1977).

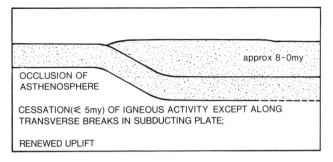

approx 8-0my

OCCLUSION OF ASTHENOSPHERE

CESSATION($\leqslant$ 5my) OF IGNEOUS ACTIVITY EXCEPT ALONG TRANSVERSE BREAKS IN SUBDUCTING PLATE;

RENEWED UPLIFT

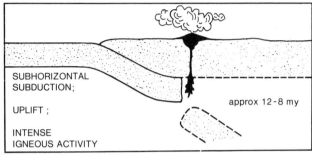

SUBHORIZONTAL SUBDUCTION;

UPLIFT ;

INTENSE IGNEOUS ACTIVITY

approx 12-8 my

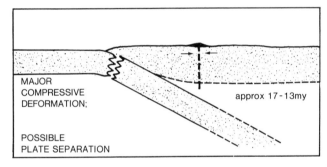

MAJOR COMPRESSIVE DEFORMATION;

POSSIBLE PLATE SEPARATION

approx 17-13my

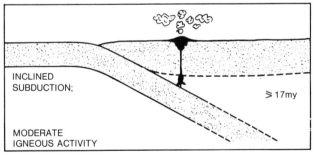

INCLINED SUBDUCTION;

MODERATE IGNEOUS ACTIVITY

$\geqslant$ 17my

trench very like that off the coast of South America. These deeps also bound a Benioff zone dipping towards the continent and marked by earthquakes reaching depths of several hundred kilometres.

Some arcs face east (Indonesian arc), some arcs have no continent behind them (South Sandwich Islands), and some lines of islands have all the attributes of arcs such as Benioff zones and trenches but lack the arcuate shape (Solomon Islands). Nevertheless the island-arc complex makes one of the distinctive landform and tectonic assemblages of the earth and, despite many complexities and variations to be discussed later (Ch. 14), island arcs mark plate junctions.

Continent–ocean collision with obduction

In a few collision zones a slab of ocean floor has overridden rather than underridden the continent. In Papua New Guinea a slab of basic rocks with petrology, layering and structures exactly like those thought to be typical of ocean crust makes the mountainous terrain of the Papuan Ultramafic Belt. Another area is the Troodos Mountains of Cyprus (Fig. 4.6) where dyke-ridden oceanic rocks override continental rocks. Obduction is the term for this overriding, in contrast to subduction when the oceanic slab goes down below the continental mass.

Fig. 4.6 Development of the Troodos Massif by obduction (after Gass and Masson-Smith, 1963): (1) Formation of an intrusive swell in the oceanic area between Africa and Eurasia. (2) With the approach of Africa and Eurasia some of the sialic layer underrides the mantle, which is obducted. (3) Isostatic rise of the sial heaves up the ultrabasic and related rocks of the mantle.

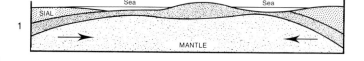

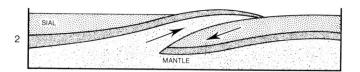

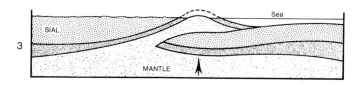

Continent–continent collisions (Himalayan type)

The most spectacular example of this collision type is that presumed to have occurred between India and continental Asia. Geological data show that India was once united with the southern continents of Gondwanaland, and until the Mesozoic shared their faunas, their ice-age and other features. Since then it has separated and drifted north at a rather high speed (16 cm/y) until it collided with the Asian mainland. This geological information is confirmed by the evidence of sea-floor spreading in the Indian Ocean. For a while part of a great sea, the Tethys Sea of Mesozoic and Early Tertiary times, lay between the two continental slabs, but this was crushed and uplifted as the Indian plate drove under the Asian plate. Further effects of this collision are the formation of a double thickness of continental crust. Isostatic compensation of this mass (see p. 31 for a discussion of isostasy) led to the formation of the Tibetan plateau, the highest in the world, fronted by the highest mountains in the world, the Himalayas. The double thickness of crust may also account for the lack of volcanoes on this plate boundary, for it is too thick for magmas to penetrate. The zone is, however, marked by many earthquakes (Fig. 4.3).

In some situations a continent–continent collision may continue as a continent–ocean collision. The Zagros–Makran line is regarded as a plate boundary of convergence by some (Farhoudi and Karig, 1977) but there is a marked change of character from the Zagros thrust, which is a continent–continent collision, to the Makros region where oceanic crust is thought to be subducted. From south to north the

Fig. 4.7 (*a*) Tectonic sketch of the
Zagros Makran area. (*b*) Schematic cross
section along line A–A'. (*c*) True scale
section along line A–A' showing possible
relationship of descending lithosphere to
tectonic units (after Farhoudi and Karig,
1977).

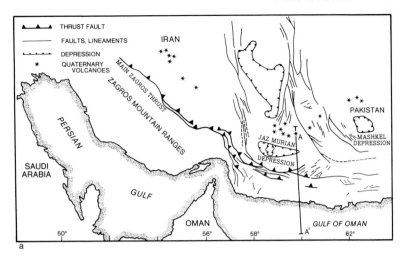

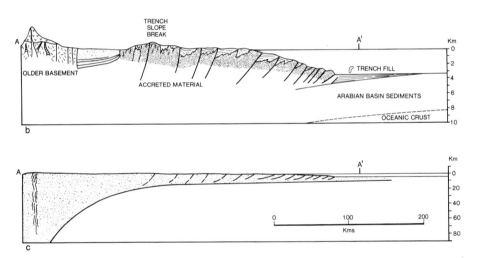

Makran ranges become higher, older, and more deformed. Further
north still are subsiding basins, and beyond them active volcanoes
(Fig. 4.7).

Ocean–ocean collision

An arc may be formed, as in the South Sandwich islands (Scotia arc)
where a strip of spreading Pacific seems to be pushing a finger of
oceanic crust into the oceanic crust of the Atlantic. The contact is
marked by a trench, earthquake zone, active volcanicity and two sets of
sea-floor spreading (Fig. 4.8). The Caribbean arc is somewhat similar,
with the Pacific pushing into the Atlantic.

The curvature of Indonesian arc may result partly from the Pacific
Ocean spreading into the Indian Ocean.

Arc–arc collision

Yet another possibility is that two arcs collide, and a possible example
is provided from the Philippines (Roeder, 1977). The chief features of
the region are the Manila Trench and the Philippine Trench, with
convergent Benioff zones, and the collision zone is interpreted by
Roeder as shown in Fig. 4.9.

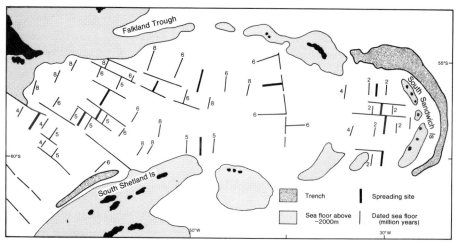

Fig. 4.8   Map of Scotia Arc showing
spreading site immediately behind the
South Sandwich Islands and older
spreading patterns with different
orientations further back (after Barker
and Dalziel, 1980).

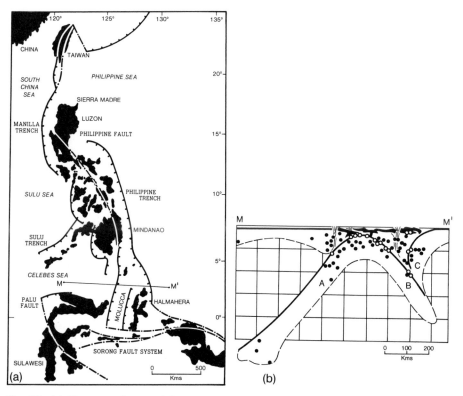

Fig. 4.9   (*a*) Plate tectonic map of the
Philippine–Molucca area. (*b*) Section
across Molucca strait and adjacent waters.
Black=crust; White=lithosphere, with
three downgoing slabs A, B, C; dots =
earthquake centres. Background grid has
100-km mesh (after Roeder, 1977).

Flake tectonics

Yet another version of collision tectonics is that of flake tectonics (Oxburgh, 1972), a model in which it is assumed that material arriving at a collision site may be sliced up into flakes, and the flakes transported across the continental plate (Fig. 4.10). Coleman (1971) proposed that some ophiolites could be emplaced onto continental edges by this kind of action; Laubscher (1971a) suggested a similar mechanism to peel off the allochthonous oceanic sediments of the northern Apennines from a downgoing slab; and Oxburgh (1972) invoked this mechanism to derive high-grade metamorphic nappes. A significant feature of flake tectonics is that the flakes appear to dip in the opposite direction to the subduction plane.

Fig. 4.10 A diagrammatic representation of flake tectonics. The continental plate on the right acts as a scraper that peels off material from the sinking slab and causes it to be emplaced on to its edge (after Smith, 1976).

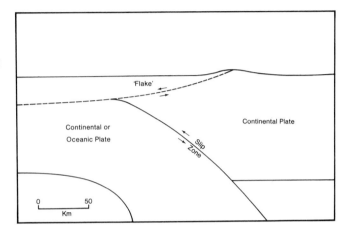

Other features of plate tectonics
Small ocean basins

Many small seas are not merely extensions of the ocean onto continental shelves, but are distinct geomorphic features (Ch. 14). They have anomalous crust, intermediate in thickness between continent and ocean, but they are simatic and have traces of spreading history which may be currently active or old and now stopped. There seem to be two main types: some are old crust preserved behind island arcs or continental fragments, and some are new sea floor. The Caspian Sea and Black Sea have old spreading sites and old crust. Seas like this have fairly thick sedimentary fills, more than the deep ocean. Back-arc basins behind island arcs (Ch. 14) are also small spreading sites, bordering continents. Some are backed by new sea floor (like that behind the Scotia arc); some are extinct spreading sites with old sea floor, like the Coral Sea which was active from about 60 to 20 m.y.; some have moved at different times, like the Solomon Sea which was active 20 m.y. ago and again at about 3 m.y.

In the Caroline Basins, between New Guinea and the Philippines, the topographic expression of spreading sites does not follow the normal pattern (Fig. 4.11). The West Caroline Trough was a spreading centre that was abandoned about 28 m.y. ago, and the Kilsguard Trough is a spreading site that was abandoned about 31 m.y. ago. The next spreading site was the Eauripik Ridge, which was active for at least 5 m.y. (Mammerickx, 1978). Although topographic features were used in this study to locate spreading axes, the nature of the topographic features is quite different. Nobody has previously postulated spreading from a trough, or inversion of an old ridge into a trough. Certainly it would be unwise to assume that troughs are all subduction sites.

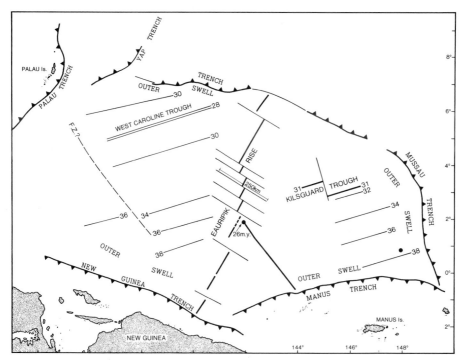

Fig. 4.11   Tectonic sketch of the
Caroline Basin. Two extinct spreading
sites correspond to troughs rather than
ridges, but the Eauripik ridge is a younger
spreading centre, active for about 5 m.y.
but also extinct.

Triple junctions

Triple junctions are the Y-shaped junctions where three plates come together, and where three spreading sites join. The junction of the Nazca, Antarctic and Pacific plates is an example. The origin of new plate boundaries seems to be commonly associated with triple junctions, but in many instances two arms become dominant and the third dies out. This 'failed arm' is a tectonically important feature, and is known as an aulacogen.

Plate tectonics and topography

In relating plate-tectonic boundaries to geomorphology it is clear that there is some correlation between topographic features and plate boundaries, but this is far from perfect. A great many mountains, plateaus and other landscape features have no apparent relationship to plate-tectonic situations. Some of the postulated explanations of mountains by plate-tectonic theory are rather tenuous and require special events such as reversals of subduction direction, changes of subduction position, and similar complications. Some mountains may also be related to subduction zones of the past, so we may have palaeoplate tectonics to add to plate tectonics in our search for the explanation for the origin of mountains. The elements of plate tectonics are vital to understanding the world, but alone will not enable us to account for the topography of the earth.

Plate tectonics in the past

There are no pre-Jurassic sea floors and all the evidence we have of continental drift is that it occurred since the break up of Pangaea starting in the Jurassic. What happened before that?

The commonest asssumption is that continents have split and reassembled many times in the history of the earth. We may imagine continents growing by accretion of mountain belts, rather like North America, and then drifting until they collide with another continent, like India with Asia, and become welded to it, perhaps with the formation of other mountain chains such as the Himalayas. The double continent may be joined by yet more continental fragments until eventually either all the continents have assembled together, or the supercontinent splits up along new lines. It does seem a remarkable thing that all the continents should have been assembled together in the Jurassic, and for much of the Palaeozoic. However there is some indication from palaeomagnetic studies to support the notion of repeated splitting and joining of continents.

The expanding earth theory of course simply has one earth-sized continent (except for the Eopacific) until expansion and sea-floor spreading start in the Jurassic.

Plate-tectonic reconstructions are now frequently attempted for Palaeozoic settings, like the Caledonian plate-tectonic model shown in Fig. 4.12 (Phillips, Stillman and Murphy, 1976). While present plate models are perhaps weak on the hidden rocks, reconstructions depend on ancient rock exposures but are very speculative on palaeo-geography and geomorphology. Better geomorphic constraints on the plate models of today will be valuable to stratigraphers and palaeographers if they ultimately enable them to model their palaeogeography in a more systematic way.

Fig. 4.12 Palaeogeographical and tectonic sketch of parts of the British Isles in Early Ordovician times (after Phillips, Stillman and Murphy, *et al* 1976).

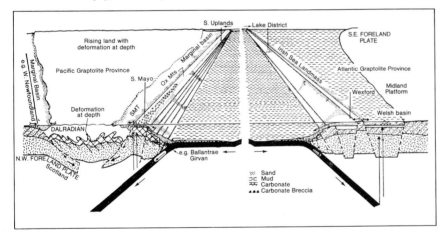

The northern Appalachians are explained by Dewey and Bird (1970) in simple plate-tectonic terms. The continental crust is rifted in Late Precambrian time, and a proto-Atlantic ocean is formed. Sedimentation occurs on the margins of the opposed continents, and then the continents move together again. At first there was 'Andean type' orogeny on the Appalachian side (though why not on the European side is not explained), and then when the opposed continents collided there was 'Himalayan type' orogeny, with disappearance of the proto-Atlantic ocean.

Dietz (1972) provides a slightly different plate-tectonic model for the Appalachians using lateral compression and plate tectonics (Fig. 4.13). Starting with geosynclinal deposits and platform deposits, deformation, vulcanism and intrusion took place about 350 m.y. ago at a plate convergence, before the opening of the present Atlantic started about 225 m.y. ago.

Fig. 4.13   Plate-tectonic interpretation of the Appalachians and the Atlantic according to Dietz (1972): *a.* Rifting in the Late Precambrian; *b.* Opening of the ancestral Atlantic Ocean, and continental margin sedimentation; *c.* Closing of the ancestral Atlantic Ocean with folding of sediments (note subduction of old spreading site) with Andean-type orogeny in the Early Palaeozoic; *d.* Disappearance of the ancestral Atlantic Ocean, with Himalayan-type orogeny as opposed continents collide; *e.* About 180 m.y. ago the Atlantic re-opens, along the old suture line; *f.* Today the Atlantic Ocean is still spreading, and sediments accumulate at the continental margins.

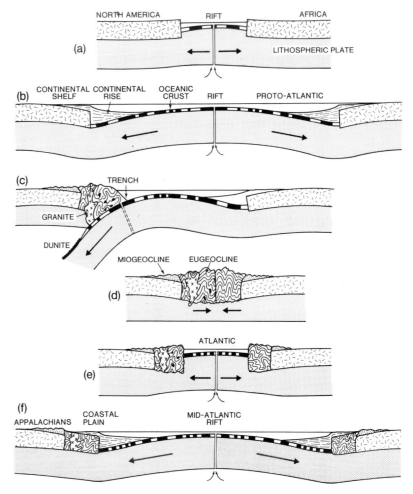

Thomas (1977) adds a complication to the Appalachian story. The thickness and facies of Late Precambrian and Palaeozoic rocks suggest an originally zig-zag continental margin, with salients and recesses along the length of the Appalachian–Ouachita structural system. He suggests that the original zig-zag rift was the expression of transform faults of Late Precambrian age, which created an orthogonally zig-zag geosyncline. Thomas suggests that later compression, not parallel to the movement during rifting, caused folding and uplift. On the promontories compression was transmitted directly through continental crust: within re-entrants compressive stress has been transmitted through the wedges of thick incompetent clastic sediments.

In both these models it must be borne in mind that the existence of the present Appalachian Mountains is *not* the result of plate tectonics. The mechanisms of Bird and Dewey, Dietz, and Thomas produce a fold belt by plate tectonics. This fold belt is presumed to have been a mountain range, but the folded rocks have been worn down to a plain, and uplift of this plain created the present mountains much later in earth history and at a time when simple plate tectonics cannot apply.

Precambrian orogeny

The whole tectonic and geomorphic system was very different in the distant past. Wynne-Edwards (1976) believes that the Proterozoic was a time of ductile flow rather than brittle fracture and failure characteristic of Phanerozoic time. This impression might derive, however, from the nature of the geological record, with only the ductile rocks of the deeper crust preserved and all the old brittle rocks eroded away. England and Richardson (1977) have pointed out that blueschists develop on the low-temperature end of the metamorphic geotherm and are succeeded in exposure at the surface by greenschists or amphibolite-facies rocks. The time scale for this process, he believes is consistent with the virtual absence of Precambrian blueschists.

Two models based on isotopic data have been proposed to explain crustal data:

(*a*) The rate of continental crust generation has been increasing exponentially from the Archaen to present. This is generally thought unlikely.

(*b*) Probably at least two-thirds of the crust formed by 2 500 m.y. ago. Condie (1976) believes that the seismically determined thicknesses of cratonic areas formed since 2 500 m.y. are independent of age and vary only between 38 and 40 km, and Wise (1974) suggests that the relative elevation of the continents relative to sea level has stayed the same throughout the last 2 500 m.y., implying that the continents have had at least 90 per cent of their present thickness since that time. Tarney and Windley (1977) conclude that a substantial proportion of the continental crust evolved rapidly during the Late Archaean in connection with major crustal thickening. Tarney and Windley suggest that before 2 500 m.y. the continental crust grew at Andes-type continental margins with underthrusting of oceanic crust, generation and underplating of extensive tonalitic-granodioritic material, and that it was associated with widespread nappe stacking and imbricate interthrusting. The continents thus thickened by tectonic and magmatic processes were metamorphosed to granulites deep in the crust. Chapman and Pollack (1977) working on the relationship between heat flow and lithospheric thickness produced a map of the thickness of the lithosphere. They predict a lithospheric thickness of a few tens of kilometres in young oceans and continental orogenic provinces, in contrast to over 300 km in shield areas. This suggests greater viscosity beneath shields, offering an explanation for what they claim is an observed retarded motion of plates with shields.

They point out that in the past the lithospheric plates were certainly thinner and the asthenosphere more well developed – conditions that should have facilitated Precambrian continental drift. In the future the lithosphere will continue to thicken, though there may be interruptions like the Mesozoic disruption of Gondwanaland. As shields continue to thicken they will act increasingly as viscous anchors, and plate motion will diminish and eventually cease, bringing to an end the plate-tectonic phase of the earth's evolution.

The idea that orogenic belts result from the collision of cratons derived from widely-separated positions can be tested for the Precambrian orogenic belts by palaeomagnetic data analysis (McElhinney and McWilliams, 1977).

For the period 2 300 to 1 900 m.y. data from the West African and Kaapvaal cratons form a coherent set with poles of similar age from each craton falling consistently on a combined apparent polar-wander path constructed for this time interval. Data from the Kalahari and Congo cratons likewise form a coherent set for the interval 1 100 to 700 m.y. These data strongly suggest that these cratons were not previously widely separated and then converged to form the Pan African or any other orogenic belts.

Data from Australia for the time interval 2 500 to 1 100 m.y. also form a coherent set, so orogenic belts such as the Musgrave belt did not arise from convergence of widely-separated cratons.

From North America data from the Superior, Churchill, Nain, Bear and Slave provinces form a coherent set, so the Judsonian orogengy cannot be the result of plate convergence.

Thus palaeomagnetic data preclude plate-tectonic models involving convergence of widely-separated cratons to explain all the Precambrian orogenic belts that have been examined. It remains possible though unlikely that the belts may result from opening and closing of small (less than 1 000 km) intercratonic basins, so long as the cratons returned to their original relative positions.

This idea that plates may split, move slightly apart to form a rift for sediment accumulation, and then move together to form a mountain chain is sometimes called the Wilson model after the man who suggested it. As discussed earlier, it has been invoked for the Caledonian and Appalachian orogenies, but seems to have limited success elsewhere. In contrast to McElhinny and McWilliams, Burke, Dewey and Kidd (1976) find some palaeomagnetic results that they claim are compatible with the operation of the Wilson cycle.

Plate-tectonic collisions cannot account for Precambrian orogeny, and the plate-tectonic mechanisms that are evident today must have originated at some time between the Late Precambrian and the Mesozoic.

The mechanism of plate tectonics

At least four mechanisms have been suggested for plate tectonics:

1. Plates are pushed apart by mantle upwelling. Magma forces its way into the ridge spreading sites, forcibly pushing the plates apart.

   One objection to this hypothesis comes from the mechanics of the system. Is it possible for a hot and liquid or viscous dyke to push aside a huge solid slab? Another objection is that it seems improbable that applied pressure at the back could push a slab only a few kilometres thick but thousands of miles wide.

   Yet another objection comes from the apparently permissive injection of dykes. In Iceland, which may be regarded as an accessible portion of a ridge, the combination of subsidence in the rift contemporaneously with spreading suggest that the dykes are permissively filling cracks that appear as the island is pulled apart (Decker and Einarsson, 1971). The dykes are not the driving force. Wherever dyke swarms associated with spreading are studied, as in Scotland (see p. 108) they are part of crustal extension and appear to be permissive.

2. Plates slide under gravity (Hager, 1978). Intrusion of magma at the

ridge creates a topographic elevation, and the trench provides a low, so there is a gradient between the two. The sea floor may slide down this gradient like a table-cloth over an inclined table. This hypothesis has the advantage of applying a body force to the sea floor rather than an applied force behind it, so movement of the large thin sheet of crust seems more plausible. The objection is that many parts of the ocean do not have a consistent direction, and features such as seamounts or ridges like the ninety east ridge which might be expected to provide a driving force quite fail to do so.

3. Pull by downgoing slabs. In models of island arcs and trenches with a downgoing slab, it is suggested that the descending slab sinks because it is colder and denser than surrounding rocks. It is further suggested that the downgoing slab pulls the sea floor behind it, so that it eventually follows into the subduction zone. The mechanism has been likened to having a table covered by a table-cloth, one corner of which dips into a bucket of water. The water makes the wetted cloth heavy, and the entire table-cloth may be pulled over the table towards the bucket.

Objections to this idea come from two sides. First there are mechanical arguments. Does the density difference of the presumed downgoing slab provide enough energy to pull half an ocean floor? Is the strength of the crust of the sea floor sufficient to enable the whole strip, thousands of kilometres across, to be pulled from one end?

The second objection comes from some of the more complicated plate-tectonic models that require flipping of subduction zones, formation of new sinking slabs, or descent of sea floor at a new location. In these models it would be impossible for the detached downgoing slab to continue to exert its influence, and no mechanism has been proposed to form new sinkers from normal sea floor.

It is also found that in some places subduction has consumed the entire sequence of sinker, half-plate and spreading site (Pitman and Hayes, 1968). A spreading site is the antithesis of a dense downgoing slab, and if it can be consumed the pull mechanism appears to be impossible.

4. Convection cells are frequently proposed as the driving force for continental drift and plate tectonics. Transform faults mark places where the convection currents would be flowing in opposite directions, and indicate that they are extraordinarily narrow compared to their length, not the broad flow we might imagine. In the convection currents we find in boiling jam or other analogues the change in pressure from top to bottom is trivial, but in the earth a convection current in going down has to go from a region of low pressure to high pressure, which is not easy to envisage. The only layer in the earth that seems capable of supporting convectional flow is the low-velocity layer or asthenosphere, so we have to have the convection cell confined to this zone, which is very thin (150 km) compared with its area.

Because there are problems with shallow convection currents several workers have suggested whole mantle convection, as much to explain earth magnetism as to solve plate-tectonic problems. The rather meagre evidence is summarized by Smith (1979).

The spreading sites themselves, migrate over the globe, so the migration of convectional cells must also be postulated. Convection is necessary if the earth is to maintain constant volume. Since the sea floors are clearly moving towards the continents, some material must be welling back to fill the space. There is no need to imagine a very direct loop and the counter-currents may be in complex and different directions, just as the counter-currents in the oceans do not all flow in direct loops beneath the surface currents. There is no necessity to think that the convection is the driving force of continental drift any more than thinking the oceanic circulation at depth drives the surface currents.

Difficulties and objections to plate tectonics

Because plate tectonics has been the ruling theory in earth science for the past two decades does not mean it provides all the answers, and there are several significant objections, and numerous places where plate tectonics fails to provide satisfactory models. Some discussion of these points will be provided later in the book, but some points may be noted now.

1. The length of spreading ridges is much longer than the total length of subduction zones. Why is this so? If the volume of the earth is constant, and assuming a continuity equation for flow, the rate of subduction must be very much faster than that of spreading. There must be some places where plate motion accelerates, but we have not found them.

2. Plate-tectonic theory does not explain why subduction is almost entirely around the Pacific, while spreading is in all oceans.

3. Since spreading is symmetrical around some continents (Africa, Antarctica) the mid-ocean ridges move over the globe. This conclusion is also inevitable when it is realized that some plates have increased in area by 50 per cent since the Mesozoic yet have no subduction sites, so their bounding spreading sites have moved away from the continent on both sides and therefore relative to each other.

   The fact that spreading ridges themselves drift is not inconsistent with plate tectonics, but it is a seldom-discussed aspect. What can make these passive, responsive ridges with their active volcanism migrate over the surface of the globe?

4. Sometimes evidence of plate movement is difficult to resolve with geological evidence. For example, Molnar and Wang-Ping (1978) present palaeomagnetic evidence that shows a Tibet–Asia convergence of 2 500 to 3 500 km, and since there is no evidence of a former sea floor in this position it seems that this large amount of north–south shortening must have taken place entirely in siallic crust. This seems hardly compatible with the little-disturbed Cretaceous limestones and sandstones that cover the Tibetan plateau.

5. Sometimes the evidence suggests that the plates are moving in two directions at the same time, which is impossible. The enormous volume of granitic rocks emplaced in western North America may suggest relative motion of the mantle, bringing in siallic rock to be remobilized. Westward drift of the continent over the Pacific (or underthrusting of the Pacific) might do this. But there is also the northwest displacement along the San Andreas fault. The two sets of movement required cannot both be attributed to the same simple cause, whether the cause be subduction, convection currents, lateral compression or whatever. Either we are dealing

with different directions at different crustal levels, or we are seeing two resultants of a very complex movement.

6. The Benioff zone is not complete in many places. In North America, for instance, the deep and even the intermediate earthquakes are missing. Although in some places the earthquake foci indicate an inclined plane, in many others the evidence is very vague.

7. The North America plate rides indiscriminately over the North Pacific (and other) plates with no regard to spreading sites, plate margins, or transform faults. Essentially the North American crust is maintaining its identity while crossing a very diverse substratum, and it does this without creation of any trench or Benioff zone. It would be simpler to say that the American plate is overriding the ocean plates to the west: subduction, with all its connotations, is a superfluous hypothesis in this situation.

8. It is not even certain that the plates do indeed act as single units. The number of minor plates or platelets seems to be increasing suggesting at least complications to the simple theory. Kahle and Werner (1975) have pointed out that gravity lows in the wake of drifting continents suggest movement of the continent over the sea floor, which cools when the continent has passed with a resulting increase in density. They liken the lithosphere passing over the asthenosphere to the movement of a flat iron over a blanket, and if this is so then the continent is *not* part of the plate it passes over.

9. Plate tectonics fails to explain mountain building. As already noted, the present Appalachian Mountains result from dissection of an uplifted planated area – the tectonic processes that folded and faulted Appalachian rocks might be explained in plate-tectonic terms, but the cause of the later uplift is not. Similarly many other mountainous areas – the Eastern Highlands of Australia, the Drakensberg of South Africa, the Sierra del Mar of Brazil – are not related to plate tectonics.

10. Plate tectonic mechanisms of mountain building depend on compressive folding. Later it will be shown that there are strong objections to compressive folding, and that most folding results from gravity sliding rather than compression. It remains possible that mountain building may result from such things as crustal thickening, underplating, and more remote manifestations, but if crustal compression is rejected, many of the more elementary concepts of mountain building by plate tectonics must also be rejected.

Plate tectonics and mineral deposits

Many attempts have been made to relate mineral deposits to plate tectonics, some with apparent success. It would be a mistake to think that this provides hard evidence with economic support for plate-tectonics ideas.

Virtually all attempts to relate plate tectonics and mineral deposits adopt a similar approach whereby major plate boundaries (spreading zones, subduction zones, transform faults, collision sites) are first described in broad terms, followed by descriptions of the mineral deposits thought to be associated with each plate-tectonic element. This produces a list of plate-related deposits which seems impressive in the number and diversity of the economic minerals. But when the list is considered against the total list of economic deposits it seems that apart from small and inconsequential deposits only porphyry coppers, volcanogenic sulphides and carbonatites have good plate-tectonic links.

On the other hand it can be shown that many deposits can *not* be related to plate-tectonic sites or processes, including red-bed Cu-U-V, Proterozoic banded iron formation, Kupferschiefer copper, Mississippi Valley Pb-Zn, various stratiform ores, and ores related to weathering, erosion and deposition. Of course, many mineral deposits are Precambrian and were perhaps formed in pre-plate-tectonic ages, so the argument should be restricted to Late Precambrian and Phanerozoic deposits. Even so, the fact that many major deposit types occur *on* the continental plate and not at the margins suggests that plate tectonics is only of limited use in understanding the origin and distribution of mineral deposits.

5 Structural and tectonic landforms

In preceding chapters we have discussed some of the major features of the earth, its rocks and its structural and tectonic features. This chapter is concerned with the details of the structures, their erosion, and what they look like as landforms. While much of the book is concerned with first- and second-order landforms, this chapter will be much concerned with third-order landforms.

Fold geometry

Stratified rocks may remain horizontal, in which case differential erosion creates landforms after possible uplift (see p. 162) or they may be folded.

The nomenclature for folds is shown in Fig. 5.1. Arch-shaped folds

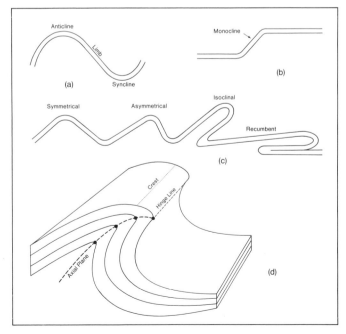

Fig. 5.1 (*a*) Terminology of a simple fold. (*b*) A monocline. (*c*) Types of folds defined by attitude. (*d*) Axial plane, hinge line and crest.

are called anticlines, downfolds are synclines. The middle part between the top of an anticline and the bottom of a syncline is the limb of a fold. An anticline may be symmetrical, the limbs having equal dip, asymmetrical or isoclinal when opposite limbs are parallel.

The line along which a change of dip takes place is known as the hinge line, and often coincides with maximum curvature. A plane which joins the hinge line of successive beds in a fold is the axial plane. Fold axis is a term used in different ways by different writers, but is roughly a line parallel to the hinge, or the intersection of the axial plane with a bed.

The crest line marks the highest points on the same bed in an anticline. With a vertical axial plane the crest is coincident with the trace of the axial plane, but with an asymmetrical fold with an inclined axial plane the axis and crest will be parallel but not coincident.

The axial plane can itself be curved.

Similar terms can be used to refer to synclines, with trough corresponding to crest.

In a plunging anticline or syncline the limbs will converge in plan (Fig. 5.2), even if the fold is symmetrical.

Fig. 5.2 (*a*)   Converging limbs in a plunging anticline. (*b*) Converging limbs in a plunging syncline. (*c*) Concentric folds diminish downwards. (*d*) Similar folding. Material must flow from the thinned limbs into the thickened crests and troughs. (*e*) Culminations and depressions.

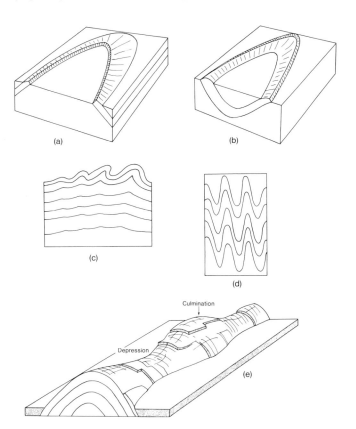

If the axial plane bisects the fold, the fold is said to be symmetrical or upright. If the axial plane has a dip, the fold is inclined or asymmetrical. If the fold axis is near horizontal, the fold is said to be recumbent: recumbent folds are usually on a large scale. Monoclines are small steep parts in generally horizontal strata. They often overlie a fault at depth and mark an uplift that has not yet turned into a fault.

Cylindrical folds are those whose profile is essentially semicircular and remains constant when traced along their axes. A parallel or concentric fold is one in which the successive folded beds have a constant centre and increasing radius. More common are similar folds, in which many successive beds have almost identical shapes. To create this kind of folding the intervening beds have to be squeezed out from the limbs of the fold to accumulate in trough and crest positions. The beds with constant thickness are said to be competent; those that flow are incompetent.

Angular folds have straight limbs and sharp hinges, often a point of fracture. The axis of a fold may vary along its length, with culminations and depressions. A fold nappe is a recumbent fold in which the middle limb has been sheared out completely.

Domes are antiform structures with dips in all directions and often no distinct fold axis. A basin is the synclinal equivalent.

Fold geomorphology

Primary fold relief appears to be rather rare. A possible example is the Rough Range in northwestern Australia, where the hills certainly correspond with anticlines, and it seems possible that the actual ground surface is being folded. Another example is the Moorhead anticline (p. 171).

In New Zealand several active anticlines warp the coastal plain near Wellington (Te Punga, 1957). The anticlines are folds in the Hawera series of Late Pleistocene age, and all have direct geomorphic expression. The Mt Stewart–Halcombe anticline is a typical example. It is 11 km long with an even crest. The western flank slopes west at 2° and the eastern flank slopes east at 6°. The surface of both flanks is parallel to the dip of the underlying beds, and the arched surface is a faithful replica of the underlying fold. Along the sides of the anticline the drainage flows in primary synclinal courses, with tributaries draining the dome, including marked radial drainage at the ends of the dome. The crest of the anticline is undissected, and there are several undissected interfluves of several square kilometres.

The larger Pohangina anticline can be traced for about 45 km and rises to 350 m, and where highest it is well dissected, illustrating that erosion will often keep pace with tectonic uplift. Each anticline is related to the edge of a fault block in the underlying greywacke. They may be termed drape folds.

Some of the valley anticlines and similar gravity structures described in Chapter 9 are also original surface folds, but they result from surface processes. Most folds that are exposed as landforms were folded under a deep cover at some time after deposition of strata, and the folds we see are exposed by erosion, usually following uplift.

Even amongst erosion-exposed folds, anticlinal hills are not common, although the Jura mountains contain many anticlinal hills that are scarcely breached. The crest of an anticline, once exposed, tends to be soon eroded and the commonest landform assemblage associated with folding is a breached anticline or dome. The strike ridges on each side have typical dip and scarp slopes (see p. 162), a typical drainage pattern, and between streams that cross the strike the dipping strata have the typical form of flatirons, triangular facets with their bases parallel to the strike and their apices pointing up the dip of the rock. With horizontal folds the strike ridges are very long; with domes the strike ridges form concentric rings. Structural basins are also surrounded by strike ridges, and the flatirons point in the opposite direction.

Individual landforms then are flatirons or facets, scarps and dipslopes, breached anticlines, infacing and outfacing escarpments. If the folded rocks have been through a period of planation there may be bevelled cuestas which shows that the landforms result from differential erosion and not from simple folding of surficial strata.

In general it seems that most folds are formed by tectonic processes affecting strata long before they reach the ground surface. Many folds are created soon after the strata are deposited. There are few processes that form folds at the ground surface. If folds *are* created at the surface, the crests are eroded first, possibly while the fold is rising, and erosion does not wait for the creation of the complete fold before destruction sets in. Domes are reasonably common as surface features.

Laccoliths, formed by igneous intrusion doming up the overlying strata, may give rise to surficial domes. Salt domes erupt at the surface, and may even be breached by a salt glacier.

Gneiss domes are generally thought of as deep-seated structures, but Ollier and Pain (1980) have shown that some domes in Papua are surface features (p. 101) and these still have the form of domes, about 3 km high and 15 km across. Although deeply dissected they still have facets like flatirons, some individual facets being 2 500 m high. The gneiss lineaments are concentric with the outcrop of the dome.

The term basins refers to structural features formed by deformation of bedded sediments, and also to major topographic depressions such as the Kalahari Basin (Fig. 6.5, p. 79).

On the broader scale the ground surface may be warped up to form major swells like those described in Chapter 6, which disrupt drainage as explained in Chapter 12. Monocline is also the name given to the warping of a planation surface into a steep zone between two fairly horizontal stretches. They seem to be especially common on continental edges, like the Natal monocline in South Africa.

Where two monoclines converge an area of extra uplift is produced. In Australia the north–south monocline of New South Wales meets the east–west monocline of the Victorian Highlands, and in the corner is Mt Kosciusko, the highest mountain in Australia. In South Africa similar conflicting monoclines run from Algoa Bay past the Tandjesberg to the Compassberg, the highest point in Cape Province. Likewise in Brazil, two similar monoclinal flexures converge to Cabo Frio and the same uplift continues to Pico da Bandeira, the highest point in Brazil. As it happens these three examples have precisely the same orientation.

Fault geometry

Faults are fractures through rocks, with differential movement of the rocks on opposite sides of the crack. There are several kinds (Fig. 5.3).

Normal faults are steep or vertical, and the blocks on opposite sides have moved mainly vertical. If the fault plane has a slope, the angle between the fault and vertical is the hade of the fault. The fault block originally above the inclined plane is the hanging block, and the trace of the fault on this block is called the hanging wall of the fault. In normal faulting the block that was above the inclined fault plane moves relatively down. Consideration of the geometry of the situation suggests that this can happen only in an extensional environment.

Thrust faults come in two kinds: high-angle thrusts and low-angle

Fig. 5.3 (a) Normal fault. (b) High-angle reverse fault. (c) Thrust fault.

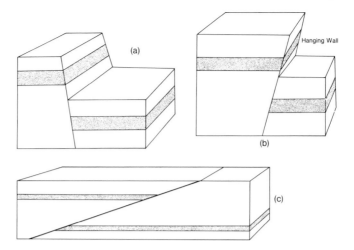

thrusts. Both are indicative of local compression. High-angle thrusts look rather like normal quilts, but the sense of movement is reversed, with the hanging block rising. They are sometimes called reverse faults. In low-angle thrust faults too, the upper block moves up relative to the lower block. These are often of very low angle, and very large size, and merge into nappe faults.

Strike-slip faults (also known as transverse, tear or transcurrent faults) are nearly vertical, faults along which the opposite sides have moved in opposite horizontal directions parallel to the fault plane. Transform faults are a special case of strike-slip faults, as described on p. 18.

Strike slips are called dextral (right lateral) or sinistral (left lateral) depending on the sense of movement. Standing on one side of the fault and looking across it, if the far block has moved right then it is dextral fault. The sense of movement should of course be consistent if one piece of country is moving past another, but sometimes the local evidence seems to conflict, and opposite directions of movement have been reported from different parts of the same major fault.

It is clear that two factors are involved in fault description: one is the attitude of the fault, whether near vertical or near horizontal, and the second is the direction of movement on the fault plane.

The directions of movement of the blocks are shown in Fig. 5.4. If the movement is down the dip of the fault plane, it is a dip-slip fault. If movement is along the strike of the fault plane it is a strike-slip fault. It is also possible for a block to move down and sideways, when it is called an oblique-slip fault. The blocks between separate faults may rise or sink relative to other blocks. A high block between two sunken blocks is a horst. A sunken block between two high blocks is a graben. A block that has been tilted rather than simply raised as a whole is called, appropriately, a tilt block. Rift valleys are a special kind of graben of large size and global implication, as described in Chapter 6. Thrust faults occur where a surficial mass is spreading, like the Uinta Dome (Fig. 5.5), as nappe faults which are many kilometres long, and as deep-seated thrust faults.

Many thrust faults are topographically inactive, and only differential erosion causes any landform effect. A few others appear to be mobile. Many of the gravity-slide nappes (see Ch. 9) are riding on a layer of easily-flowing material such as salt, dolomite or suitable clays.

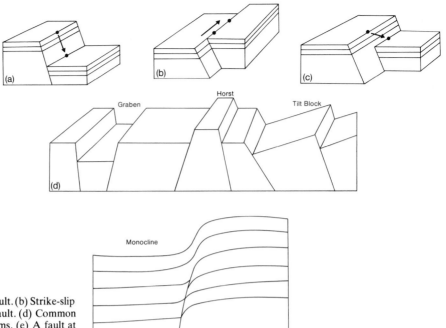

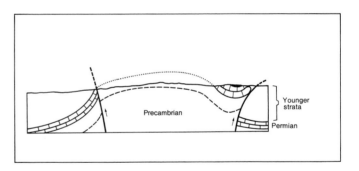

Fig. 5.4 (a) Dip-slip fault. (b) Strike-slip fault. (c) Oblique-slip fault. (d) Common fault-generated landforms. (e) A fault at depth giving way to a fold at the surface.

Fig. 5.5 Diagrammatic section across the central Uinta uplift, Utah, showing the outward spread of the uplifted Precambrian along thrust faults.

The rate of movement of nappes, estimated at 10 000 B, suggests that they might still be moving at the surface but rates of erosion could probably keep up with them. Deep thrusts in the earth grind rock along the fault plane into a powder that is fused together by frictional heat to produce a dense rock called mylonite. Thick mylonites – hundreds of metres – are particularly common in Precambrian rocks. Such faults have no marked geomorphic effect, other than guiding differential erosion to some extent. The Aswa River of northern Uganda follows a particularly thick mylonite bed and has a very straight course.

Fault geomorphology
Normal and reverse faults

Normal faults frequently break the ground surface, displacing parts of the landscape such as erosion surfaces and valleys, and on a small scale breaking river terrace deposits and turf.

Examples of small-scale faulting have been described from Utah, where the Wasatch Fault had produced minor tectonic landforms in young glacial and post-glacial deposits. Fresh fault scarps are up to 20 m high, and other surface manifestations are multiple-fault

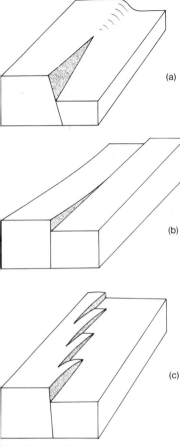

Fig. 5.6   (a) Hinge-fault passing laterally into a monocline. (b) Scissors fault. (c) En echelon faults.

scarplets, graben, rotation of blocks, triangular facets along scarplets, lines of springs where ground water utilizes fault planes, and slickensides (parallel scratches and polish) on fault surfaces. The most obvious features produced by faults are fault scarps – steep slopes parallel to the fault trace on the ground, on the upthrow side. These tend to be dissected by valleys, and divided into triangular scarp facets. After an earthquake fresh scarps, called scarplets if they are especially small, may be seen, but these weather in a few years or less and lose their freshness. A fault scarp hundreds of metres high would have taken a long time to form, and would be increasingly weathered towards the top, unless erosion has exposed fresh rock.

In plan normal faults are usually fairly straight, though they may be curved or sinuous. There may be several faults associated together, generally roughly parallel since they formed in response to the same stresses, and they may split and join to some extent.

Faults eventually die out at the ends, so they have maximum throw somewhere in the middle. At the ends they may merge into folds, especially monoclines. In this case there is said to be a hinge where the fault joins the monocline. In yet other cases the fault may show opposite sense of uplift on opposite sides of a pivot, to produce scissors faulting (Fig. 5.6). If faults die out to be replaced by parallel ones that are slightly offset, it is possible to produce en echelon faults.

Of course a fault formed at the ground surface will have a scarp on the upthrown block, facing towards the downthrown block. This is a fault scarp. Eventually erosion may remove all trace of the original scarp, but so long as rocks of different hardness are present on opposite sides the position of the faults is likely to be preserved by differential erosion. It may happen that a new scarp will be produced, but if it is formed by erosion rather than direct tectonic action this should be called a fault-line scarp.

Fault scarps tend to be destroyed by weathering and erosion on the upper slope, and the lower part tends to be buried by debris derived from upslope. Because of this it is quite often difficult to find an actual outcrop of the fault plane itself, even in a place like a rift valley.

The Sierra Nevada, California, is a gigantic tilt block, with a gentle slope to the west and a dramatic fault scarp on the east. In rift valleys the emphasis is on the grabens that commonly run along the crests of continental arches, though in some instances the rise to the rift is some distance from the faults (see p. 175). Rift valleys are commonly bounded by high land, as the Vosges and Black Forest border the Rhine graben, and in places there are great topographic contrasts, like the Albert Rift Valley which has a base (below lake sediments) about 1 000 m below sea level and is bordered by Ruwenzori reaching 5 118 m. In the Tien Shan, the Turfan Depression is about 200 m below sea level, and is bordered by a peak nearly 7 000 m high.

Although dominated by normal faults, rift valley faults include strike-slip faults, and a few high-angle reverse faults, and in some instances chasmic faults (see p. 71). The rift system may be explained by vertical tectonics, with uplift forming a broad arch, with the rift giving more localized compensating depression; simple tension, which accounts for the rifts but not the bordering arches or high-fault blocks; compression (based on an improbable explanation of gravity data); and strike-slip faulting, which undoubtedly occurs in some rifts such as the Dead Sea but does not itself explain the rift form.

In some places there is a broad area of fault blocks rather than a

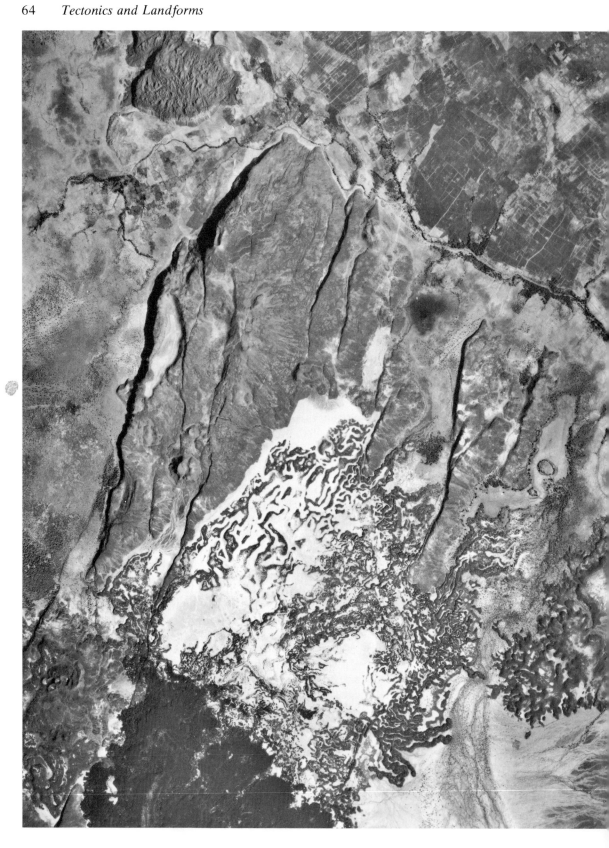

Fig. 5.7 Air photo of part of the Afar
triangle, showing faults and associated
lava flows. (Supplied by the State
Rivers and Water Supply
Commission, Victoria – part of an
aerial survey carried out for the Food
and Agricultural Organization's
Awash Valley Development Project –
Ethiopia, 1972).

linear system of rifts, and this is called fault-block topography or fault-block landscape.

Fault-block landscapes have been classified into types, using particular geographic examples to provide the name (Bloom, 1978).

Oregon-type fault-block landscapes This consists of originally flat-lying rocks, which in the type area are lava flows. There the faulting is so recent, the erosional and depositional modification so slight, that the region displays almost block-diagram simplicity.

Steens Mountain, Oregon, is a single fault-block of Pliocene lavas with gentle tilt, bounded by fault scarps to east and west, and represents the simplest and most obvious fault-block mountain of the horst type. The Grand Teton mountains are an example of a deeply eroded horst.

The Afar triangle in Ethiopia consists of fault blocks of young lavas which could be classified as Oregon type (Fig. 5.7).

Basin and range fault-block landscape In this type the bedrock has been folded and metamorphosed, though planated and perhaps capped by lava before faulting produced the present tilt blocks. This is associated with considerable shearing, but more importantly with crustal extension, a feature common in areas of block faulting and rift-valley swells. Very good evidence of extension is provided by several elliptical to circular fault-bounded depressions in Oregon such as Summer Lake and the Upper Alvord playa, described by Fuller and Waters (1929). If they were due to compressional faulting then the forces must have acted centripetally like the closing of a camera shutter. Seemingly a dome would have been a more logical structure under these conditions. Upper Alvord playa is less than 5 km across yet the bounding fault scarps have been thrust up over 300 m; the exact mechanism is not clear, but some sort of downfaulting of the depression during general extension seems the most plausible. The geomorphology of the classical Basin and Range province of the United States is complicated by aridity and the deep accumulation of sediments in the basins, so another type has been defined where this factor is absent.

New Zealand type fault-block landscape This differs from the previous types in having through-flowing rivers for the most part, so a fluvial geomorphic system affects the fault block which does not accumulate debris. The block-fault topography of the Melbourne region and the Adelaide region in Australia could be regarded as this type.

In the Wellington area of New Zealand Fig. 5.8 shows some of the

Fig. 5.8 Diagram to illustrate movement of tilt blocks in the Wellington region, New Zealand, that accompanied the 1855 earthquake. Note the dimensions are in feet (after Stevens, 1974).

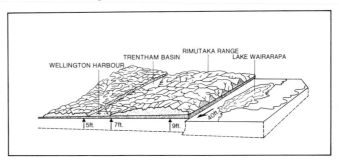

fault blocks, and the differential movement that accompanied the 1855 earthquake. These faults moved obliquely, actually moving laterally more than vertically.

The range of possible types is great, and further examples might be needed to account for fault-block topography in the humid tropics as in parts of New Guinea or the New Hebrides, in glacial areas and so on. Suffice it to say that the geomorphic features depend on the nature of the bedrock, the nature of the faulting, and the relative accumulation or removal of sedimentary debris.

Crustal extension

The normal faulting associated with most fault-block mountains is generally taken to be a sign of crustal extension, and indeed it is possible to calculate the amount of extension in some instances. For the Great Basin between the Wasatch Range and the Sierra Nevada Eardley estimated that the faults imply an extension of 50 km during the last 15 m.y. (300 B). The Rhine graben has extended 4.8 km in 45 m.y. (about 100 B) according to Illies (1972). Crustal extension in the Rio Grande rift near Albuquerque amounted to 83 km in 26 m.y. according to Woodward (1977), a rate of 300 B, an estimate based on interpretation of a decrease in the dip of faults. It should be remembered that these extensions may be confined to the uppermost 5 or 10 km of the crust and may not be associated with extension at deeper levels. Fig. 5.9 shows four possible models relating block faulting to regional displacement, and although they all indicate extension at the surface only one involves extension throughout the crust. Such tensional areas with normal faults that flatten at depth are rather like landslide faults, and indeed landslide slip planes grade in scale to faults as described in Chapter 9. On the edge of fault blocks it is not always clear whether a given fracture should be regarded as a fault or the slip plane of a landslide.

Fig. 5.9 Possible relationships between block faulting and deeper displacement: (*a*) Block faulting in response to crustal extension and thinning of the lower crust. (*b*) Normal faulting in response to crustal uplift, by large-scale landsliding. (*c*) Rift valley formation along the crest of a regional upwarp. (*d*) Rift valley formation in response to downward vertical movement at depth (after Hobbs *et al.*, 1976).

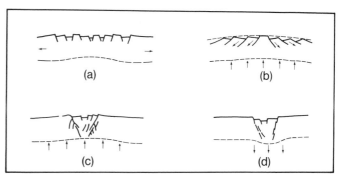

Elsewhere vertical uplift may not be accompanied by any regional extension. The Colorado Plateau is a vast area that was uplifted about 2 000 m, bounded by faults and monoclines, and too large to be explained in the manner of the tilt blocks of the Basin and Range province. To the north are the Uinta Mountains, a plateau-like uplift about 70 km across, bounded by steep *reversed* faults and deep sedimentary troughs (Fig. 5.5). The total vertical movement is about 10 000 m, accompanied by very considerable spreading of the upper part of the plateau, which must be distinguished from regional extension.

Fig. 5.10 shows the separation of collapse blocks at the front of a rising thrust fault. It has been argued that the development of

Fig. 5.10 Collapse blocks at the front of a rising thrust fault partly supported by an apron of smaller debris (after Fuller and Waters, 1929).

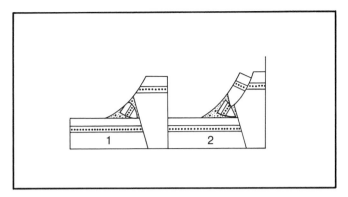

step-faulted blocks of large size requires that a great overhanging scarp shall remain suspended and unsupported during emergence until a large-scale landslide breaks away. This seems improbable, and it is more likely that support is provided by an apron of debris which is overridden by the thrust block until landslide faulting takes place.

Fig. 5.11   Tertiary evolution of the Niger Delta, illustrating the formation of faults during deposition (after Evamy et al., 1979).

The Niger Delta is much faulted, with some major faults that define distinct provinces within the delta region and control sedimentary deposition to a large extent, and many smaller faults (Fig. 5.11). These

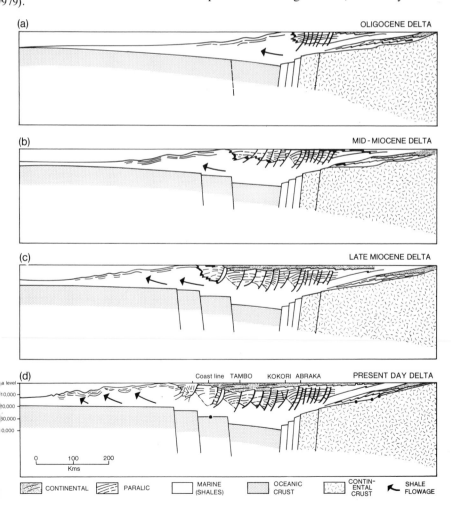

are significant because they demonstrate that not all faults are associated with topographic features on the land – some major faults never did produce fault scarps on the ground surface and by analogy wherever we see faults in old delta sediments in older rocks we should not think of faulting as being related to younger tectonics and uplift for ridge crests may be penecontemporaneous with the deposition as in the Niger Delta.

Strike-slip faults

If a strike-slip fault affects rugged country, many fault scarps will result. All the ridge crests on one side of a fault will be displaced in the same direction. When they move about half the distance between ridges they effectively block off the valleys (Fig. 5.12), and have been termed shutterridges (Buwalda, 1936).

Fig. 5.12   Formation of shutterridges by strike-slip faulting across the grain of the topography.

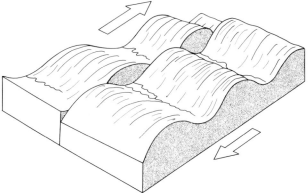

Transcurrent faults offset streams, ridges, moraines, and terrace escarpments, and if the topographic features are sufficiently sharp the amount of movement, both lateral and vertical, can be determined. For example, the Branch River fault in New Zealand (Fig. 5.13) is clearly a major transcurrent fault, but it also has a vertical component so is really an oblique fault (Lensen, 1968).

Strike-slip faults are often in very long complexes. They tend to be followed by rivers so the faults themselves may be obscure, but some, like the San Andreas fault, make very clear surface indications of their presence by faulting highways, fences, and orange groves. There are many splinter faults, and along the zones of transverse faults there may be blocks of highland and lowland, with some slices of rock raised as part of the space accommodation problem that arises when faults that are not perfectly straight slide past one another.

Strike-slip faults have often moved many hundreds of kilometres and to get the amount of movement it is necessary to look at major geological outcrops rather than geomorphic features to work out the amount of movement.

The Great Glen of Scotland was interpreted by Kennedy (1946) as a strike-slip fault, mainly on the evidence of the offset of the Strontian and Foyers granites (Fig. 5.14). This has been disputed, but if correct indicates a lateral movement of 105 km. If it is a transverse fault, it is an old one, and there is no modern movement. It is a marked topographic feature, but the topography is brought about by differential erosion.

The Alpine fault of New Zealand (Fig. 5.14) is a dextral fault or

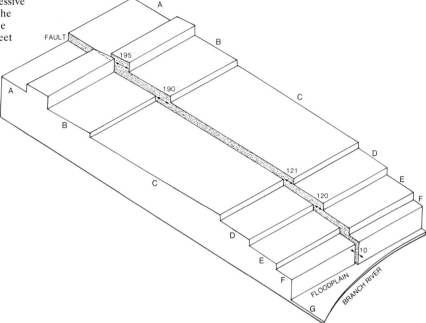

Fig. 5.13   Diagram to show progressive offset of terraces and channels of the Branch River, New Zealand, by the Wairau Fault. Dimensions are in feet (after Lensen, 1968).

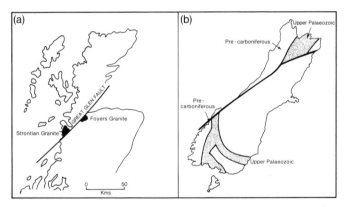

Fig. 5.14   Major strike-slip faults. (*a*) The Great Glen Fault of Scotland. (*b*) The Alpine Fault of New Zealand.

which major structures have been displaced by about 500 km (Wellman, 1955). The main Alpine fault is virtually a line, but at the south and north it splits into a splayed mass of faults bounding fault blocks. The New Zealand Alpine fault gives an average rate of movement since the Jurassic of 3 000 B and in recent times is moving at 9 000 B.

The San Andreas fault is a major fault zone, and relates to sea-floor spreading and transform faults in the Gulf of California. The fault is still very active. The total length is over 1 000 km and a displacement rate of about 3–5 cm/y (40 000 B) has been determined by geodetic surveying since 1970. To move 1 000 km at this rate would take only 25 m.y.

When the San Andreas fault system of California and the Alpine fault system of New Zealand are compared (Fig. 5.15) the main features are remarkably similar, suggesting a genetic relationship. Scholz (1977) suggests that in both situations the normal mode of slip is of relatively frequent large earthquakes. Great earthquakes occur

Fig. 5.15 Fault maps of southern California and South Island, New Zealand rotated so that the slip vectors are parallel for the two regions (after Scholz, 1977).

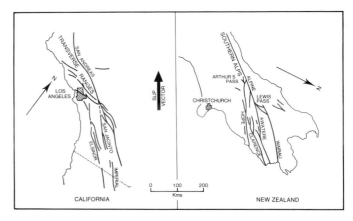

only where the effective normal stress on the fault is unusually high because the fault is oblique to the regional slip vector, and that fault creep occurs only when the effective normal stress is very low.

Strike-slip faulting may also lead to vertical uplift of local areas, a model that Kingma (1958) has suggested for the Southern Alps of New Zealand (Fig. 5.16). Strike-slip movement may lead to alternating areas of tension and compression, and the compressed areas may reach equilibrium by vertical movement. Likewise, Ruwenzori, Uganda, is bounded on almost all sides by fault, but on the southeast side the erosion surface of the neighbouring plain is simply warped up.

Fig. 5.16 Strike-slip faulting may cause alternating areas of tension and compression, compensated by vertical movement (after Kingma, 1958).

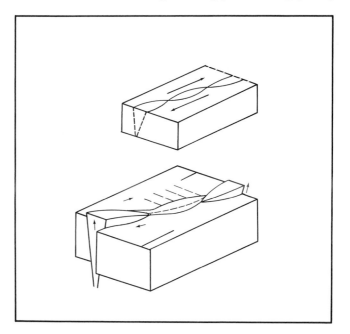

Surface low-angle thrusts

Low-angle thrust faults are usually less conspicuously expressed in the topography than are high-angle thrusts. The front of the overthrust block tends to be jagged and irregular, with embayments brought about by initial topographic irregularities and later slumping of unsupported masses and subsequent erosion.

The Tawonga Fault (Fig. 5.17) is the only proven post-Mesozoic thrust fault in the Eastern Highlands of Australia. The southern block has moved from southwest to northwest along this fault in Tertiary or Quaternary times, thrusting a block of Palaeozoic rocks over a boulder conglomerate, which is interpreted as alluvium (Beavis, 1960). Since the alluvium under the fault is continuous with that currently being deposited the latest movement cannot be older than Pleistocene. Any suggestion that the boulder conglomerate might be a Permian glacial tillite can be discounted by the presence in the conglomerate of boulders of Tertiary basalt, the rounded nature of the boulders, and the virtual absence of clay-size particles from the matrix.

Fig. 5.17   Map and tunnel section across the Tawonga Fault, Victoria (after Beavis, 1960).

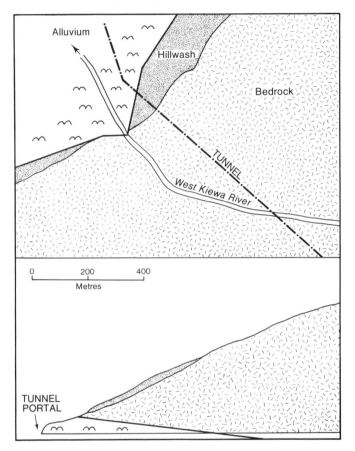

According to Beavis the movements along this low-angle thrust fault have resulted in the elevation of the upthrust block by 600 m, the difference in average elevation of the country on opposite sides of the fault – 1 800 m to the southeast and 1 200 m to the northwest.

If ideas of gravity tectonics are correct it would seem that low-angle thrusts should be fairly common, but they appear to be rather rare. This suggests that many thrust faults may be created in basins of deposition rather than as terrestrial landforms.

Chasmic faults

When Carey (1958) proposed the formation of sphenochasms and rhombochasms by crustal separation, he did not coin a name for the faults that bound the crustal fragments. It is useful to have a term for such features, and Osmaston (1973) proposed chasmic fault, 'to define any major age discontinuity extending through the lithosphere'. Each such 'fault' might be single or a complex fault-zone depending upon the tectonic details of early crustal separation. The search for fits between continental shelf outlines has as its purpose the delineation of corresponding pairs of chasmic faults. The jigsaw pieces that were created by continental drift are bounded by chasmic faults.

The Dead Sea appears to be a site of crustal spreading as well as strike-slip motion (Fig. 5.18). A neat demonstration of the sideways movement is provided by the absence of deltas at the mouths of the Wadi Zarqa and the Wadi Mujil, and yet the presence of a delta with no adequate source in the El Lisan peninsula. If the delta was built by these wadis in former times, then there has been about 40 km movement since they parted.

Fig. 5.18 The Dead Sea rhombochasm, diagrammatic and more realistic.

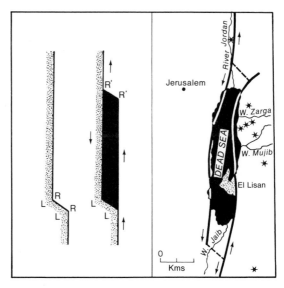

Using this and other evidence Quennell (1959) found that lateral movement in the region amounted to 107 km, of which 62 km was from Early Miocene to Early Pliocene, and the second phase of 45 km movement was from Middle Pleistocene to the present, and is still going on.

The Dead Sea rhombochasm was not filled with sea-floor material, and there is no associated vulcanicity.

The best known place where splitting happened recently is the Red Sea–Gulf of Aden region which split about 25 m.y. ago. The continent on each side is raised about 1 000 m above normal, with deposition of much sediment in the graben.

As the splitting starts, the continental crust is heaved up by the hot mantle coming up beneath it, not because it is pushing upwards but because, being hot, it is less dense and so buoys up the crust. Oceanic crust begins to form in the crack, but because it is narrow, and because of the elevation of the surrounding land, this is soon covered with sediment. If the climate is warm, any sea that enters the basin may evaporate to leave salt. This is especially so if the basin is closed or nearly closed at both ends like the Red Sea, not so good if one end is open like the Gulf of Aden. Erosion of the elevated continental margins will spread sediment on top of the salt. Thus we find a model for the break-up of Pangaea in the Mesozoic.

6 Cratons, fold belts, rifts and lineaments

To explain some of the classical terminology in tectonic geology we shall first go through a simplified account of the history of a mountain belt. Figure 6.1 shows the basic division between continent and ocean. The ocean crust bounds a thicker slab of siallic material known as a 'craton', 'shield', or sometimes simply as 'basement'. Erosion of the continent produced sediments which are deposited in the sea, mainly as a wedge or prism of sediments against the continental slope, but extending to some extent onto the craton as platform sediments.

The continental wedge is that part of the sediments on the continent, but big deltas like the Mississippi and Niger can build out onto oceanic crust. The sedimentary prism may be complicated in various ways, but approximates to a geosyncline (see Ch. 13), and is the site of a future 'fold belt' or 'mobile belt'.

Compression from the ocean side of the prism folds the sedimentary rocks in a belt of *foreland folding,* and oceanwards of this is a belt of *thrust faulting* and the formation of *nappes*. Somewhere at the back of the thrusts and nappes is a 'core zone' or 'root zone' of highly metamorphosed rock, thought to represent the most squashed and altered part of the mobile zone. The sediments that were derived from the continent are now apparently pushed back onto it. At the same time there is intrusion of granite deep in the folded sediments, and uplift of the folded rocks to form a range of mountains or an orogenic belt.

In the later stages the lower part becomes very hot, and the prolonged period of unrest culminated in one or more violent episodes of mountain building or orogenesis during which the narrow mobile belt, already distorted by long subsidence and beginning to be softened by heat, is squeezed between the stronger and more stable forelands. The geosynclinal filling and its basement are folded, thrusted and may begin to undergo metamorphic changes and become partly molten. Eventually the whole deformed belt is heaved up to produce a range of mountains. The molten material generated deep in the crust is squeezed up through the folded rocks to produce intrusions of granitic composition (Read and Watson, 1966). After this deformation (and supposed crustal shortening) and uplift comes a period of rapid

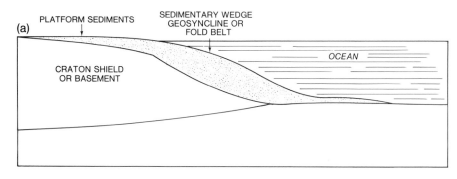

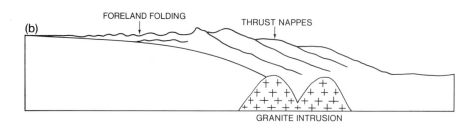

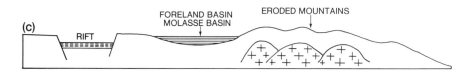

Fig. 6.1 Some tectonic terms illustrated.
(*a*) The relations of craton, ocean,
platform and sedimentary wedge.
(*b*) Foreland folding and nappes result
from a push towards the continent, and
folding may be accompanied by intrusion
of granite. (*c*) The mountains are eroded,
and sediment deposited in basins at the
foot of mountains. The foreland may start
to break up by faulting.

erosion. The derived sediments, coarse grained and occasionally
accompanied by volcanic contributions, are deposited in depressions
next to the main orogenic belt. The depressions are known as 'foreland
basins' or 'molasse basins', and may extend almost as a moat in front of
the mountains. Eventually the mountains become thoroughly worn
down, and renewed platform sedimentation takes place.

In this late stage there may be a development of faults, rifts, graben,
ring dykes, and features associated with late granites (post-orogenic
granites). In Australia this phase is called transitional tectonism. The
whole process leads to the welding of a belt of altered sediments to the
craton, a process known as 'cratonization'. Some people use the term
orogenesis to mean the adding of a welt of new rock to a craton. In the
jargon of tectonics 'orogenic provinces' become cratonized and turn
into 'platform provinces'.

The Precambrian basement of most Phanerozoic fold belts consists
largely of highly metamorphic rocks, with gneiss being very common.
The structures have long been thought comparable to core zones of
younger orogenic belts, and so the Precambrian structures were
thought to be the worn down stumps of old 'orogenies'.

Eventually the cratons or shields break up, and drift apart. This
process may be heralded by doming, but certainly the continents
develop rifts, which grow wider and lead to separation. The present
continental rifts mainly take the form of rift valleys. When continents
drift apart they have a leading edge and a trailing edge which have very
different properties that affect their geological history, geophysics and
their landforms.

Shields and cratons

The term shield is applied to those areas which have not been folded or complexly deformed since the end of the Precambrian. Exact margins are not easily determined as the edges are usually covered by thick, often undeformed sequences of sedimentary rocks.

Shields are subject to erosion, but contain very few mountains. Most are well above sea level, but physiography is gently rolling, with some differential erosion but no volcanoes, earthquakes or active deformation.

Craton is the name given to the central stable portion of a continent. Cratonization is the conversion of mobile belts around the edge of cratons into part of the craton.

The Archaean (pre-2 500 m.y. old) regions consist of two types of rock: high-grade metamorphic rocks such as granulite and gneiss, and low-grade metamorphic rocks, the greenstones, derived from volcano–sedimentary rocks. On the eroded stumps of the Archaean, the Proterozoic rocks were deposited. These consist of vast thicknesses of volcanics and sediments. Cratons were by now stabilized, and some of the great fracture systems that might affect future crustal development had already appeared. Asymmetrical fold belts were developed in some areas, and may be interpreted in terms of plate tectonics, or other mechanisms that prevail in later times. The major break in tectonic processes is between the Archaean and the Proterozoic rather than between the Proterozoic and the Phanerozoic.

A shield may consist of several cratons, together with other ancient rocks. Geologically Africa consists of three old (over 1 100 m.y.) cratons surrounded by non-cratonic regions with a dominant age of 540–650 m.y. known as the Pan African crust. The Pan African crust is about 35 km thick with typical siallic density, and consists mainly of granite, granodiorite, sediments and volcanics. It has been unaffected by compression throughout the Phanerozoic and is typical continental crust.

Miyashiro (1972) has made a case for supposing there should be no significant difference between the depth of erosion between Precambrian shields and Phanerozoic fold belts. In this case the absence of blue schists from the Precambrian cannot be an erosional feature, and the restriction of granulite exposures to Precambrian shields suggests that metamorphism was more intense and geothermal gradients steeper in the earlier parts of the earth's history.

Fold belts

A very obvious second-order feature of the globe is the mountain range. Although as we shall see these ranges were not formed by the folding of rocks, the old idea that crustal shortening and folding created mountains has not been dispelled, and these ranges are still referred to as fold belts.

Fold belts of different age are known, and are often referred to as 'orogenies', with the implication that the orogeny or folding can be related to a specific time, the date of which is later than the sedimentation and younger than intrusions that follow the folding (noting however that some folding may be almost contemporaneous with sedimentation, as may some intrusion).

What is clear from any map is that mountain ranges, or fold belts, tend to be long and thin. These may be related to continental borders, to geosynclines, or to island arcs, which are topics discussed later. (Ch. 13, 14). Mechanisms of folding the fold belts will also be discussed later (Ch. 9 and 19), so the subject will not be pursued further here.

Cratonization

Cratonization is the name given to the process whereby sediments and volcanic rocks deposited at the edge of a craton become welded to the craton, becoming part of it. Cratonization may take place in various ways, such as subduction, accretion, or compression and granitic intrusion. With subduction, material is added to the base of the crust so is not evident on maps of rock distribution, but if it is accreted in some way the newly-added border to the continent should be apparent. If this process were repeated, cratons should grow by the addition of new concentric rings of crust, and the idea arises of growing continents. Only North America really lends much support to this idea (Fig. 6.2).

Fig. 6.2 North America with bands of younger rock accreted around a nucleus of older rocks, a possible example of cratonization (after Gastil, 1960).

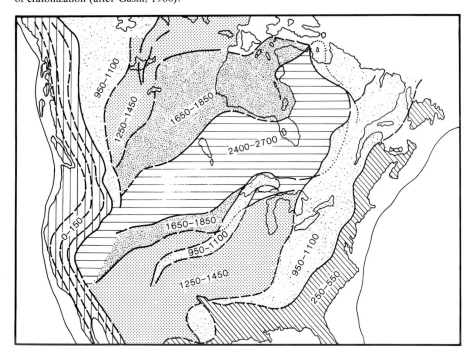

Successive accretion on the western margin of the northwest United States may account for the distribution of the five ophiolite belts (Fig. 6.3) which range from Early Palaeozoic to Jurassic in a belt less than 200 km wide (Coleman and Irwin, 1974).

Cratonization provides the answer to one geological dilemma, the conservation of continents. If continents were merely eroded and erosion products spread in the sea, the continents would become thinner and broader until they disappeared. They do not do this – indeed there is some evidence that they are growing steadily thicker (see p. 51) so some process is required to put the eroded sediment back on (or under) the continents. Cratonization does this. It is clear that continents have not simply accreted rims of younger material, but it is possible that after periods of growth continents will split into smaller fragments, which after drifting for a while may once again start growing by cratonization of younger belts of rock.

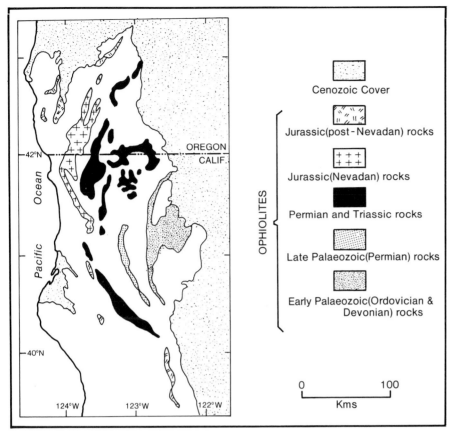

OPHIOLITES

Cenozoic Cover

Jurassic(post-Nevadan) rocks

Jurassic(Nevadan) rocks

Permian and Triassic rocks

Late Palaeozoic(Permian) rocks

Early Palaeozoic(Ordovician & Devonian) rocks

0   100
Kms

Fig. 6.3   Distribution of five ophiolite belts in the northwest United States that become younger towards the Pacific Ocean, possibly as a result of continental accretion (after Coleman and Irwin, 1974).

Gass (1977) has attempted to interpret the Precambrian of Africa in terms of cratonization. Geologically Africa consists of three old (>1 100 m.y.) cratons separated by non-cratonic regions. The term Pan Africa has been coined to identify the non-cratonic part of the African basement, dated 1 100–500 m.y. The Pan-African crust averages 35 km thick, average density 2.7–2.9, average heat flow 50–100 mWm. Granitic-granodioritic composition predominates and the terrain has been unaffected by compressive deformation for most of the Phanerozoic. On usually accepted geological and geophysical grounds the area is undoubtedly continental. In northeast Africa and Arabia 60 per cent of the ground is covered by granite plutons, the host rocks are volcanic products of calc-alkaline affinity and volcanoclastic sediments, all metamorphosed to greenschist facies.

Gass explains the origin of this crust as starting with production of ocean lithosphere at spreading sites in marginal basins adjacent to island arcs. The arcs are now represented by ultramafic belts. Then progressive cratonization by plutonism, vulcanism and sedimentation prevailed. Several belts of ultrabasic rocks in the Pan African are interpreted as ophiolites, and intervening areas as fragments of sea floor that existed as island arcs.

The small back-arc basins were probably of the type that now characterizes the southwest Pacific. However, by the end of the Pan African (450 m.y.) the mature island arcs had been swept together and the process of cratonization was complete.

Continental re-assembly

Continents do not merely develop new fold belts around their edges, which become firmly fixed by cratonization. They also drift, and sometimes collide. The collision site between continents is another favoured place for the formation of fold belts and mountain ranges – the continent–continent collision mountains of plate tectonics. It is rather remarkable that these mountains and fold belts are so similar to the mountain ranges at the edges of continents which are restricted in many ways to one-sided development.

Sutures are zones along which continental blocks have collided. Most geologists agree that the Indus suture north of the Himalayas marks the site of closing of a substantial ocean, and there is similar agreement that the Urals mark the site of suturing of Europe and Asia in Permian times. Many other sutures have been suggested, but without general agreement. It must be stressed that the amount of ophiolite in the sutures is quite trivial compared with the amount of sea floor that has been consumed, so there is no quantitative argument for regarding ophiolite bands in suture zones as remnants of old sea floor. It is possible that some other mechanism might account for the presence of ophiolites in long thin zones.

Basins and swells

One of the striking features of second-order landforms is the way in which some land masses appear to be divided into basins or lowlands by a series of gentle ridges or swells. Africa is perhaps the best example, and the major features are shown in Fig. 1.4 (p. 7).

Africa has a very long history as a land mass, and the swells vary in age but go back a long way. Figure 6.4 shows the range of ages, which go back to the Upper Palaeozoic. Nevertheless many of the swells and rises are sufficiently young to have direct geomorphic effects such as warping erosion surfaces and diverting rivers (see p. 175).

Fig 6.4   The swells and major rift faults of Africa (after Le Bas, 1971).

Swell province	Approximate age
1 Ethiopa–Aden	Tertiary
2 Kenya	Tertiary–Recent
3 Western rift (Uganda–Congo)	Late Tertiary–Recent
4 East African	Late Cretaceous–Tertiary
5 Rungwe	Late Tertiary–Recent
6 Chilwa–Zambesi	Cretaceous
7 Rhodesian	Early Mesozoic
8 Transvaal	Precambrian
9 Kimberley–Pretoria	Mesozoic
10 Bushmanland–Sutherland	Late Cret–Early Tertiary
11 S.W. Africa–Uruguay	Mesozoic
12 Angola–S. Brazil	Cretaceous
13 Cameroon	Tertiary– Recent
14 Nigeria–N.E. Brazil	Late Pal.–Early Mesozoic
15 Guinea–Guiana	Cretaceous
16 Hoggar	Tertiary– Recent
17 Tibesti	Tertiary– Recent
18 Suez–Sinai	Tertiary

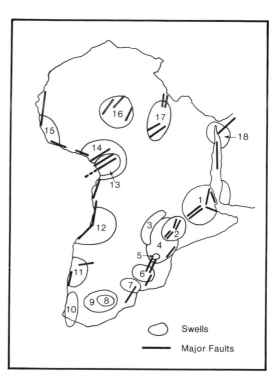

Many of the swells are followed by faults, and in Africa some of the swells and rifts make up the great rift valley system of East Africa. The swells are regarded as plume-generated uplifts by Burke and Dewey (1973) who suggest they are typically crested by volcanoes and rupture into three rifts – the triple junction – with axial dykes in the rifts, and one arm commonly failing to develop and so becoming an aulacogen.

Swells on other continents are also followed by rifts, or at least by faults. The Rhine graben in Germany is at the crest of a swell 190 km wide, and is associated with crustal thinning and a small amount of vulcanicity. In Australia a line of swells along the Great Divide makes a reasonable imitation of the African situation and is followed by numerous faults and volcanic eruptions but no rift valleys.

Swells seem to be the precursor of splitting, which starts with the appearance of faults, then graben as the crust extends, and finally there may be a splitting of the continent. One may therefore expect to see half-swells – the remnants of old swells – along the edges of continents that have drifted apart. Le Bas (1971) has mapped such features on the Africa–South America junction (Fig. 6.4).

Large basins include such features as the Congo Basin and the Kalahari Basin. Basins of this size may be created by the withdrawal of material from below or by uplift of the rims around the edges, and the two processes may be connected – sub-crustal material moving from basin to swell sites. It is also possible that the edges may rise as a result of isostatic compensation related to erosion of a continental edge and deposition of thick sediments offshore. The high rim that runs around southern Africa appears to be of this origin, and the interior basin is an inevitable consequence, even if sub-crustal material did not flow from the basin (Fig. 6.5).

Fig. 6.5 The Great Escarpment of southern Africa and its relationship to the Kalahari Basin internal drainage (stippled) (after Brock, 1972).

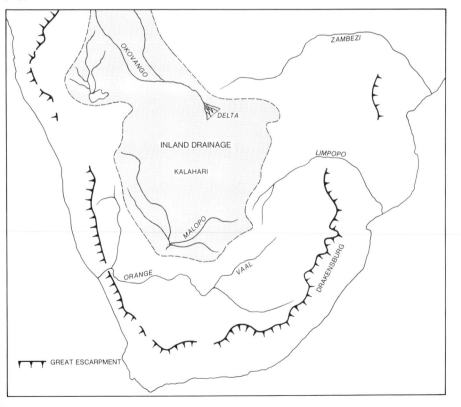

Rift valleys

Rift valleys are elongate grabens – the African ones have a total length equal to one-sixth of the world's circumference – commonly but not always located on the crests of swells. They have remarkably uniform widths, as shown below:

Western Rift (L. Albert)	35–45
Tanzania	40–50
Malawi (Nyasa)	40–60
Dead Sea	35
Turkana (Rudolf)	55

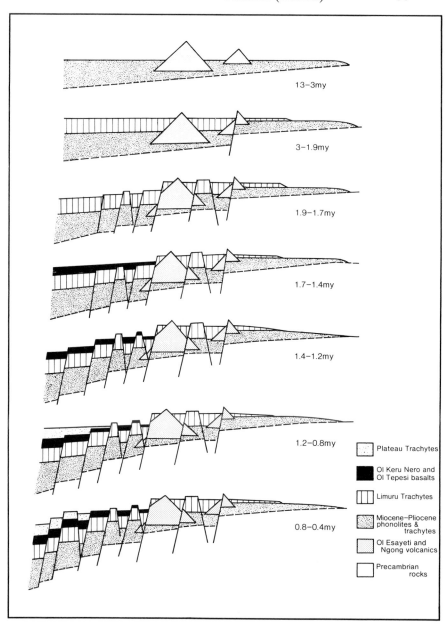

13–3my

3–1.9my

1.9–1.7my

1.7–1.4my

1.4–1.2my

1.2–0.8my

0.8–0.4my

Plateau Trachytes

Ol Keru Nero and Ol Tepesi basalts

Limuru Trachytes

Miocene–Pliocene phonolites & trachytes

Ol Esayeti and Ngong volcanics

Precambrian rocks

Fig. 6.6 Diagrammatic sections showing the evolution of the east margin of the south Kenya rift (after Baker and Mitchell, 1976).

They are bounded by normal faults with huge throw, so that the base of the graben may be below sea level. The basement in the Dead Sea graben is from −2 600 to −1 229 m; Tanzania varies from 2 150 to 2 560 m; Malawi varies from 1 565 to −1 005 m. Opposite walls of the rift may be at different height, so clearly the outer blocks have risen, by different amounts as well as the centre block sinking. The continental rift valleys have marked negative anomalies associated with their thick sedimentary fills.

Many writers believe that rift valleys, especially the African rifts, make use of older structures – they are suggesting resurgent tectonics and reactivation of old faults. In some places the grain of the basement rocks is parallel to the rifts, but in others it is oblique. In places undisturbed Precambrian rocks, virtually horizontal, overlie complex older rocks and are not folded, though cut by the later rift valleys. There is a danger that some older structures such as thrust faults might be attributed to the present rift activity when in fact they have no connection.

Vulcanicity is associated with many rift valleys, but not all. The huge Lake Baikal rift, for instance, 2 500 km long and 50–60 km wide, has no associated volcanicity at all. In contrast, parts of some rift valleys consists wholly of volcanic products, but here the chronology is relatively easy to work out, and the tectonic evolution can be followed in some detail. Part of the South Kenya rift valley is of this type (Baker and Mitchell, 1976). The tectonic evolution of the eastern margin followed pre-rift eruption of lavas from shield volcanoes starting about 13 m.y. ago. An eastern boundary fault formed 3.3 m.y. ago and was followed by formation of an inner graben from 1.9 m.y. ago. This was followed by an almost continuous succession of block-fault movements (Fig. 6.6).

Several rifts are known to be extending at measurable rates. The Rio Grande rift is about 850 km long, running roughly north–south from Colorado to New Mexico. There has probably been about 8 km of crustal extension in a northwest–southeast direction near Albuquerque since about 26 m.y. ago (Woodward, 1977), an estimate based on the assumption that the normal faults bounding the graben decrease in dip downwards and extend to a depth of about 10 to 15 km, below which the crust undergoes plastic deformation (Fig. 6.7). The rate is 300 B, which compares with a rate of 400 to 1 000 B for the eastern rift system of Africa (Baker, Mohr and Williams, 1972, p. 54) and 100 B for the Rhine graben (Illies, 1972, p. 42).

Fig. 6.7 Diagrammatic section across the Rio Grande rift near Albuquerque, New Mexico. Extension of 8 km calculated between points x and y. S=sedimentary graben fill; P=Palaeozoic to Oligocene rocks; PC=Precambrian rocks (after Woodward, 1977).

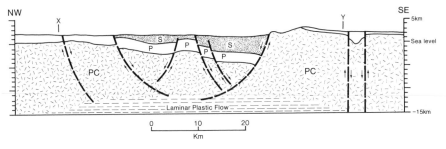

As noted in a previous section, the formation of swells seems to initiate faulting, rifting and extension, and it is interesting that the rift-valley system of Africa can be traced continuously to the Red Sea, and thence to the Carlsberg sub-oceanic ridge. The Red Sea has major differences from rift valleys, as it has a positive gravity anomaly and a presumed simatic sea floor. It is in fact an example of the earliest stage of sea-floor spreading, and it is easy to imagine the spreading site working along the rift system and isolating East Africa from the rest of the continent, just as in the past a similar break-up presumably separated Madagascar.

The Baikal system of rift valleys is remarkably similar to that of East Africa in many ways, including scale (King, 1976). It extends about 2 500 km and there are about twelve major depressions ranging from 100 to 700 km long and 15 to 18 km wide. Rift structures are graben along arched uplifts. They are usually asymmetric, with a steep northern flank and the other margin more gently sloping, either downwarped or with minor faults. The depressions are mainly dry, but there are lakes in places, such as Lake Baikal, 670 m long. Sediment thicknesses are around 1 500–2 500 m, but the exceptional Baikal depression probably has up to 6 000 m.

The Baikal system is essentially Cenozoic, but tends to follow earlier structural trends. Before rifting, in Cretaceous and Early Cenozoic times, the area was one of tectonic stability with subaerial weathering and erosion. The graben fill of continental sediments and volcanics can be divided into two groups. A lower group is mainly lake, swamp and river deposits of Oligocene to Pliocene age, with thicknesses from 1 000 to 3 000 m: an upper group consists of 500 to 1 200 m of very diverse sediments accumulated under cold conditions from later Pliocene to Holocene times. They were produced from mountain relief which increased through time. Volcanism, mainly basaltic, is largely independent of the rifts and is mostly on the uplifted blocks and arches. Intense gravity lows are found over Lake Baikal, resulting mainly from the great thickness of light sediment.

The later stages of rift evolution are best studied so far on the Atlantic coasts, when each side of the rift valley evolves into an aseismic continental margin.

The details of landforms associated with rift valleys and discussion of their origin are given in Chapter 4.

Aulacogens

Related to rift valleys and to ocean-ridge spreading sites are the failed arms of triple-junction spreading sites known as 'aulacogens'. These have already been introduced (p. 48) and Fig. 6.8 shows the position of some postulated aulocogens on the Atlantic coast. There are many other possibilities which constitute tectonic landforms, or landform assemblages.

A very fine example of an aulacogen is found in southern Africa, followed in part by the Limpopo River (Fig. 6.9). Reeves (1978) describes how the aulacogen is followed by a dyke swarm, slightly wedge shaped and focusing on Estosha Pan in Namibia. It is followed by several lowland features and towards the coast it separates two monoclines of post-Karoo age, the Lebombo monocline and the Sabi monocline, which both relate to the break up of Gondwanaland if not actually marking old continental boundaries. The three axes intersect at angles of almost precisely 120°.

Continental break-up

According to Kent (1977) Mesozoic sedimentation on a world scale took place in basins on the margins of old Palaeozoic massifs. Vertical faulting prevailed from the end of the Palaeozoic, and basin formation was characterized by intermittent marginal faulting and graben formation, in Permian, Triassic and Jurassic times.

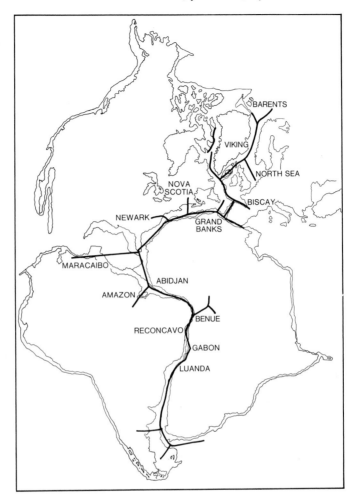

Fig. 6.8 Triple junctions around the Atlantic. Most of the rifts are part of the rift between America and Africa, but the failed arms are preserved on the continents as aulacogens.

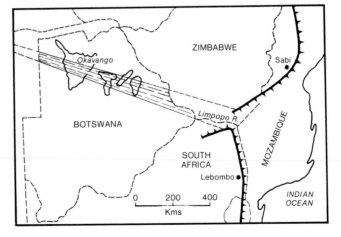

Fig. 6.9 The Limpopo aulacogen, followed by a post-Karroo dyke swarm, showing its relationship to the Lebombo and Sabi monoclines (after Reeves, 1978).

This basin and graben formation preceded the break up of Gondwanaland and the opening of the North Atlantic, but the future sutures were already defined and followed by marine transgressions in Late Palaeozoic time.

Faulted basin development along what were to become the world's aseismic or Atlantic coasts (see Fig. 15.4, p. 225) became widespread by the Permian, and continued intermittently until the Middle Cretaceous.

Some basins are not directly related to continental margins, like the North Sea basin which is 500–1 000 km away yet has great thicknesses of sediment. The North Sea basin has a history of subsidence apparently typical of many basins on Atlantic-type coasts of the world (Kent, 1977). It was initially in two parts, a northern basin with a Devonian ancestor, and a southern basin that dates back to the Carboniferous at least. Broad subsidence was initiated in the Permian and fault-controlled subsidence through the Early and Middle Mesozoic. In the Upper Cretaceous simple subsidence resumed, and continued through the Tertiary to the present day. We still do not know why this pattern of events was nearly synchronous on a world scale, nor why there was a sudden end to the phase of major faulting in the Lower Cretaceous followed by simple unfractured subsidence since then (Fig. 6.10).

Fig. 6.10 Diagrammatic section of the North Sea, illustrating the change from faulting to unfractured subsidence in the Lower Cretaceous (after Kent, 1977).

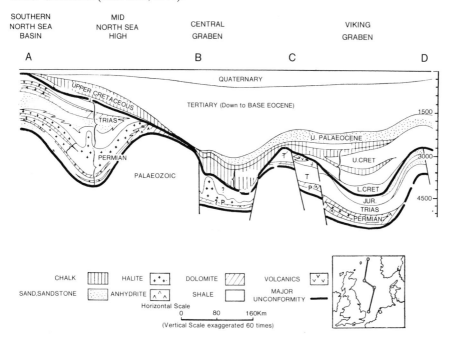

In places as far apart as Greenland, Canada, Gabon and Australia the Mesozoic regime of tectonic graben and basin formation was replaced in the Aptian by a simple system of subsidence with prograding marginal or basin sedimentation with only minor faulting. The change from Permian and Early Cretaceous fault-controlled subsidence was almost, but not precisely synchronous around the world. It is about Aptian though sometimes a little earlier or later, but the small range of variation in date suggests that the change in sedimentation style is fundamental. It is not likely to be brought about by eustatic effects but is likely to be tectonic. Kent thinks it may be related to start of drift in many areas, though drift commenced in different places over a much longer time span.

The rifting that preceded continental break-up was accompanied by two other features: the formation of salt deposits in marginal basins, and vulcanicity. Some closed salt basins may have anaerobic black sediments, the precursor of oil. The rifting of a craton is best documented by studies around the Atlantic. In North Africa, for instance, Van Houten (1976) finds the development of evaporite basins in the Middle Trias, rifting sets during the Late Trias, and basalt flows appear in the Early Lias.

Montadert *et al* (1974) have presented detailed stratigraphic, structural and geophysical information to produce a detailed history of the Bay of Biscay, which has the following main features:

1. Rifting of the platform during the Triassic and Early Lias, with thick evaporite and volcanic deposits.
2. Following a quiet period, subsidence began and intensified until the Albian, and the Bay of Biscay began to open, with an oceanic floor.
3. Opening continues with creation of the central part of the Bay at the end of the Cretaceous; volcanic activity until the Eocene; the continental platform tilted and subsided over 3 000 km.
4. Subduction possibly took place under the northern edge of the Iberian plate until the Oligocene or Miocene.
5. The abyssal plain deepened during the Upper Eocene, except on sea mounts.

The opening of the Bay of Biscay was hampered from the start by the convergence of Africa and Europe causing compression in the whole Tethyan area, and the northern margin of the Iberian plate became a subduction zone probably as early as the Middle Cretaceous. The north coast of Spain is quite abrupt, without the sedimentary wedges found off North America, as one might expect from its sphenochasm origin. Africa has a number of peripheral evaporite basins varying in age from Triassic to the present day. In the Red Sea marine Miocene rocks have an early clastic phase, smoothing the pre-Miocene relief, and an evaporite phase. The lower group was filling a subsiding graben. The evaporites have associated gypsum, dolomite and marl. Basins off Tanzania have Mesozoic and Tertiary sediments in which evaporites reach a thickness of 2 500 m, of Middle Jurassic to Triassic age.

The Gabon salt basin has little salt-tectonics deformation so is easier to study. Before salt deposition sedimentation was essentially continental, indicating a basin cut off from the sea. Post-evaporite deposits are marine. The evaporites are of Lower Cretaceous age, like those of the Brazilian Reconcavo and Sergipe–Algoas basins, and the assemblage clearly marks a line of continental evaporite basins forming in graben that were precursors to the drift apart of South Africa and South America and the intrusion of the sea.

The Congo basin is similar. The Senegal salt basin appears to be of the same Cretaceous age, but is complicated by extensive salt tectonics and consists of about ten salt domes.

In the eastern Algerian Sahara is an extensive salt basin of Triassic age, 250 000 km² in area. Evaporites overlie continental beds and are overlain by Liassic anhydrite and then thick continental red beds of Jurassic and Cretaceous age. Individual salt units are not more than 300 m thick, and there is little salt tectonics.

In Western Australia events since the Late Carboniferous are interpreted as reflecting a typical sequence of intracontinental rifting,

rupture, and sea-floor spreading, forming a continental margin th[
faced first a juvenile and later a mature ocean (Fig. 6.11). Th
cessation of rifting, the outpouring of basalt, the onset of marin
transgression and the generation by spreading of the adjacent sea flo[
coincide and indicate the time of rupture. This was Late Jurassic alon
the northwest margin and Early Cretaceous along the southwest. Sin[
rupture the older northwest margin has subsided further than th
southwest margin.

These observations support what has become the classic type [
rifted margin, with the following general phases:

1. A phase of arching and rifting, with fill that varies from marine [
   the north to fluvial in the south, with alkaline volcanics. Th
   phase lasted from Late Carboniferous to Late Jurassic or Ear[
   Cretaceous.

2. Rupture of the arch in the Late Jurassic (north) and Ear[
   Cretaceous (south) with local basalt flows. Marginal subsiden[
   and marine transgression.

3. Since the Upper Cretaceous there has been progradation [
   carbonates on a slowly subsiding margin.

Fig. 6.11 Diagrammatic evolution of the Western Australian continental margin (after Veevers, 1974).

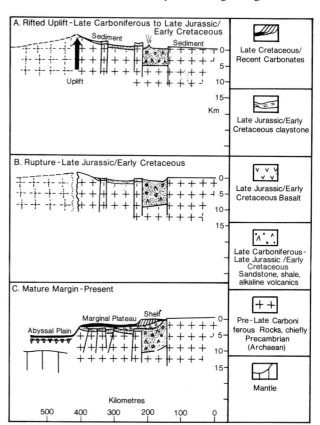

The orocline concept

Carey (1954) proposed a set of terms that are very useful in describin
the movement of relatively rigid fragments of the earth's surface an
the associated bending in neighbouring pliable zones. These term
were needed because although there were adequate terms for th
deformation and dislocation of strata in the vertical plane there wer

no terms for deformation of structures in plan, still less for the bending of a whole orogenic belt.

Carey reasoned that a rifting region where stable shields were separating could not be wholly obliterated by later events, and the pliable zones must still record the stretching and bending induced in them by the motion of the shield.

The 'orocline concept' is the name given to the whole set of ideas associated with such movements, a fundamental concept in tectonics related to lateral movements. Sea-floor spreading, continental drift and plate tectonics are all concerned with lateral movements, but the orocline concept adds to these the interrelationship of the spreading

Fig. 6.12   The Baluchistan orocline, the Arabian sphenochasm and the Punjab orocline. (a) Before spreading. (b) After spreading (after Carey, 1958).

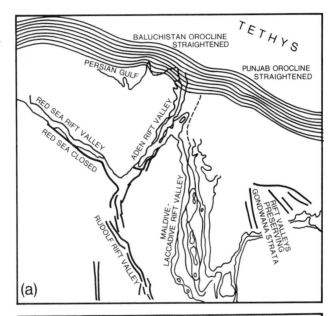

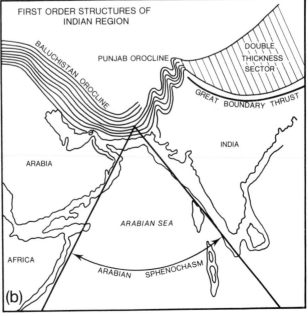

sites and bending sites, rather than the spreading sites and subductio
sites of plate tectonics.

The main terms are as follows:

Orocline   An orocline is an orogenic belt with a change in tren
which is interpreted as an impressed strain. The definition refers t
'orogenic belt,' and may indicate either a topographic mountain chair
or a belt of folded rocks. The Baluchistan orocline (Fig. 6.12) serves a
an example. There is a tacit assumption that orogenic belts are initiall
straight and that any bend must be imposed later. It remains possibl
that some mountain ranges or fold belts may be curved from thei
inception, like island arcs.

Coupled oroclines   Two oroclines forming an S shape are said to
be coupled. The Baluchistan and Punjab oroclines are examples (Fig
6.12).

Sphenochasm   A sphenochasm is the triangular gap of oceani
crust separating two cratonic blocks with fault margins converging to a
point, and interpreted as having originated by the rotation of one of the
blocks with respect to the other. Examples are the Bisca
sphenochasm (Fig. 6.13), the Arabian sphenochasm, the Arcti
sphenochasm and the Coral Sea sphenochasm.

Fig. 6.13   Formation of the Bay of
Biscay and the Pyrenees by rotation of the
Iberian Peninsula (simplified from Carey,
1976).

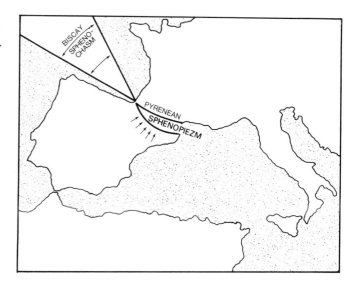

Rhombochasm   A rhombochasm is a parallel-sided gap in the
sialic crust occupied by simatic crust and interpreted as a dilation.
Much of the Atlantic Ocean is a gigantic rhombochasm. The Red Sea
(Fig. 6.14) is a smaller scale rhombochasm.

Sphenopiezm   A sphenopiezm is a wedge-shaped squeezing
together, opposed to a sphenochasm. The Pyrenean sphenopiezm,
opposed to the Biscay sphenochasm, is an example (Fig. 6.13).

Of course these tectonic features do not occur singly but in
interrelated groups, so that the rift oceans, for instance, consist of an
assemblage of rhombochasms and sphenochasms, and oroclines are
found in related land masses.

Fig. 6.14 The Red Sea and Gulf of Aqaba as a rhombochasm. Arrows show the dominant direction of movement.

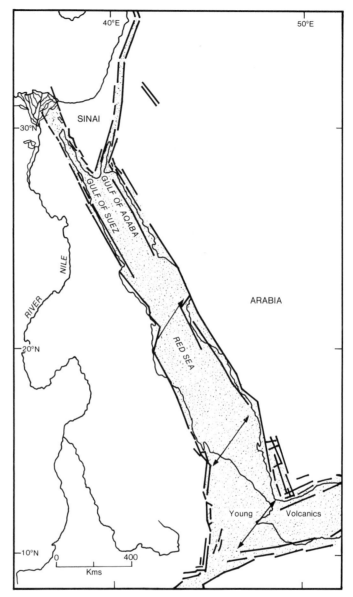

The major rotations associated with the opening of sphenochasms should be testable by palaeomagnetic methods. This has been done in many instances, and Carey has provided data and references (1976, pp. 4 and 5) to show that the predicted angles of rotation match well the rotations determined by palaeomagnetism. The data are summarized in Table 6.1.

Table 6.1 Tectonic rotation subsequently confirmed palaeomagnetically (after Carey, 1976).

Rotated block	Rotation predicted by Carey	Palaeomagnetic rotation observed
N. America to Europe	30°	30°
Africa to S. America	45°	45°
Newfoundland	25°	25°
Spain	35°	35°
Italy	110°	107°
Corsica and Sardinia	90°	50°
Sicily to Africa	0°	0°
Arabia to Africa	3½°	7°
India to Africa	70°	70°
New Guinea	35°	40°
Honshu N. and S.	40°	58°
Mendocino orocline	60°	63°
Puerto Rico to S. America	45°	53°
Jamaica to S. America	42°	50°
Hispaniola to S. America	39°	40°
Colombia	large	80°
Appalachian Arcs	20°–40°	29°
Malay Peninsula	Ca 70°	70°

Lineaments

Many features, on all scales, suggest a geometrical pattern. Coasts mountain ranges, faults, drainage patterns, fold axes and joints have all been used to define geometrical properties of the earth's surface. Any linear feature of the earth's surface may be termed a lineament. Brock defines lineament as a geological or topographical alignment too precise to be fortuitous. A tectonic origin is not implicit in the word, although in a great many instances the tectonic implications are plain. The study of lineaments involves two stages – first the recognition and definition of the lineament, and secondly its interpretation.

Most lineaments are straight lines (Fig. 6.15), but some are curves. The most obvious pattern of curves is that of the island arcs. The question of scale is important. A long straight-line lineament may appear curved on some map projections. Some lineaments that appear curved at a small scale break down into a series of lines when examined in more detail.

At times the search for lineaments verges on numerology, and their alleged significance can take on almost magical properties. One of the early giants of geology, Werner, believed that parallel mineral veins were of the same age. The notion that gold is only found in north–south aligned rocks in Victoria, Australia, is apparently true.

Several authors are convinced that there are two sets of lineaments which are fundamental to patterns of structure and physiography all over the world – a meridional set and an orthogonal set, and a diagonal set.

Vening Meinesz (1947) worked out the shear pattern that might result from a somewhat improbable shift of the north pole over 70° of latitude along the 90° meridian. Hills (1956) found that the Vening Meinesz net, combined with trends that bisect the angles of the net, accounted for virtually all major Australian lineaments (Fig. 6.16).

Examples of north–south trends include the Malvern axis and the Pennines, in England, the Heathcote axis and Palaeozoic fold axes in eastern Australia (Fig. 6.15), and the Rhine Graben in Germany. East–west lineaments include the Hercynian axes of Europe, the equatorial shear zone, and some of the Pacific fracture zones. Diagonal lineaments include the rift valleys of East Africa, the drainage pattern

Fig. 6.15 Lineaments in eastern
Australia. These are meridional except on
the shield, where diagonal trends
predominate (after Hills, 1956).

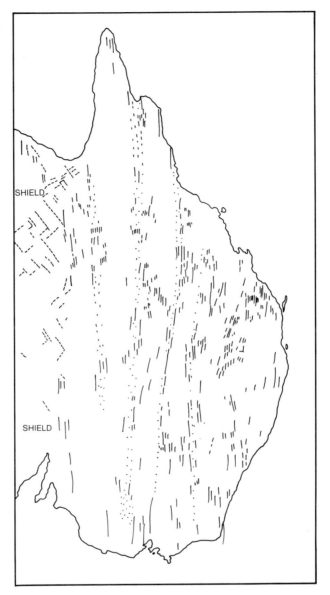

of northern Uganda, Caledonian and Charnoid trends in England and
the Darling lineament in Australia.

There seems little doubt that some of these features are real; in
other words some lineaments exist as observable features. It is not so
clear that the world-wide patterns exist. For one thing, how accurately
can the lineament be measured, and what sort of tolerance will be
allowed in putting it into one of the regular sets? If 10° either way is
allowed, then 40° in each 90° will be included in one or other of the
favoured sets, which is almost half of a random set of lineaments.

Another problem with finding world-wide sets of lineaments is that
the patterns on different continents should *not* fit after continental
drift, unless each continent kept its orientation despite drift. But we
know from palaeomagnetic results that the continents have rotated as
well as drifted, so the lineaments should also be rotated if they were of

Fig. 6.16   Major lineaments in Australia compared with Vening Meinesz's network. Solid lines = trends parallel with the network. Broken lines = intermediate trends (after Hills, 1956).

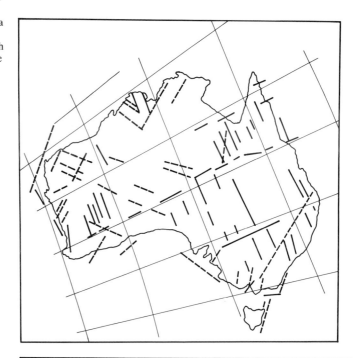

Fig. 6.17   Rosette of magnetic lineaments of northwest Europe, based on 611 lineaments with a total length of 39 733 km (after Affleck, 1970).

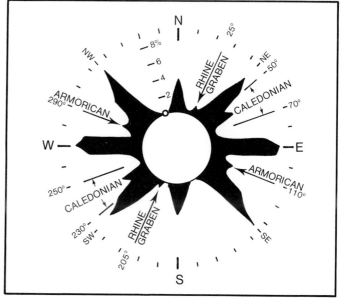

pre-drift origin. Since many of the lineaments are indeed pre-drift many of the apparent matchings must be put down to laxity in pattern matching, or to coincidence.

Brock (1972) has mapped countless lineations and mapped vertices, triangles, rhombs, spokes, spigots, A-frames, asterisks, antipodal relationships, comparable shapes, dimensions and radii, and a host of features that convinced him of the reality of a global mosaic. He was also aware that his conclusions needed a condition of fixity in the earth, and that his concept and even the facts on which they were based were incompatible with the theory of continental drift. He therefore devoted a chapter to refute drift and make a case for fixity, even in 1972, but although he concludes 'the mosaic hierarchy . . . provides the global view and a stable one, and without any gross and unprovable assumptions' the idea of fixity has very few adherents at the present time.

If there really is a world pattern of lineaments, it suggests a world-wide system of stress and strain and one or more of the many suggestions for the pattern must be considered. These include differential movement of the north and south hemispheres (producing a shear pattern), and the sliding of the lithosphere as a whole over the inner earth. Plate tectonics does not seem able to produce any such pattern, and for the present the search for world-wide lineament patterns seems to be out of fashion, though as recently as 1977 O'Driscoll presented evidence for yet another pattern – a double helix. The structure of the earth may rival that of DNA in complexity, but this is a surprising coincidence indeed.

Air photography makes lineament work much easier; satellite work such as the ERTS and LANDSAT imagery adds another dimension to lineament work. Haman (1975) presents a lineament analysis of the United States based on ERTS imagery. A total of 1 585 lineaments were analysed, ranging in length from 50 km to 600 km. The average lineament density is 0.04 km/km², with high values of 0.08 km/km² in several places and a low of 0.002 km/km² on the Idaho batholith. A very high density covers the interior plains with platform deposits over either Precambrian or Palaeozoic basement. The Cordillera has a marked low density. The lineaments were analysed for orientation, using 15° sectors. The WNW and ENE lineaments are mostly confined to the interior platform areas and are conspicuously absent from the Cordillera, and are therefore of pre-Mesozoic age. The NNW- and NNE-trending lineaments are more widely scattered throughout the United States and are interpreted to have been formed in post-Palaeozoic time. If the WNW–ENE and the NNW–NN lineament systems represent conjugate shear fractures, we may conclude that the United States was subjected to east–west compression in pre-Mesozoic time, and to north–south compression in Late Palaeozoic and younger time. The change in stress orientation appears to have occurred in Pennsylvanian-Permian time.

Lineaments can also be detected from geophysical data, such as magnetic data. Fig. 6.17 shows the rosette for magnetic lineaments of northwest Europe, and once again the dominant features are the equatorial, meridional and diagonal lineaments, with minor peaks that can be attributed to Caledonian, Armorican, and Rhine graben directions. As Affleck (1970) points out 'These patterns all suggest the permanency of the continents, and that the relation of the earth's spin axis to the crust has been constant since Precambrian time'.

7 Granite and metamorphism

Igneous rocks are those thought to be derived from the cooling of molten rock or magma, the word being derived from the Latin word for 'fire'. Of the many kinds of igneous rocks distinguished by petrologists, the two that dominate at the earth's surface are basalt, with many related rocks, and granite, also one of a family. What is more the basalts seem to be mainly of volcanic origin, but the granites appear to have a sub-crustal origin.

In classical geology igneous rocks were divided into plutonic, hypabyssal and volcanic classes. The plutonic rocks were formed at depth, the hypabyssal rocks were formed at intermediate levels and were the feeder connections between the plutonic and the volcanic rocks, which erupted at the ground surface. Unfortunately it is not possible to get mainly basaltic volcanoes from a granite pluton, so this simple story is untenable. What is a better story?

Basic intrusive rocks

Volcanic rocks are varied, but for simplicity we shall reduce them to basaltic volcanoes and andesitic volcanoes. The former which are relatively poor in silica are called basic volcanics, and the latter intermediate to acid volcanics. Basic volcanics can be derived by melting the sima layer beneath the oceans, or possibly beneath the continents. The intermediate-to-acid rocks could be derived by various mechanisms of fractionating a basalt magma into an extra-basic fraction and an acid fraction, but large amounts of fractionation would be required to make a significant amount of acid magma. They might also be made by melting of continental crust, a process called anatexis, and some might be made by contamination of basaltic magma by various amounts of continental crust. There is a vast literature devoted to the petrology, geochemistry and mineralogy associated with arguments for various methods of magma formation, mostly not related to the theme of this book and far too technical to pursue any further here.

What we can note is that the intrusion of feeders to volcanoes (Ch. 8) are very different from the intrusion of granites and similar plutonic rocks.

From detailed petrological and physico-chemical considerations it is possible to put constraints on what rocks can originate in various ways, and although the history of petrology shows that views may change, some mechanisms are apparently impossible. Wyllie *et al.* (1976) provide some conclusions along these lines.

The calc-alkalic rocks of batholiths or their precursors (roughly granodiorites) may be generated in deep continental crust, in subducted oceanic crust, in the overlying mantle wedge, or in processes involving material from all three sources. Primary granite magmas cannot be derived from the mantle or subducted ocean crust. Primary granite magmas with low H_2O content are generated in the crust and erupted as rhyolites. Primary andesite is not generated from mantle peridotite, nor from the crust unless temperatures are very high. Batholiths are produced from crustal rocks as a normal consequence of regional metamorphism, and some batholiths receive contributions of material and heat from mantle and subducted ocean crust.

Granite intrusions

Granitic batholiths fall into two distinct types, which we shall refer to (though begging the question) by two oldfashioned names.
1. Plutonic granites. These are intrusive bodies that cut across pre-existing rocks, with sharp contacts, limited aureoles of thermal metamorphism around the pluton, and indications that they are intruded at high level. They have been derived from some deeper level and risen to their present position by various mechanisms to be considered later.
2. Granitization granites. These are often vast bodies of usually foliated rocks, with the composition of granites but with signs of gneissic structure, that gradually merge into the surrounding rocks. 'Ghost stratigraphy' of pre-existing rocks may be traced through the granites, suggesting that the older rocks were converted to granite by some kind of 'soaking' or metasomatic alteration.

The obvious conclusion is that type 2 granites are formed by conversion of other rocks to magma by matasomatism at great depth and if they should become sufficiently mobilized they might rise diapirically to form the type 1 granites, cross-cutting plutons.

Unfortunately this simple picture does not seem to be quite adequate in the light of later findings.

Chappel and White (1974) showed that there are two contrasting types of granitoid in southeastern Australia. These are called the I- and S- types to stress their different origins, the S-types being derived from the partial melting of metasedimentary source rocks and the I-types from source rocks of igneous composition. S-types are usually foliated and either contain cordierite or white mica secondary after cordierite. I-type granitoids are generally massive and frequently contain hornblende. Geochemically I-types have higher Ca, Al, Na_2O/K_2O and lower Fe, and Mg than S-types of comparable SiO_2 values.

The differences between the two groups are not the result of differences in the melt-forming process but reflect differences in the nature of the material. Thus the geochemical features of the S-type granitoids are indicative of their source rocks having been through a process of chemical weathering in a sedimentary cycle. Conversely the I-type granitoids were derived from fractionated rocks that had not been involved in weathering processes (Hine *et al.*, 1978).

This seems to imply that the I-type granites cannot be derived from the S-type granites, and if these can be roughly equated with the plutonic and granitization granites it seems that the plutonic cannot be derived from the granitization types. Two mineralogically similar rocks have been derived from quite different sources, one from sediments and one by direct igneous processes, presumably from the mantle.

It is hard, however, to see how bodies the size of the Great Peruvian batholith – the largest in the world – could originate from the mantle for the main chamber before differentiation would have to be even bigger. High-level granites are thought to be emplaced mainly by stoping, a process in which blocks of the enclosing rock are detached and sink into the lighter magma, becoming xenoliths (foreign rocks) within the magma and eventually perhaps being entirely assimilated.

Some granites simply push aside the overlying rocks. In the heart of the Rand Basin in South Africa lies the Vredefort dome, a cylinder of granite 40 km across surrounded by a collar of near vertical and sometimes overturned sediments (Fig. 7.1), situated at the deepest part of an unfolded sedimentary basin. The uplift could not be formed

Fig. 7.1 Vredefort dome, and postulated mantle intrusion (after Brock, 1972).

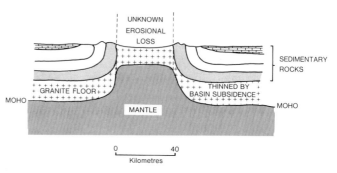

by any kind of lateral compression, but results from simple vertical tectonics in the middle of the basin. Evidently the pluton has pushed up the overlying strata, with some faulting, and has also spread to cause overturning of the sediments. The rise, pushing up unconformable sediments above, is reminiscent of some gneiss domes such as the Rincon Mountain dome of Arizona, which has shed overlying sediments by gravity sliding (Davis, 1975). The Vredefort Dome is associated with a puzzling gravity high, and Brock (1972) suggests that the plug of granite has been pushed up by an underlying plug of mantle material. Some plutons are rather regularly spaced (Fig. 7.2) suggesting some fundamental control on the nature of intrusion.

Fig. 7.2 A large batholith breaking into a number of evenly-spaced plutons towards the surface near Walcha, New South Wales (after Korsch, 1977).

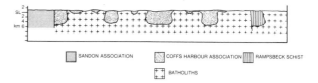

Deep granites may be formed by gradual alteration to granite, but this process appears to be unpopular among some experts, even in instances where ghost stratigraphy is preserved and traces of beds in confining rocks can be traced into the granite itself. Pitcher (1970) reviewed the ways in which ghost stratigraphy and structure can arise in granite plutons, and concluded that rarely is this relict structure itself proof of a metasomatic replacement origin. Explanations involving stoping by magmas are often adequate, at least for plutons in low- and middle-energy environments. True granitization ghost structure may still be a true explanation for high-energy environments deeper in the crust.

In the Andes the intrusion of the massive batholiths through the Mesozoic was controlled by major fault lines in the basement (Pitcher and Bussell, 1977). Individual plutons were controlled by transcurrent faults and by smaller-scale joint patterns. Intrusion took place over a long time, and the same structures were repeatedly exploited. Generally subsidence of large fracture-bound crustal blocks provided the space for the emplacement of magma in bell-jar and cauldron subsidence structures, and stoping along closely-spaced regular joints produced rectilinear patterns of contacts. Exceptionally the forceful intrusion of plutons lifted the roof blocks, with movement along pre-existing faults.

The lineaments of Andean trend have existed throughout the Phanerozoic, and it is thought that the fault blocks express resurgent tectonics along deep-seated fractures in the crystalline basement which everywhere underlies the Peruvian Andes.

It is generally a mistake to think of granite 'magma' as a liquid, like lava at the earth's surface, and granites are emplaced in a condition that approximates to a solid. The mechanism of flow in solid state is discussed in Chapter 10, and discussion of the rise of plutons will be delayed until then. Suffice for the present to say that gravity is generally the driving force in the ascent of plutons.

It is worth mentioning here, however that some granite may have been rigid in response to short-time deformation, while being capable of flow on a long time-scale, and even supporting internal reaction as heat was gradually released by continuing crystallization.

If a rising granitic pluton could be liquefied, by stress release or any other mechanism, it could give rise to lavas of intermediate and acid composition, and this certainly seems to be the source of some. In some instances the pluton continues to rise until it intrudes its own earlier-formed volcanics.

Associated with granitic intrusions are various late-stage emanations that may cause hydrothermal alteration of the surrounding rock or the intruded rock itself, and may be related to the emplacement of minerals of commercial value. There is some evidence that the role of these rather mysterous emanations may have been overplayed. Some ore deposits formerly thought to be of hydrothermal origin are now thought, or in some instances known, to be the product of reactions in a weathering or sedimentary environment rather than a plutonic one (Ollier, 1977a; Amstutz and Bernard, 1973).

The genesis of kaolinite in altered granites has long been a source of controversy, and both deep weathering and hydrothermal alteration being proposed. The latter had most general support, but Sheppard (1977) has produced convincing evidence in support of the former. There is a well-defined relationship between the D/H and $^{18}O/^{16}O$ isotope ratios in kaolinites formed in a weathering environment, a ratio which is different from that found in plutonic environments, and Sheppard found that the kaolinites of the Cornubian batholith in southwest England are of weathering origin. There is no evidence that the china clay kaolinites were first formed hydrothermally and subsequently underwent isotopic exchange.

Metamorphic rocks and metamorphism

Metamorphism is the process of change in the mineral assemblages, structures and textures of pre-existing rock. It occurs when the rock is subjected to physical and chemical conditions considerably different from those in which the rock originated. The two main processes are chemical alteration and recrystallization, and mechanical reorganization.

Rocks produced by mechanical breakdown or cataclasis are called cataclastic rocks. Breccias (consisting of fragments of broken rock) in fault zones may be considered the least-altered type, but pressure and frictional heat lead to the formation of friction breccias, mylonites which are more sheared and granulated so that grains are less than 0.1 mm, and eventually pseudotachylite which looks glassy.

Contact metamorphic rocks are produced by increase in temperature without any increase in pressure, and are found in aureoles around granites and other igneous rocks. A tough fine-grained rock called hornfels is the commonest type. There may be chemical reactions, such as the alteration of calcium carbonate to the silicate (limestone to wollastonite), and some metamorphic aureoles around plutons may be divided into several zones characterized by different mineral assemblages.

Deformation of rock by pressure tends to make flaky minerals assume a parallel orientation, perpendicular to the stress, and the rock develops a tendency to split in this direction – it has developed a foliation (*folia* is Latin for 'leaves'). This exists in metamorphosed shales and slates as slaty cleavage, cleavage being a direction in which the rock is easily cleaved or split. These are very fine-grained rocks.

With more intense metamorphism the grains recrystallize into larger grains, so that mica flakes can be seen by eye. At the same time there is a tendency for different minerals to segregate into bands. These rocks are schists, and the foliation is called schistosity. Coarse-grained metamorphic rocks have very distinct bands of different minerals but usually few flaky minerals. These are gneisses and the foliation is called gneissosity. The foliation may be complicated by various corrugations and folds, and elongated minerals that can also produce lineations in the metamorphic rock. The microstructures are studied in structural petrology.

The most widespread and important metamorphic rocks are the regional metamorphic rocks, which are the schists and gneisses. Metamorphosed shales were studied in Scotland by George Barrow, who discovered a series of zones characterized by a critical or index mineral. The least metamorphosed rocks retain mica; further metamorphism leads to chlorite; then biotite, almandine garnet, staurolite, kyanite and finally sillimanite. These are called Barrovian zones. The zones were found to be inconsistently developed around the world, so later workers developed ways of dealing with metamorphism under different pressure and temperature regimes. It was also found better to use groups of minerals, or mineral facies, which indicate the phase (pressure-temperature relationships) better than individual minerals (Fig. 7.3).

The lowest-grade regional metamorphic rocks contain zeolites. The next stage contains chlorite, epidote and actinolite – all green minerals that give rise (with feldspars and other minerals) to the greenschist facies. Intermediate-grade metamorphic rocks contain the amphibole mineral hornblende, which with feldspar and garnet characterize the amphibolite facies. The highest-grade rocks are granulites, consisting

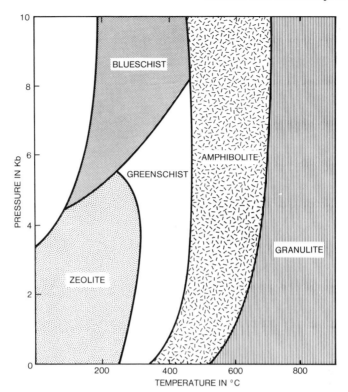

Fig. 7.3 Simplified diagram of the pressure–temperature fields of common metamorphic facies.

of feldspar, pyroxene and sometimes garnet. The mineral assemblage is like the igneous rock gabbro, but granulites have more sugary, granular textures.

Where high pressure is combined with low temperature in metamorphism the blue mineral glaucophane is found, and rocks formed in these conditions are of the blueschist facies. These have assumed particular importance in plate-tectonic theory as they might be indicators of plate collisions, subduction zones and so forth.

During metamorphism the chemical composition of the rock may be changed by the migration of elements into or out of the rock, a process called metasomatism. During low-grade metamorphism, for instance, when a shale is converted into a slate, water is driven off so that the rock is dehydrated and the water migrates elsewhere. If ferric iron phases are present they may be reduced, driving off oxygen, which may combine to form CO_2 or SO_2 if carbon or sulphur are available. Laboratory experiments show that at high pressures such hot aqueous fluids can dissolve large amounts of alkalis and silica. It seems, therefore, that highly-metamorphosed rocks at depth will be dehydrated, and give off fluids charged with alkalis which migrate towards the surface. They react with cooler rocks and by adding alkalis and silica convert them to a composition nearer to that of granite, a process called granitization. With ultrametamorphism these rocks may partially melt and become indistinguishable from an igneous magma. In other words, at great depths magmatic and metamorphic processes merge, and we have to try to conceive rock behaviour in conditions quite unlike those at the surface of the earth.

In general greatest metamorphism will result from deepest burial, so the deepest part of a geosyncline is where we might expect

conversion to gneisses, granitization and the intrusion of granit
magmas. By examining the distribution of former pressure an
temperature relationships, as in the double belts of metamorphic rock
associated with some island arcs (p. 206). The directions of foliation
in metamorphic rocks may indicate the directions of applied forces a
the time of metamorphism, but interpretation is not always easy. Th
vertical foliation in root zones of mountains may be associated wit
lateral compression, but some regard it as an indication of vertic
movement of the gneiss belt.

Metamorphic domains

Metamorphic domains with parallel foliation may terminate abruptl
against another lot of metamorphic rocks with a foliation in
completely different direction, even perpendicular to the first set. It i
commonly assumed that the foliated metamorphics are the roots of ol
orogenies, and if two lots cross, one orogeny must be younger than th
other. This in itself present problems: how could the younger rocks b
compressed to give a new foliation without changing the underlyin
rock? Was the younger set in a sort of trough with the older rock
plunging beneath it, or was some other special relationship possible

In some instances it seems possible that this is a non-problem, wher
investigators have been asking the wrong question. The metamorphi
domains represent zones where the foliation has been changed, but th
rocks are not really younger.

The Grenville Province in the Precambrian of Canada is a classi
example, consisting of high-grade gneiss 1 000 to 800 m.y. old. Marbl
belts can be mapped from the neighbouring Superior Province into th
Grenville, so the Grenville is simply a metamorphosed version of th
Superior. The line where one changes into the other is a metamorphi
front, not an unconformity between two great orogenies.

In the Precambrian of Scotland the Laxford rocks have a foliatio
almost perpendicular to the apparently older Scourian rocks. Th
Scourian rocks include charnockite, magmatite and much altere
rocks that have been intruded by dykes after deformation. Toward
the Laxford 'boundary' the dykes become first sheared, then folded i
a complex way, and finally develop axial plane structures with biotit
growth in axial planes. The boundary is a metamorphic front, and ther
was no new sedimentation in Laxfordian time.

Although some people have made detailed scenarios for th
evolution of the Pan African metamorphic rocks by plate tectonics o
other orogenic means, there are some who regard the Pan African as a
metamorphic event rather than an orogeny involving earlie
deposition.

Gneiss domes

Some granites have pushed up overlying rocks into domes, and a
gneissosity is found to be concentric with the granite. This may be i
the granite or in the overlying metasediments, or in both. In som
Scandinavian examples fragments of the granite are found in th
sediments that overlie the granite, yet granite also intrudes some of th
sediments; the granite has not only been domed up after a period o
erosion and sedimentation, but has been remobilized to the extent tha
it can intrude the younger sediments (Eskola, 1949).

Many, but not all gneiss-mantled domes show this unconformabl
relation between granite and overlying metasediments. To account fo
the simple types it is suggested that after a period of erosion the granit
is buried by sediments, and later becomes sufficiently mobile (but no

liquid as early writers sometimes thought) to dome up, arching the overlying sediments and imparting a gneissic composition and foliation.

In some instances the material inside a gneiss dome is itself gneiss, again with a foliation parallel to the dome surface. The foliation was presumably imparted during the doming, though some writers have suggested that prior thrust faulting created the foliation and this original planar gneissosity was domed up by later events. Either way, the kind of intrusion by 'doming up' rather than clear-cut plutonic intrusion is accepted as a fact.

Generally the evidence points to deep metamorphism, and bathydermal intrusion of the gneiss domes. Indeed Pitcher (1970) wrote: 'such domes are invariably found in high-grade metamorphic rocks so that a replacement origin of the granite component is entirely possible and a true relict ghost stratigraphy only to be expected.'

There are a few examples, however, that are formed at the surface. Ollier and Pain (1980) describe a gneiss dome from Goodenough Island, Papua New Guinea, which consists of amphibolite facies gneiss, is cored by granodiorite only 2.9 m.y. old, and which appears to have risen to be a surface landform – a dome 2 500 m high – by shouldering aside surrounding rocks. The main evidence is that the preservation of facets of the dome is too good to be produced by differential erosion, and if the whole region were uplifted since the granite intrusion it would be impossible to create the dome by selective erosion in the time available.

Another example is the Dayman Dome on mainland Papua New Guinea. This consists mainly of greenschist facies metamorphics, with a foliation everywhere parallel to the surface of a remarkably-preserved dome. The shape of the dome is indicated by generalized contours (Fig. 7.4) and part of the flank is shown by detailed contours

Fig. 7.4 Generalized contour map of the Dayman Dome, Papua New Guinea.

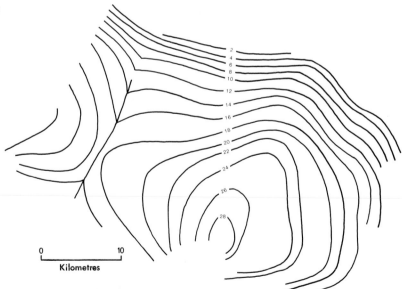

(Fig. 7.5) and an air photograph (Fig. 7.6). The small section of the north flank looks like a typical but exceptionally well-preserved fault scarp, which it is. Furthermore the increasing perfection of the slope

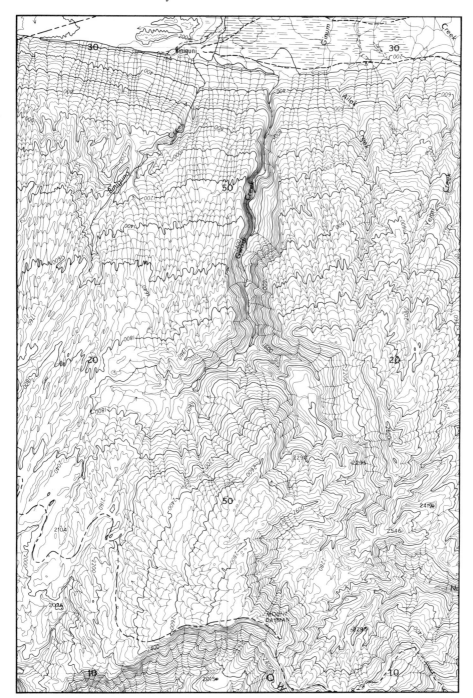

Fig. 7.5 Contour map of part of the
Dayman Dome.

and the reduction in dissection towards the base suggests that the fau
scarp is still emerging. The air photo even shows bands parallel to th
base of the scarp with different drainage density and tone, suggestin
bands of different age of emergence. The gneissosity is parallel t
the topographic surface or fault plane. The remarkable thing is th
this linear fault scarp can be traced continuously around the dome, an
onto the flatter upper surface, so the entire dome surface can b
regarded as a fault plane. As with Goodenough Dome it would b

Fig. 7.6 Air photograph of part of the Dayman Dome.

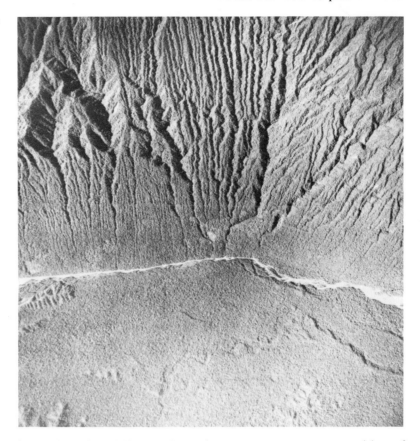

impossible for differential erosion to remove vast quantities of overburden to produce this dome, and it is regarded as a primary tectonic landform. The foliation is produced by shearing and perhaps plastic flow in a zone that defines the shape of the dome.

The Rincon Mountain dome of Arizona may be similar. It is a granitic gneiss mass bounded by a dome surface known as the Catalina Fault, which parallels the attitude of the foliation in the gneiss. Around the base of the dome are gravity slide structures that have a radial structure about the dome (Davis, 1975), so presumably the dome was pushed up as a topographic feature beneath the beds that slid from it.

The foliation parallel to the surface of gneiss domes may have formed by thrust faulting and been later domed (Davies, 1978), or it may represent the shear directions near the dome surface associated with vertical uplift of the dome (Ollier and Pain, 1980).

8　Volcanoes

Volcanoes and magma

Volcanic activity produces a distinctive set of landforms built from the products of molten rock or lava that erupts from beneath the earth's crust. Magma, which differs slightly from simple molten rock because of its dissolved gases or volatiles, is derived from somewhere near the base of the crust. It is thought that the sub-crustal material which is normally solid becomes liquefied for some reason such as sudden reduction of pressure and rises as magma through the crust by way of fissures or pipes until it reaches the ground surface as lava where it erupts in some form of volcanic activity.

In the ocean basins the dominant magma is of basaltic composition. In chemical composition this is almost identical with the mantle, and could be derived from it by melting. In the Pacific an area of entirely basaltic volcanoes is separated from those areas around the edge of the Pacific by an 'andesite line', on the shoreward side of which andesite magmas are commonly erupted, though basalt magmas are also erupted. Andesitic magmas have more silica than basalt magmas and usually erupt with greater violence. Even more silica-rich (more 'acid') magmas are known, with the approximate composition of granite, and these give rise to volcanoes of rhyolitic composition or to ignimbrite pyroclastics. Small amounts of andesitic and rhyolitic magma may be produced by various differentiation processes from basaltic magma, but the production of large volumes of acid magmas probably requires the melting of siallic continental rock in some way, and these rocks are confined to continents and their margins.

The fact that andesites are absent from the Pacific Ocean within the 'andesite line' (Fig. 8.1) shows that in the Pacific at least andesites are not created by fractionation of basalt magmas, and there seems little reason to suppose they are elsewhere. Plate motion rates may be relevant to the genesis of magmas, and Smith (1976) says that slip rates of about 10 cm/y (100 000 B) are commonly regarded as typical values needed to generate andesites. Liquid lava flows away from its point of eruption for distances that depend on local topography and the viscosity of the lava. Acid lavas are the most viscous and never flow freely: basalt lavas flow more readily and can form very long thin flows. The surface of a lava flow has distinctive features which have been

Fig. 8.1   The Andesite Line.

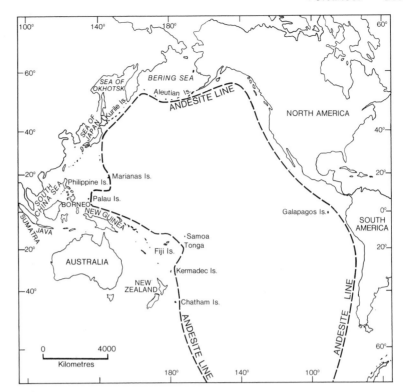

given names. Aa lava is blocky and slow moving, with a thick skin, broken into blocks, that ride on the massive, pasty lava underneath. Pahoehoe flows are the most liquid type. Cooling forms a thin skin that may be dragged into folds by movement of the still fluid lava beneath to form ropy lava.

Volcanic eruptions that contain large amounts of gas are explosive and produce pyroclastic deposits, fragments of volcanic rock ranging from dust size to huge blocks. Lavas may be innately explosive, as are most very acid lavas. Other pyroclastics are produced when lava hits the sea or groundwater, and production of scoria seems to be common in the last stage of some basaltic eruptions, presumably when the supply of lava is running down but gas is still abundant.

Some other volcanic products indicate the environment in which they were erupted. Pillow lavas, as the name implies, look like a stack of pillows, each with a glassy edge and more crystalline interior. They are a sure sign of eruption in water. In some places, such as Sicily, the level of pillow lava can be used to determine old sea levels. Elsewhere it is clear that the pillow lavas were deposited in the deep sea. These are often rich in sodium and are commonly associated with chert and serpentine – the so-called Steinmann trinity. Pyroclastic rocks produced by eruption in shallow water (called hyaloclastites) may be found associated with some pillow lavas and volcanic breccias in volcanic deposits resulting from an eruption into an ice cap. It is the recognition of such an origin for datable volcanics in Antarctica that gives the age for the appearance of the Antarctic ice sheet.

Major volcanic landforms

Volcanoes produce a number of well-known landforms, mainly of constructional origin.

Volcanic cones   The roughly conical hill or mountain is the most obvious feature of volcanicity. They may be built of lava, of pyroclastics, or a mixture of the two. Shield volcanoes consist mainly of lava and are usually very large, with gentle slopes like the major volcanoes of Hawaii.

Scoria cones consist of generally coarse pyroclastics are fairly steep sided, and generally small, many less than 300 m. Ash cones are similar but have finer pyroclastics and more gentle slopes.

Strato volcanoes consist of alternate layers of lavas and pyroclastics. Most of the world's great volcanoes such as Fuji, Vesuvius, Egmont or Rainier are of this type. They are the typical volcanoes of popular imagination and art.

Volcanic cones may also be divided into monogenetic, those volcanoes with a single style of eruption like scoria cones or lava cones, often resulting from a single series of eruptions from a single centre, and complex volcanoes, including both shields and strato volcanoes, where eruptions of various kinds occur, several eruptive centres may operate and parasitic volcanoes are common, dykes and pipes intrude the volcanic pile, and eruptions may be repeated near the same site for thousands of years.

Craters   Most volcanoes have a crater at the top, the vent from which lava is erupted. In the maar-type volcano the crater extends below the general ground surface into sub-volcanic bed rock and is surrounded by a rim of pyroclastics. Such craters must be explosively formed, in contrast to the more normal type which is built up of volcanic ejecta.

Lava flows, plateaus or plains   These features result from the eruption of rather fluid lava, and are generally confined to basaltic eruptions.

Ignimbrite plateaus   These are major igneous landforms resulting from highly explosive eruption of intermediate or acid lavas.

Calderas   These are depressions in volcanic areas or on volcanic centres resulting from either vast explosions or tectonic sinking, perhaps following eruption. Fig. 8.2 shows an example from Rabaul, New Guinea, where the caldera is occupied by the sea, making a great harbour surrounded by volcanoes on the caldera rim.

For further details of volcanic landforms readers may refer to Cotton (1944), Ollier (1969), Green and Short (1971) and Green (in

Fig. 8.2   Caldera at Rabaul, Papua New Guinea (photo by John Barrie).

press). In this chapter the interest is not so much on constructional and erosion volcanic geomorphology as on the tectonic relationships of volcanic features.

Basalt volcanics

Basaltic magmas are generally intruded through cracks which open through tension and become filled with magma to form dykes. These not only provide the feeder channels for surface volcanoes but they themselves add considerably to the bulk of the crust they traverse. Iceland has been spreading at a rate of 5 cm/y by dyke intrusion. In Arran, Scotland, Tertiary volcanic intrusion accounts for an extension of the crust of 1 in every 15 kilometres.

Crustal spreading on Iceland takes place along a series of rift zones, broken by several transverse fracture zones. Each lava pile appears to have built up as a series of lenticular lava units, each representing a single eruptive episode and related to its own feeder dyke swarm. Iceland has thick crust and deep seismic activity. Some workers postulate a primordial mantle plume magma source beneath Iceland.

At the ground surface eruptions from dykes give rise to fissure eruptions. These commonly erupt liquid lava which flows readily, covering wide areas and modifying the surface topography by flows, but not building great volcanic mountains, though small scoria cones may grow along the line in the last stages of activity.

Towards the surfaces many dykes break up, and the ascending sheet of magma turns into a series of distinct pipes. Each pipe gives rise to a volcano when it reaches the surface.

Some volcanic activity is dominantly around a single vent in contrast to that along fissures. These are the central volcanoes and include the low-angle shields of Hawaii, and steep oceanic islands like Tristan da Cunha. Activity continues at the one centre for periods of thousands or even millions of years. Sometimes lava is produced, sometimes volcanic ash and other fragmental pyroclastic deposits, but flows predominate in most basaltic volcanicity. Scoria volcanoes tend to be late-stage eruptions and rather small.

Areal volcanism (also known as polyorifice volcanism) is characterized by monogenetic volcanoes and the absence of any tendency for eruption centres to be localized at definite points for any length of time. Individual volcanoes are short lived, ranging from a few weeks to perhaps 12 years, and seldom grow bigger than 450 m. Scoria cones, lava cones and maars are the dominant volcanic types and strato-volcanoes are absent or rare. The spatial distribution of the volcanoes is irregular, with some clustering and occasional linear groups. The petrographic composition is basaltic and remains fairly constant within an areal province. Examples of areal volcanic provinces include Victoria (Australia) and the volcanic regions of Auvergne, Armenia and Mexico.

The type of volcano, the nature of the magma and the nature of the eruption are very much affected by the tectonic setting. Mid-ocean vulcanicity is largely of fissure type but may be confined at times to a few central volcanoes. The island of St Helena is now about 250 km east of the ridge, and probably had an origin like Ascension but it is now extinct and has been carried away from the active site by sea-floor spreading. It was, however, active from 15 to 8 m.y. ago, though not continuously.

Other volcanoes of the oceans appear to be related not to sub-ocean ridges but to hot spots under the crust. The hot spot periodically erupts and forms a volcano. After a period of quiescence the hot spot is reactivated, but meanwhile the crust has drifted over the hot spot and so the newer volcano is in a different position relative to the crust, as described in Chapter 2.

Hot spots can also lead to eruption on continents. Wellman and McDougall (1974) plotted the age of central volcanoes of eastern Australia against latitude (Fig. 8.3) and found a remarkable correlation. They suggest that the Australian continent has moved in a northern direction over two hot spots. The apparent southward migration rate of the centres of vulcanism in eastern Australia, 665 mm/1 000 y, then becomes the rate of northerly movement of the Australian plate over the asthenosphere.

Fig. 8.3 Relation between latitude and age of Cenozoic volcanoes of Australia. Dots = central volcanoes; rings = lava field eruptions; triangles = areal provinces (data from Wellman and McDougall, 1974).

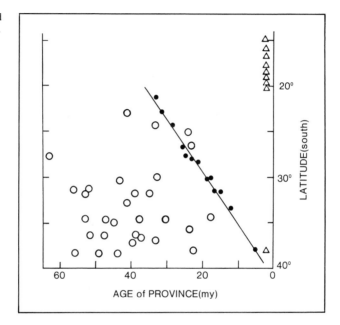

Basaltic lavas are generally the most fluid, so they can flow for great distances, like the 120 km Frambuni flow from Trolldaungja in Iceland. On level ground lavas can form a volcanic plain of great extent with little volume, like the 150 000 km² volcanic plain of Western Victoria. The Laki flow of 1783 in Iceland covered an area of 565 km² in seven months, with an average thickness of about 25 m.

On irregular topography lava flows down valleys, sometimes filling them and spilling over interfluves, and generally disrupting earlier drainage systems. Sometimes lava production is on such a vast scale that even the valleys of originally mountainous regions are completely filled and a lava plateau is produced. The Columbia River Plateau, for instance, which is of Miocene age, completely buried an original topography with a relief of over 1 500 m – deeper than the Grand Canyon. The Columbia Plateau basalts average 1 000 m thick, made up of many individual flows averaging 10 m thick and they cover an area of about 130 000 km².

Rather similar, and frequently treated together with the Columbia River basalts are the Snake River basalts of Quaternary age covering 50 000 km². The Deccan traps of India are of Cretaceous to Eocene age with a present area of 500 000 km² and thickness up to 2 000 m. The Cretaceous Parana plateau basalts of Brazil cover an area of over 750 000 km² and the Karroo basalts of South Africa, which range from Late Triassic to Early Cretaceous, cover 140 000 km² but were once probably ten times as extensive (Cox, 1972).

Such vast deposits could only be derived from the mantle, with minor additions from other sources, and must relate to very major tectonic events. Cox (1972) considered that the Karroo basalts were related to a single thermo-tectonic event related to the establishment of the continental margin of southeastern Africa during the fragmentation of Gondwanaland. The waning stages of this cycle overlap with the main period of activity in South West Africa which is presumed to be associated with the development of the South Atlantic coastal margin. It is postulated that a rise of mantle material approximately equidimensional in plan, took place beneath the central part of Gondwanaland, and caused the observed surface volcanic and tectonic effects.

Windley (1977) considered that the Jurassic lavas of Gondwana-land lie on the foreland of the Mesozoic fold belt, and represents an abortive attempt at fragmentation, while the Cretaceous lavas are located along or near actual separated continental margins.

Intermediate and acid volcanics

A completely different setting for volcanoes is the Pacific margin of the Americas. Here andesitic volcanoes are dominant, and they occur along distinct lines or chains. Why the single chain of southern South America gives way to a double chain at latitude 30°S is not clear, but there appear to be definite lines at depth governing the location of these huge central volcanoes. It is significant that this area is also intruded by many batholiths of andesitic composition. The line continues into North America, but is less distinct. The Cascade Range of the United States is almost entirely volcanic. It is contemporaneous with the Columbia River basalts to the east, but whereas the latter are basaltic the Cascades are andesites and granodiorite intrusives. Later stages produced the many large composite cones such as Mt Rainier, Mount Hood, Lassen Peak and Mt Mazama.

On the western side of the Pacific island arcs predominate, and here again andesitic volcanoes are present, similar in most respects to those of the Americas. One interesting feature discovered in island-arc volcanism is that the potassium content seems to vary systematically with distance from the front of the arc (Fig. 8.4). Some have related

Fig. 8.4 Variation in potassium content of andesites across an island arc (after Wright, 1971).

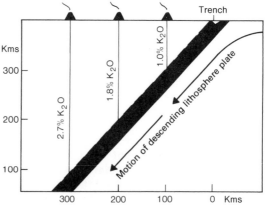

this to depth of the subducted plate, but the correlation of potassium content with the assumed depth of the plate, or the depth of the Benioff zone as determined by earthquakes, does not actually provide an explanation for chemical variation.

Island arcs may have double bands of contrasting volcanism, with eruption of basic volcanic rocks, pillow lavas and breccias on the sea floor in the trench, and intermediate to acid volcanics making andesitic volcanoes or ignimbrite deposits on the volcanic arc. The same relationship, if found in the past, suggests an arc situation, as in Wales where throughout the Ordovician acid volcanics appear as sub-aerial ignimbrites, while basic volcanics always occur as submarine pillow lavas and pillow-lava breccia.

A mixture of solid particles suspended in a gas can act as a liquid, a principle used in industrial 'fluidization' for the transport of such materials as cement and coal dust through pipes. The process occurs naturally in some volcanic eruptions, when pyroclastics suspended in volcanic gases flow like extremely mobile liquid. They look like rapidly projected dust clouds, and at night they glow, and they are generally known as *núees ardentes,* translated as 'glowing clouds'.

An eruption of this type deposits a layer of volcanic ash known as an ignimbrite (Cook, 1966) which is a rock unit term, not a petrological term. Rhyolite and andesite composition is commonest, and many ignimbrites are welded, that is gases are expelled and the hot ash sticks together, to form welded tuff. Welded tuffs often look very like rhyolite, and many rocks that were identified as rhyolites in the past are now known to be ignimbrites. In Cotton's classic book of 1944 the ideas of ignimbrite had not been incorporated, and he writes of rhyolite plateaus.

Pyroclastic flows are very mobile. They flow downhill and fill depressions rapidly, and when they come to rest they have a remarkably flat top. Flows have been traced for 70 km from Mt Mazama, Oregon. The Lake Toba flows of Sumatra cover 25 000 km^2 and the Taupo–Rotorua ignimbrites have an area of 26 000 km^2.

Calculated volumes of ignimbrites are very great. The New Zealand field has a volume of 8 300 km^3, the San Juan ignimbrites of California 9 500 km^3. Individual calderas can produce large amounts such as 90 km^3 from Aira calderas and 80 km^3 of Aso, both in Japan.

Ignimbrites erupt very quickly. The Katmai eruption in Alaska produced 28 km^3 in 60 hours, and it is estimated that the 2 800 km^3 of northern Queensland ignimbrites were erupted in a few days (Branch, 1966). On the other hand some ignimbrite areas had many pauses during construction, with sufficient time for forests to be established. In the Lamar River area of Yellowstone Park there are the remains of 27 distinct forests, with trees in an upright position, each buried in its turn by pyroclastics.

Other volcanic types

The volcanoes of the Mediterranean area are sometimes classified as a separate group. They tend to be relatively rich in alkalis, especially potassium, but they show considerable variety. However, central volcanoes tend to be dominant. The Mediterranean is a complex tectonic area, and it would be surprising if one simple association of volcanic types could be found in this area.

In rift valley areas some volcanoes are basaltic, but there are no andesites or rhyolites. Many lavas are very rich in alkalis, sodium and potassium, even when highly basic, and some are rich in carbonate including the remarkable Oldoinyo Lengai in Kenya. Carbonatite is a volcanic rock consisting largely of igneous calcite, and suggests vast accumulations of carbonate at the base of the crust.

Alkaline/carbonatite volcanism recurs at continental-rift intersections over a long time period during which plates have moved long distances. This indicates that fixed plumes with sources below the lithosphere cannot be responsible, but rather the structural fabric of the lithosphere is determining the site of the magmatism and its character (Bailey, 1977). The distribution and origin of carbonatites show they are somehow related to continental interiors, but further tectonic correlations are not yet available, although it has been suggested that the alkaline lava and carbonatite association of West Africa and Brazil may be related to aulacogens.

Kimberlite is a variety of peridotite, rich in olivine and notable because some kimberlite pipes contain diamonds. Geologically they are important because they appear to come from deep-mantle material. The depth of origin of kimberlites has been determined from the presence in some kimberlite inclusions of coesite, a high-pressure form of quartz, combined with the absence of its even higher-pressure counterpart, stishovite. This defines the point of origin of kimberlite as being within the upper mantle between 150 and 300 km, five to ten times deeper than the base of the crust. The fast speed of emplacement of kimberlites (estimated as about the speed of sound) is one of several features that together give rise to a unique form of intrusion that randomly samples material through its entire passage to the surface, and kimberlites contain abundant foreign inclusions (Fergusson, 1978).

Kimberlite contains rounded masses of eclogite, and presumably comes from or through an eclogite layer deep in the earth of a pressure equivalent to 110 km. It may be possible to produce ecologite by the high-pressure metamorphism of basalt, and it may be possible to produce basalt by pressure release on eclogite. The volume change between basalt and eclogite is about 100 to 85. Since eclogite is of basaltic composition, its fusion at great depths could account for the great floods of plateau basalts or the voluminous sills associated with continental break-up, and moreover such a transformation would inevitably cause vertical crustal movements on a considerable scale, because of the density changes.

About 100 ± 20 m.y. ago there was a peak in the intrusion of carbonatites, kimberlites, alkali and ultrabasic rocks possibly related to a particularly active phase of sea-floor spreading, the opening of the Tethys Sea, or similar tectonic activity.

The majority of kimberlites are of Mesozoic–Cenozoic age and are restricted to the interior and margins of stable cratons (Dawson, 1977), and rarely in rift valleys. Some are on swells, others in fault lineaments or graben. Some, such as the 94 Middle Cretaceous kimberlite pipes of Angola, line up with transform faults in the Atlantic.

Williams, H. R. and Williams, R. A. (1977) have pointed out that in West Africa kimberlite locations appear to fall on continuations of oceanic fracture zones under the continents, and Australian kimberlite shows the same relationship (Fergusson, 1978).

Subsidence and caldera collapse

To erupt vast quantities of rock in very short time, as in ignimbrite eruptions, requires very special tectonic conditions, and the likeliest explanation is that the magma chambers that fed the eruptions were very close to the surface. This idea is supported by geophysical data and by the discovery of old ignimbrites that have been intruded by a related pluton (Branch, 1966). Some mechanism of rapid release of lava causing expansion and vesiculation of neighbouring magma, which releases pressure beneath leading to more vesiculation and eruption, and so on is required – a kind of chain reaction that enables the magma to 'boil over' very rapidly. It is also necessary that the magma had abundant volatiles to provide the gas.

Such rapid release of magma, following the 'boiling over' analogy, might empty the magma chamber beneath a volcano and lead to caldera collapse. This seems to be borne out by the very frequent association of calderas with ignimbrites. The so-called Katmai eruption of 1912 led to Mt Katmai losing 240 m from its top and producing a caldera 5 km across, but subsequent investigation showed that an altogether new mountain, Novarupta was the source of the main eruption. Although the main eruption took place from Novarupta it was Katmai, 8 km distant and 1 500 m higher, that collapsed. Presumably they shared a common magma chamber, and the magma moved 8 km underground from beneath Katmai to be erupted at Novarupta.

Branch (1966) described subsidence of enormous areas in north Queensland, including one of 120 km × 55 km, and he suggested a distinction between cauldrons formed by collapse resulting from the subterranean withdrawal of magma from a deep-magma chamber, and calderas caused by collapse as a result of colossal eruptions of pumice derived from high-level magma chambers.

Volcano-tectonic depressions as defined by Van Bemmelen (1930) are surface depressions caused by the collapse of roofs of magma chambers brought about by rapid emptying of the magma. The type area was Lake Toba basin, Sumatra, and the idea was also applied to the Rotorua–Taupo depression in new Zealand. In both places it has been shown that the depressions are tectonic and pre-date the eruptions, and that subsidence took place intermittently over a long time, inconsistent with a single paroxysm by collapse of the arched roof of a magma chamber.

Settlement of volcanic cones

One of the fundamental differences between non-volcanic mountains and volcanoes is that the former are part of the earth's crust but the latter may be regarded as a heavy load placed upon the crust. In this respect a large volcanic cone resembles an ice sheet, a large lake, or a huge sedimentary pile. As with these other loads, volcanoes are subject to isostatic forces and settle under their own weight.

Large volcanoes contribute, by their own mass, to tectonic adjustment around them. The ocean crust is rather thin, and when a submarine volcano erupts it is presumed that it draws its magma from a considerable area around the vent. When activity ceases a heavy volcano is resting on a fairly limited area of thin crust, and the volcano then starts to sink. It may depress the surrounding crust to form an annulus of lowland around the base. Quite often the volcanoes are truncated by marine erosion to give them a flat top, and many flat-topped volcanoes have now subsided to depths well below the zone of wave activity. They are known as guyots. A volcano that subsides without flattening is called a seamount. Some volcanoes in tropical areas provide a base for the growth of coral reefs.

The eruption of large quantities of basalt magma might eventually lead to an emptying of a magma chamber leading to collapse of the overlying volcanic structures. In basaltic regions the mechanism of cauldron subsidence seems to be commonest, in which a cylinder of material falls as a single unit giving rise to a caldera known as the Glencoe type, after an example in Scotland.

A modern example of such a caldera is Askja in Iceland which has a rim of volcanoes pouring lava into the central hollow. Another example is Niuafo'ou near Tonga, where a complex of fissures forms a ring 5 km diameter around the central caldera. Ambrym, New Hebrides, has a caldera which appears to have subsided quietly, and is a Glencoe-type volcano although not a basaltic one. Silali caldera in the Rift Valley of Kenya is another Glencoe-type structure 8 km × 5 km, the floor of which has subsided 300 m without any significant eruption.

The settlement of volcanic cones causes various deformations at the foot of the cone such as ring faults, thrusts, and anticlinally uplifted ridges, all of which tend to encircle the volcanic cone. The effects may be divided into three kinds – fault type, fold type and mixed (Fig. 8.5), and which type actually occurs seems to depend on the nature of the sub-volcanic rock. Suzuki (1968) studied these effects on nine strato volcanoes in Japan and Indonesia and compiled information on calderas and guyots. The fault-type settlement occurs in cases where Tertiary sediments are less than 200 m thick, and where sedimentary

Fig. 8.5 Sections showing various types of deformation of volcanoes. (*a*) fault type; (*b*) fold type; (*c*) mixed type. P = Pliocene; M = Miocene sedimentary rocks (after Suzuki, 1968).

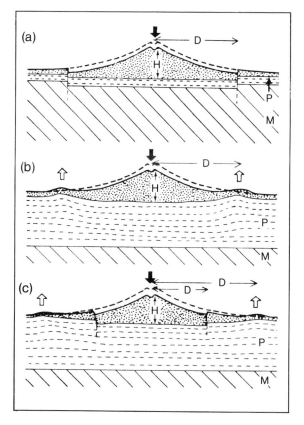

rocks are thicker – usually several hundreds to thousands of metres thick – the fold type or mixed type results.

The magnitude of settlement is about one or two hundred metres subsidence. The rate of subsidence has been determined from Iizuna, a fold type, and is about 4 mm/y or 4 000 B.

The distance (D) from the centre of the volcanic cone to the circular deformation feature is proportional to the relative height of the volcanic cone, which can be taken as a substitute for the weight. The same relationship between D and H is also found for guyots which are surrounded by circular moats or ridges, but is not found in the case of collapse calderas (Fig. 8.6, Table 8.1).

If a single large volcano can sink under its own weight it is surprising

Fig. 8.6 Relation between H (height of volcanic cone or guyot) and D (distance from the centre of volcanic cone or guyot to the deformed edge). I, settlement of volcanic cone: IA, fault type; IB, fold type; IC, mixed type (symbols a and b of IC show cases of fault and fold of the mixed type respectively); III, collapse calderas; IIIA, Krakatau type; IIIB, non-Krakatau type; IV, guyot (after Suzuki, 1968).

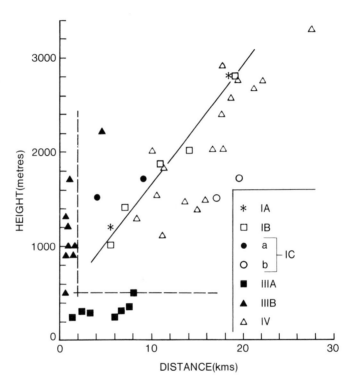

Table 8.1 Relative height of guyot (H) and distance from the centre of guyot to the circular moat or ridge at the foot (D) (after Suzuki, 1968)

Name of guyot	H (m)	D (km)
GA–4	3 290	27.6
GA–12	2 900	17.6
GA–23	2 750	19.4
Faris (a)	2 740	22.2
Faris (b)	2 655	21.2
Welker	2 560	18.6
Dickens	2 380	17.6
GA–7	2 015	17.6
GA–3	2 010	16.6
GA–15	2 000	10.0
GA–5	1 830	11.1
G22–128	1 520	10.5
GA–11	1 465	15.7
GA–14	1 460	13.6
135° 10′ W, 46° 05′ N.	1 370	14.8
GA–8	1 280	8.3
137° 10′ W, 45° N.	1 100	11.1

that the vast lava plateaus remain high, but they are high and have been so since Mesozoic times. It seems unlikely that they would have a thick siallic root – indeed areas of such basalt production are just where one might expect a lack of sial, and the reason for their continued existence as highlands is not readily explained.

Global distribution patterns

The distribution of volcanoes (Fig. 4.4) shows that most volcanoes fit on plate boundaries, but there are also some such as the central Sahara mountains of Tibesti, Hoggar and Jebel Marra, where some volcanoes are over 3 000 m high but located in the middle of a continental plate. In the oceans, the Hawaiian islands are in the middle of a plate and not related to plate edges, though as we have already seen they are related to a 'hot spot'.

The global distribution of volcanic types has been studied by Suzuki (1977). He used the List of the World Active Volcanoes of Katsu (1971) which lists 830 volcanoes. Of these 699 were sufficiently well described to be classified into volcano type. Suzuki devised a volcano-morphologic formula defined as an arrangement of the abbreviations of simple volcanic landforms in order of their formation. He found 57 varieties which were then grouped into six volcano series according to the type of the oldest simple volcano (usually the main cone) and the existence of a caldera.

The six types and their global population percentages are shown below:

Strato volcano	62
Strato volcano with caldera	10
Shield volcano	11
Shield volcano with caldera	3
Caldera volcano	7
Monogenetic volcano	6

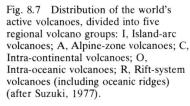

Fig. 8.7 Distribution of the world's active volcanoes, divided into five regional volcano groups: I, Island-arc volcanoes; A, Alpine-zone volcanoes; C, Intra-continental volcanoes; O, Intra-oceanic volcanoes; R, Rift-system volcanoes (including oceanic ridges) (after Suzuki, 1977).

Suzuki further divided the world's volcanoes into five regional groups: island-arc volcanoes (I), Alpine zone volcanoes (A), Continental volcanoes (C), oceanic volcanoes (O), and rift system and ridge volcanoes (R) (Fig. 8.7).

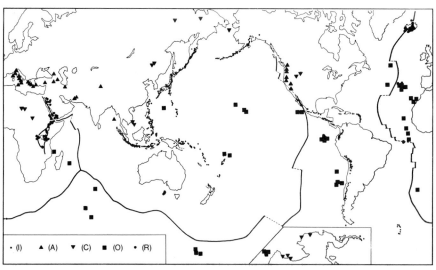

It is then possible to rank the regional groups according to the percentage of strato volcanoes, calderas or any combination of the volcanic types that they contain.

The most remarkable fact, according to Suzuki, is that the strato volcano series decreases in almost the same order as the shield series:

Strato volcano series	I	A	C	R	O
Shield series	I	A	C	O	R

This correspondence may reflect the regional difference in degree of the explosivity of erupted magmas. To make this clear quantitatively the morphologic explosion index, defined as the population percentage of the pyroclastic-rich volcanoes (including strato volcanoes, stratovolcanoes with caldera and caldera volcanoes) in the total volcanoes except the monogenetic type was calculated for each regional volcano group.

The results suggest three major volcano groups based on the morphological explosion index:

Morphological explosion index (%)

1. Island arcs	97
2. Alpine	
Continental	82–77
3. Oceanic	
Rift and ridge	41–38

The MEI may be compared with Rittmann's explosion index (1962) which is defined as the volume percentage of fragmentary material in total material produced and is calculated for the data between 1500 and 1914. Good agreement between the two suggests that within each regional volcano group the explosivity of erupted magmas remains unchanged for a long time – Suzuki suggests from the late Pleistocene to historic times.

Other results from Suzuki's analysis are:

1. Crater Lake (Krakatau) type calderas are about nine times as common as Kilauea-type calderas.
2. The percentage of volcanoes with parasitic or incidental volcanoes decreases in the order: caldera volcano; strato volcano with caldera; shield volcano; shield volcano with caldera; monogenetic volcano.
3. Lava domes rarely co-exist with other kinds of parasitic volcanoes.
4. It is expected that in the volcanoes to be formed in the future about half of the monogenetic volcanoes will be lava domes, while there will be hardly any lava domes amongst the strato volcanoes with caldera.

Aspects of the global distribution of volcanoes have been noted by many other workers, and some have been pointed out in various parts of this book. Island-arc volcanicity, hot spots, and the association of carbonatites with continental volcanicity are some of the topics noted, and a great deal of current work amongst petrologists is devoted to regional tectonic aspects of volcanicity. I have emphasized Suzuki's work here because it starts from morphology, while most others' work is based on theoretical petrology or tectonics.

Regional distribution patterns

In Guatemala there is a line of large composite andesite cones in the west, parallel to the coast. In the east most volcanoes are smaller, basaltic, and more scattered (Williams and McBirney, 1964).

Volcanoes of Mexico fall onto two lineaments. One is east–west possibly an extension of the Clarion Fracture Zone (a transform fault that can be traced for 3 000 km into the Pacific. Popacatapetl, Colima and Barcena (born 1952) are on this line. The second lineament is NNW, parallel to the structural grain of the Mexican plateau, and includes Jorullo and Paricutin.

The spacing of volcanoes may be related to the crustal thickness, as postulated by Shand (1938) for the Galapagos Islands (Fig. 8.8). As in breaking a slab of chocolate or toffee there is a smallest size that can be broken, roughly equal to the thickness of the slab, so the block defined by volcanic lineaments may indicate the thickness of the ocean crust beneath the Galapagos.

Fig. 8.8   Spacing of volcanoes in the Galapagos Islands (after Ollier, 1969).

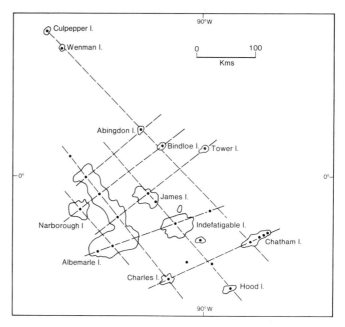

Early attempts at relating distribution were rather poor but with modern geophysical methods of determining crustal thickness a good relationship has been demonstrated (Smith, 1978).

A detailed account of structural relations of volcanicity in a small region is given by Joyce (1975) for Western Victoria, Australia where there are about 400 points of eruption.

Nearly two-thirds of these are in the Western Highlands, a fairly low plateau consisting of Palaeozoic rocks. The rest are on the Western Plains and overlie Tertiary sediments (Fig. 8.9). Lineament analysis has generally been unsuccessful, although some volcanoes occur in lines of three or four points. In some areas fault lines can be related to the distribution of vents, but only in a broad way, like the volcanoes northwest of Mt Gambier which show a general northwest alignment on a fault block bounded by a northwest trending fault zone. On the plains a northwest trend is clearly shown in rivers, valleys and lakes but is not followed by the volcanoes. The southern edge of the volcanic field is remarkably straight and is probably controlled by faulting.

Uplift appears to have preceded the main volcanic activity in the Highlands, perhaps by a substantial length of time as the pre-volcanic drainage pattern has suffered no major disruption. The volcanic

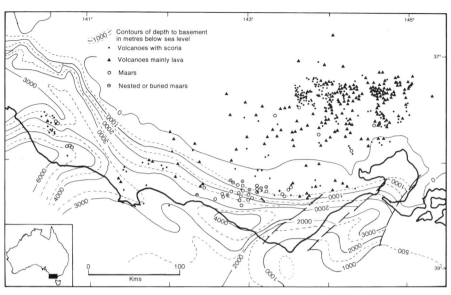

Fig. 8.9 Distribution by type of volcanoes in southeastern Australia, showing relation to basement contours (from Joyce, 1975).

activity appears to have begun on the edges of the Highlands, adjacent to areas of uplift, and then migrated into the uplifted area of the Highlands and out on to the Plains. The Western Plains show volcanicity related to a tectonic trench, with the activity beginning away from the trench and then spreading across the Plains to culminate in maar activity over the faulted edge of the sedimentary basin. This supports the view of Karapetian (1964) who thought that areal volcanoes gravitate towards tectonic trenches or to zones of deep faults. The Highland activity is like that of Shirinian (1968) who described areal volcanics as volcanic activity of monogenetic type represented by numerous independent centres of eruption that shift towards arched parts of volcanic uplands.

Intrusive basic rocks

Intrusive basic rocks are commonly feeders to volcanoes, and they occur in a few fairly well-defined forms (Fig. 8.10). Intrusive igneous rocks cool more slowly than their volcanic equivalents so are more coarsely crystalline and the crystals can be seen with a hand lens. The commonest basic intrusive rock is dolerite (diabase in America).

Dykes Most volcanoes are fed through vertical fissures, which eventually become vertical sheets of igneous rock called dykes. These vary in thickness from a few decimetres to hundreds of metres, but 1–10 m dykes are the commonest. They often occur in swarms, sometimes radially disposed around a volcano, and sometimes in roughly parallel bands perpendicular to a direction of regional tension. Parallel swarms are probably formed deep in the crust, radial ones nearer the surface. As it approaches the ground surface a dyke often splits into a number of distinct pipes, but in other instances a dyke may intrude right through a volcanic pile to the ground surface.

Necks and pipes These are the cylindrical feeders of volcanoes, and seem to be confined to a zone fairly close to the ground surface. They seem to be more common in more acid rocks than basalts, and complex pipes are found in various rocks such as the carbonate plugs of Africa.

Fig. 8.10 Diagrammatic representation
of igneous intrusive bodies (after Ollier,
1969).

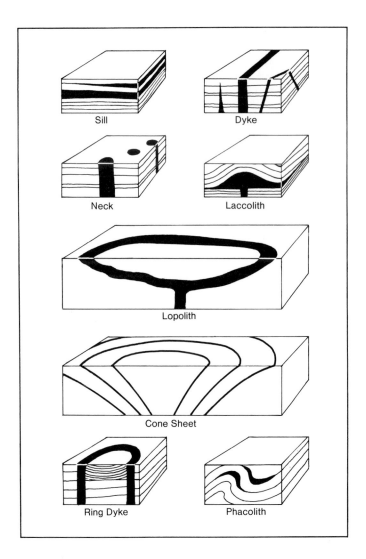

In rare instances where a volcanic neck can be descended, as at the
Shaft, Mt Eccles, Victoria, and the Hill With A Hole In on Tristan da
Cunha, the vertical pipe widens rapidly at a depth of only some tens of
metres, suggesting that the pipe is only the very last stage of what was
mainly a dyke eruption.

Some 'blind' volcanic pipes are known from Western Victoria which
fail to reach the surface, and they appear to have been emplaced by
fluidization.

Stocks or plugs   Batholiths of basalt are unknown, and the largest
intrusive bodies are known as stocks when exposed at the ground
surface. These are the solidified remains of magma chambers at fairly
shallow depths. In Hawaii shallow sub-volcanic magma chambers have
been identified by geophysical means. The largest plug is 20×12 km
and extends from 1 km below ground surface to 23 km deep. One plug
at Kilauea has a very weak magnetic anomaly so it is still at a
temperature above the Curie point.

Cylindrical or ellipsoidal magma chambers seem to be the norm, but

Etna has a plate-like magma chamber and Reunion has an extensive flat-based, centrally-domed sheet.

Sills   Intrusive rocks may form nearly horizontal tabular bodies of rock called sills, which usually follow the bedding of enclosing sedimentary rocks, and are said to be concordant. They eventually transgress or cut across the beds and spread along other bedding planes. Of course to get sills it is necessary to have horizontal bedrock, and sills are especially abundant in basins of thick unfolded sediments such as are found in South Africa, Tasmania, Antarctica and Brazil, and which were laid down not long before continental drift.

Sills can be very thick, like those of Tasmania which attain 700 m. They can also make up a considerable amount of the total rock pile after intrusion, as in South Africa where the sills in the Karroo make up 15–25 per cent of the geological column in the area. It seems that intrusion of sills lifts the overlying sediments, which are floated up to make room for the ascending rock, and considerable uplift of the ground surface should result.

Ring dykes   These structures are circular in plan and dip vertically or outwards at high angles. They seem to fill the space around a subsided cylinder or slightly conical mass. The material found inside is often volcanic with inward dips, derived from eruptions from the ring dyke. Subsidence of this kind is called cauldron subsidence and is often associated with caldera collapse. A present-day example of the surface expression of a ring dyke is provided by the island of Niuafo'ou near Tonga, which is a basalt dome with a caldera surrounded by a complex of fissures 5 km across which has erupted in the present century. Many more ring dykes are revealed by considerable erosion, and it seems probable that most ring dykes fail to reach the ground surface.

Cone sheets   These are concentric assemblages of curved dykes inclined inwards towards a common centre. They are inverted cones that become flatter with distance from the centre (Fig. 8.10), and the outermost cones also flatten to the surface like the flared bell of an inverted trumpet. Individual cone sheets are about 10 m thick, but presumably the intrusion of a mass of cones causes uplift of the bedrock within the cones.

From the dip of the cones it is possible to work out the depth of the apex of the cones, which turns out to be about 3 or 4 km.

Laccoliths   An intrusion in horizontal rocks that has a flat base but pushed up the overlying strata is called a laccolith. This intrusion domes the overlying rock and can make elevations of considerable size. The classic example is the laccolith of the Henry Mountains. It has been calculated that this was formed beneath 800 m of overlying sediment, all of which was forced up and would have made a mountainous landform. The present Henry Mountains are merely the erosional remains of this supposed dome.

Lopoliths   These are enormous saucer-shaped intrusions.

In Tasmania dolerite magma was intruded into horizontal Permian and Triassic sediments, lifting and floating them as a roof. In doing so the dolerite formed a number of very large and shallow saucers, each cradling a central raft of sediments, structures that could be regarded as lopoliths.

Apart from these fairly geometrical considerations, there ar
problems with the mechanism of intrusion of basaltic rocks, and th
source of the magma. Basaltic magmas seem to originate well in th
upper mantle, from depths up to 250 km. The rise through the earth'
crust of basaltic magmas occurs because the density of liquid basa
(about 2.6 g/cm³) is less than the mean density of rocks of the crus
After cooling, crystalline basalt that forms has a higher density (2.9 t
3.0 g/cm³). Therefore the final height of rise of an asthenolith i
determined by the relationship between the rate of rise and the rate c
cooling of the basalt (Beloussov, 1971). This may apply to some of th
deeper injections, but others move very much faster (see Ch. 17) an
cooling has little effect on the rising column. The figures ar
interesting, however, in partly explaining why a volcano tends to sin
after eruption.

A number of volcanoes have been erupted on domes pushed up i
the underlying topography. Various kinds of intrusion – sills, con
sheets, or ring dykes – occupy space which is created by lifting up th
overburden and so the ground surface is domed before the volcan
erupts. In some instances it is possible that the dome is created an
eroded into irregular topography before the volcanic deposits appea
Some examples are shown in Fig. 8.11.

Fig. 8.11   Examples of doming up of the
basement beneath volcanoes of East
Africa. (*a*) Napak. (*b*) Kisingiri volcano.
(*c*) South Ruri Hills. (*d*) Homa Mountain
(Napak section after King, 1949; the
others after King, Le Bas and Sutherland,
1972).

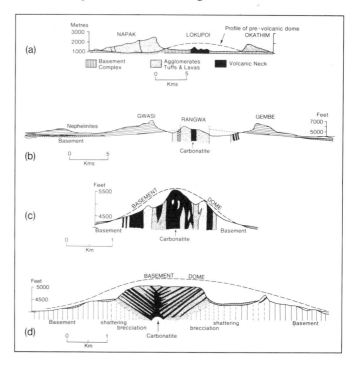

It is often assumed that the development of a volcanic centre an
concurrent intrusive activity takes place in an entirely passiv
basement which does not suffer any appreciable movement. Yet Kin
(1949) has shown that Napak volcano in Uganda is on a distinctl
dome basement. The centre of Mull, Scotland, is located on a dome
yet it is also associated with a caldera, which Rast (1970) explains b
the mechanism shown in Fig. 8.12.

The idea that stress release in the mantle can cause the initia
production of basaltic magma in the mantle was first suggested b

Fig. 8.12 The growth of magmatic domes and resultant calderas, based on deductions made in Mull and Snowdonia (after Rast, 1970).

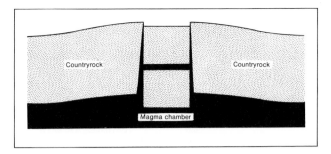

Yoder (1952) and advanced by Uffen and Jessop (1963). The release of stress associated with major earthquakes may reduce the melting temperature in parts of the mantle so that melting starts. Following this idea Kuno (1966) suggested that basalt was released from various depths on the Benioff zone. He suggested olivine basalts originate at depths of 250–400 km, high alumina basalts at depths of 250–150 km, and tholeiites at 150–100 km. Bailey (1964) drew attention to the coincidence of positions of crustal stress and volatile-rich alkaline magmatism, where stress release may act as a focus for any volatiles.

The ascent of primary magmas is conditioned by their increase in volume at the site of initiation. Lighter acid magmas may rise as a result of their buoyancy but heavier magmas move because of their inner pressure generated as a result of conversion from solid to liquid.

Fluidization

The study of volcanic pipes leads to new ideas on the nature of eruptions, especially of maar eruptions, and suggests a new process, fluidization. A mixture of solid particles suspended in a turbulent gas can behave in many ways like a liquid. Fluidization is a mechanism that accounts for the great mobility of nuée ardente pyroclastics flows, and it could possibly take place in igneous intrusions. Modern ideas of intrusion mechanism derive mainly from the work of Cloos (1941), who studied the Swabian pipes east of the Black Forest. These are Tertiary vents intruding Jurassic strata, and are funnel shaped as shown in Fig. 8.13. It was found that the larger blocks within the pipes

Fig. 8.13 Cross-section of one of the Swabian pipes east of the Black Forest. During intrusion of a pipe of tuffisite, a highly mobile mixture of gas and particles, the large blocks of bedrock are not thrown out but actually subside (after Cloos, 1941).

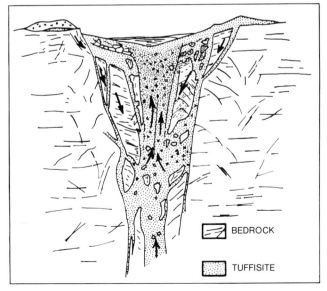

BEDROCK

TUFFISITE

could be stratigraphically matched with the wall rocks not far above indicating that they had subsided slightly rather than being thrown out Blocks in all stages of detachment from the walls were found, and were separated by veins or masses of 'intrusive tuff', which Cloos termed 'tuffisite', to distinguish it from the tuffs normally deposited on the ground surface.

Evidently the eruption took place by streaming of material around the larger blocks. Further details of the mechanism of fluidization were brought to the notice of geologists by Holmes (1965), who postulates the following:

Energy liberated at great depths is not able to throw out shattered rock like an explosion in a quarry, but may give rise to blast waves that shatter adjoining rocks. Wherever a head of gas is concentrated as, for instance, on some irregularity on the top of an ascending dyke, it will penetrate the tiniest cracks, extending them in all direction. Gas flow may be somewhat irregular, at first; but will soon become concentrated in a vertical direction, that is in the direction of lowest pressure. Rising streams of expanding gas thus force their way to the surface preparing channels for further intrusions. Along any channel of swiftly-streaming gas, irregularities are subject to thermal stress and mechanical battering, and so are abraded, adding more dust and particles to abrade and widen even more the passage. However irregular the initial passageway may be, its widening will tend to produce a cylindrical form, flaring into a funnel shape towards the surface, where lower lateral pressures allow a rapid expansion of gases and a more rapid erosion.

This mechanism for the initial drilling of a volcanic pipe would be effective even if the only gas present was derived from the magma itself, but it would be much more effective in the presence of groundwater which will flash into steam, and so co-operate in the work of disintegration.

Diamond pipes are also thought to be emplaced by fluidization and some reached the surface to produce maar-type volcanoes. The rate of ascent is very great, reaching almost the speed of sound at the ground surface, after originating at a depth of 150–300 km some minutes earlier (Fergusson, 1978).

9 Landslides to gravity tectonics

A number of types of landslide are shown in Fig. 9.1. One simple kind is the rockfall that occurs on very steep cliffs when they are undermined by erosion. Other slides depend on rock structure as in 9.1(*b*): if strata are dipping towards space a slide is possible or even probable; if strata dip towards the rock mass sliding is unlikely. Some slides occur on homogenous material in slopes that are not particularly steep as shown in 9.1(*c*), where a curved plane of failure develops and the upper material rotates (so this is called rotational slump). In slides mentioned so far the rock remained fairly coherent. If it is highly fragmental the landslide takes the form of a debris flow (if the debris is varied) or a mudflow is the debris is dominantly mud.

The rotational-slip type of landslide is especially significant as it grades upwards in size to tectonic landforms, and generates folds and faults that are exactly like those of 'tectonic' origin: indeed many features once thought to be caused by tectonic compression are now known to be caused by gravity tectonics. Fig. 9.2 shows the extensional and compressional features associated with a large rotational slump.

To get a landslide there has to be a change in equilibrium from that in which the original valley side or cliff was formed to a new equilibrium reached only after landsliding. A reduction in stability since the formation of the initial slope may be brought about by increasing the load (as by forest growth, building a town, or simply wetting the soil); by reducing the support to the slope, as by undercutting of a river bank; or by reducing the internal friction of the rock material by weathering, for example.

The sketches in Fig. 9.1 relate to fairly small landslides, but sometimes vast slides do occur.

In the year 373/2 B.C., during a disastrous winter night, a strange thing happened in central Greece. Helice, a great and prosperous town on the north coast of the Peloponnesus, was engulfed by the waves after being levelled by a great earthquake. Not a single soul survived. The next day two thousand men hastened to the spot to bury the dead, but they found none, for the people of Helice had been buried under the ruins and subsequenly carried to the bottom of the sea, where they now lie (Marinatos, 1960).

Fig. 9.1 Landslide types. (*a*) Simple fall.
(*b*) Sliding along structure. (*c*) Rotational
slump in homogenous material.

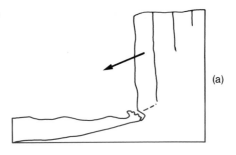

(a)

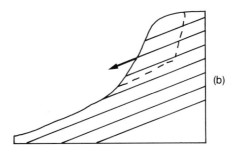

(b)

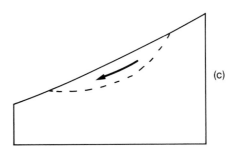

(c)

Fig. 9.2 Effects of large-scale rotational
slumping movements, with tensional
faults at the top and thrusting at the toe
(after Engelen, 1963).

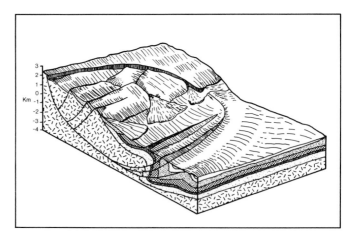

According to Seed (1968) this may record the earliest known case of a major landslide resulting from soil liquefaction induced by an earthquake. Helice was located on deltaic deposits of alluvium about 2 km from the coast. No trace of the city exists today, either on the ground or on the bottom of the sea.

Some clays are thixotropic and become fluid when subject to shock. A thin band of thixotropic (or sensitive) clay in a thick sequence of other sediments may become unstable and liquefy, leading to deformation of them all. Thixotropic clays have a structure rather like a jerry-built scaffolding, and if the struts are broken by addition of a lot of water or by a sudden shock the scaffolding collapses suddenly and the material behaves like a liquid. Similar behaviour is seen in tomato ketchup which may remain stiff and solid if tipped gently, but a sharp knock may cause it to flow very readily. Fragments in the ketchup build a framework that collapses when shaken. 'When you shake the ketchup bottle, first none will come and then a lott'l.' Thixotrophy is a common, but not essential factor in landslides.

Some very large slides take place beneath the sea from the eastern coast of North America. Heezen and Drake (1963) describe a submarine slide of over 10 000 km² approaching 1 km thick, which travelled many kilometres and came to rest on sediments of the same age as the sheet. The Agulhas Slump (Fig. 9.3) is a large submarine slump on the continental margin of southeast Africa 750 km long, 106 km wide, with a volume of over 20 000 km³. It is post-Pliocene in age (Dingle, 1977). An up-to-date review of rockslides and related phenomena is provided by Voight (1978).

Fig. 9.3 Sections of the Agulhas Slump, offshore southeast Africa. (*a*) Diagram of main morphological and structural features of submarine slumps. (*b*) An example of a trace of a seismic profile (after Dingle, 1977).

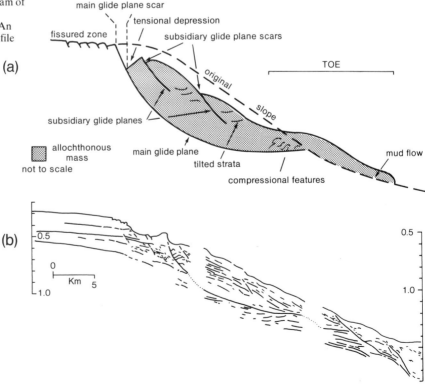

Gravity structures

The most significant feature of landslides from our point of view is that they are driven by gravity, not by an external force, although an external shock may trigger the movement. The weight of the sliding block provides the force for the slide, and the force is a body force acting on all parts of the sliding block. Gravity can produce various kinds of structure in sedimentary rocks besides landslides.

Some classical and informative structures were first reported from the Northamptonshire area of England (Hollingsworth, Taylor and Kellaway, 1944). Essentially the Oolitic limestone cap rock made a series of plateaus separated by valleys cut through to soft clays beneath. However, the cap rock did not remain horizontal, but was bent in two main ways (Fig. 9.4).

Fig. 9.4 Large-scale superficial structures in Northamptonshire. (*a*) Cambering. (*b*) and (*c*) Valley bulges (after Hollingsworth, Taylor and Kellaway, 1944).

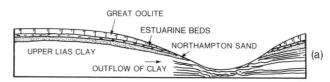

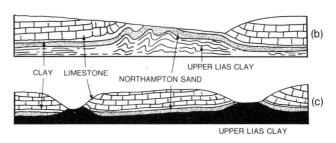

In one the plateau edges sagged down, extruding the soft clay from beneath the edge and it was subsequently washed away. This left the cap rock draped in a gentle arc on the interfluve, a structure known as a camber. The river appears to follow a syncline, but in reality the syncline followed the river.

In other situations the weight of the limestone apparently squeezed up the clay along valleys, where is was eroded, and the edges of the limestone were here bent upwards. The valleys are thus flowing along anticlines – but the rivers were initiated on flat strata and the anticlines developed only after the rivers cut down to the soft underlying clays. These features are known as valley bulges. Rocks which behave in a fairly rigid manner, like the Oolitic limestone in the former example, are known as competent beds; the rocks that flow, like the Lias clay in the last example, are said to be incompetent.

On a larger scale valley anticlines have been described from Utah that follow the sinuosities of the Colorado River for at least 35 km (Potter and McGill, 1978). The correspondence leaves no doubt that the anticlines formed as a response to valley cutting, but arching is evident as much as 600 m above the valley floor. To some extent the upward flow of evaporites in the underlying rocks is reponsible for these valley anticlines.

On an even larger scale are possible isostatic effects associated with the great Himalayan valleys such as that of the Arun (Wager, 1937). These huge valleys are sufficiently large to induce isostatic compensation, which has two effects. Firstly it causes the Himalayan belt as a whole to be elevated to a greater height than it would attain if undissected. Secondly it causes differential uplift along the length of the Himalayas, with greatest uplift at major valleys, least uplift where the weight of the mountains counters uplift. Most of the folds of the Himalayas have axes parallel to the ranges, but there are culminations – broad anticlines – that run across this trend, apparently followed by major rivers. As in the earlier, smaller-scale examples, it is unlikely that a river would follow an anticline, and more probable that the anticline formed as a consequence of the lack of mass along the valley.

A series of gravity structures in sedimentary rocks reported from Iran were perhaps the first such structures to illustrate the great importance of gravity in folding rocks (Harrison and Falcon, 1934, 1936). The structures are shown in Fig. 9.5, and they range in complexity from simple visor and knee folds to structures as complex as the cascades. The present vertical relief may be over 2 000 m, but

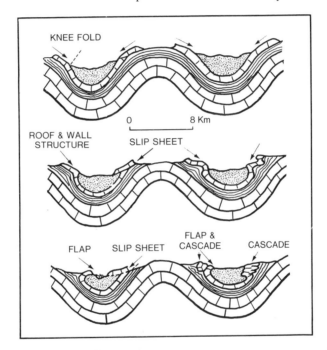

Fig. 9.5 Diagrams of gravity collapse structures of Iran (after Harrison and Falcon, 1934).

the secondary gravity folds start to form when the relief is no more than 600 m. These structures could easily have been interpreted as 'normal' geological folds, but the lack of folding in the underlying rock clearly shows that they are surficial features not related to compression of the entire earth's crust. The lack of folding at depth is a very important feature of gravity folding. It has been observed in many other areas, but the plain conclusion has not always been grasped.

Another important demonstration of gravity sliding comes from the sedimentary rocks bounding the Colorado Plateau. The plateau has been uplifted, and the stratified rocks have been shed in both directions. Clearly the uplifted block could not apply a significant lateral thrust in both opposed directions, but gravity slides could move in both directions from a dome.

Davis (1975) described gravity folding in another situation, around a gneiss dome, where no other explanation seems feasible. Most of the Rincon Mountains consist of granitic gneiss. It is bounded by a surface known as the Catalina fault, which parallels the foliation in the gneiss.

The gneiss is in the form of a double dome, which Davis interprets as due to folding of the Catalina fault and gneiss. Around the base of the dome, detached isoclinal folds, overturned asymmetric folds, and unbroken cascades of recumbent folds pervade sheets of sedimentary and metasedimentary rocks of Palaeozoic and Mesozoic age, resting on the Catalina fault, that is, on the dome surface. The slip-line directions define a radial pattern centred on the Rincon Mountains, and Davis concludes that the folding was brought about by local gravitational tectonics. This was thought to accompany the 28–24 m.y uplift that ended the metamorphism of gneiss in the Rincon Mountains, with décollement along the Catalina fault.

Erosional tectonics of the northwest Dolomites

The northwest Dolomites of northern Italy were isolated fairly early from more regional effects and from the stress gradients of the East Alpine geoanticline. In this protected position the sediments were virtually unaffected by regional stress fields and any movements were induced by their own potential energy.

In general the sedimentary column of affected areas consisted of three parts:

- *(a)*   An upper, more or less plastic zone.
- *(b)*   An intermediate zone of reefs, which are heavy and rigid, surrounded by incompetent strata.
- *(c)*   A lower zone with plastic strata and gypsum.

This arrangement is unstable because the heavy reefs (specific gravity 2.5 to 2.7) overly gypsum with a specific gravity of 2.3 and because the lower zone may move sideways under the load pressure.

The overburden pressure is maximal beneath the dense reefs and causes the lower zone to flow towards the areas with lower load between the reefs. Erosion of the incompetent rocks accentuates the differences and causes the reefs to founder. The extruded rock bulges up in domes between the reefs, and may push up the edge of the reef

Fig. 9.6   Diagrammatic representation of the foundering circuit in the northwest Dolomites. (*a*) Relationships of plastic and rigid rock masses. (*b*) Structural effects of foundering (after Engelen, 1963).

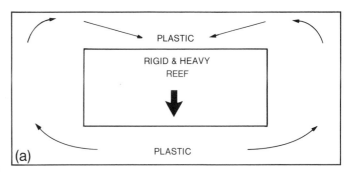

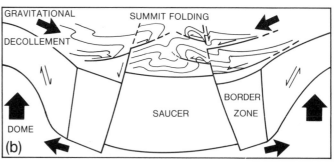

As the reefs are bent into basin shape, the upper zone may flow towards the depressions, causing summit folding. This is shown diagrammatically in Fig. 9.6.

An exceptional situation occurs with large reefs of Cargeranea, where the reef sags at the edges, but the major shape is a dome. With increasing internal pressure the top of the dome may be pierced, and the plastic rock may rise diapirically through it.

There is a relationship between the size of the rigid body (reef) and the type of deformation it undergoes when foundering. Three types were observed in the Dolomites:

(a) small rigid units fall apart into blocks (e.g. Sasso Lungo, 3 × 3 km, Fig. 9.7);

(b) rigid units of intermediate area assume a saucer shape (e.g. Sella, 7 × 6 km);

(c) the largest rigid unit assumes a dome shape and is diapirically pierced later on (e.g. Marmolada complex, 12 × 8 km).

Fig. 9.7 Break-up of a reef following outflow of plastic beds beneath (after Engelen, 1963).

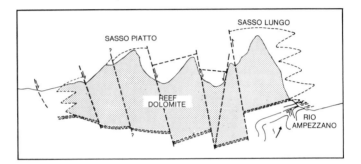

It is also possible for the rigid plate to be breached on a valley side. Once breached, the inner plastic rocks flow through, making a horizontal diapir that looks like a broken-up recumbent fold (Fig. 9.8).

All these structures have been described in detail in an excellent treatise by Engelen (1963). Presumably many other examples of erosional tectonics in the world await discovery.

Fig. 9.8 Lateral diapir on a valley side. The shaded bed is rigid, other rocks are plastic (after Engelen, 1963).

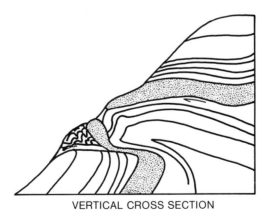

VERTICAL CROSS SECTION

Jura-type folding

Jura-type folding (Fig.9.9) refers to folding that does not involve the underlying rock. The folded rocks of the Jura could not be folded by any mechanism involving the whole crust; the folding is surficial, though the folded layer is several kilometres thick. The unfolded unconformity beneath the folded rocks clearly shows that the Jura folds were not pushed up from below, but that some sort of lateral force is responsible for the folding.

Fig. 9.9 Diagram of idealized Jura-type folding.

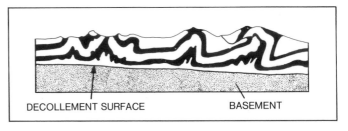

DECOLLEMENT SURFACE   BASEMENT

The Jura are a crescentic range assemblage. On the outer side are the tabular Jura, with the folded Jura behind, and behind this the almost unfolded sediments of the wide Swiss Plain. The Jura rocks may have slid from the direction of the Alps, but the old idea that they were pushed by lateral pressure from the Alps is untenable. The folding was assisted by the layer of Triassic anhydrite and salt along which the overlying beds could glide and fold. This zone of low fraction acts as a zone of detachment or, in French, *décollement* ('coming unstuck'), and this type of folding is sometimes known as décollement folding.

In the classical picture of gravity gliding the basal décollement dip towards the foreland. However, in the Jura it has been known for about 50 years that the basal décollement dipped towards the hinterland during deformation and this is also true for the central and southern Appalachians, the Canadian Rockies and the Idaho–Wyoming thrust belt. The classical gravity theory also requires the décollement eventually to cut up section at the back leaving a region of tectonic denudation rather like the scar produced by a gigantic landslide. Yet no case can be made for the existence of this type of tectonic denudation in the thrust belts mentioned above (Elliot and Johnson, 1978). They propose a more realistic theory of gravity *spreading*, rather than sliding, in which the décollement dips towards the hinterland, there is no tectonic denudation and the rock mass moves in the direction of the surface.

The idea of gravitational spreading had been conceived much earlier by Jeffreys (1931) who, although looking to a shrinking earth as a basic mechanism of mountain building, calculated the inevitability of gravity structures from the strength of materials and wrote that whenever an elevation over 3 km has been produced, either fracture or flow must set in; whichever it is, its general effect will be the spreading out of the elevated rocks over the surrounding country.

It must be admitted here, that although the Jura have a historical role in the development of gravity-tectonic notions, and that décollement, aided by evaporite beds is a reality, more recent work shows that the basement *was* itself compressed, and Laubscher (1973) claims that the stratigraphic situation unambiguously demands an uphill push at the time of the folding, and that if gravity sliding indeed played an important role it will have to be sought in the rotation of the entire sedimentary sequence away from the Alps.

Leaving aside the Jura themselves, there are many other examples of décollement folding from around the world. The folding of the Cambro–Silurian in the Oslo graben, for example, is a typical epidermis folding. The rigid Precambrian basement did not move. The zone of décollement is made of alum shales, about 50 m above the base, and the lower Cambrian still adheres to the basement.

Alpine folding and nappes

The structure of the European Alps is dominated by great nappes, huge sheets of rock that have clearly moved over fault planes at low angles, commonly bringing old rocks to lie over younger rocks. Early interpretations suggested that a great squeezing of a sedimentary basin could produce the observed folds and faults, but gravity tectonics is now more acceptable (Rutten, 1969).

The nappes show several features that provide convincing support for the gravity hypothesis, including the following:

1. Breakers at the nappe front (Branden der Deckenstirne). The nappes of the Helvetide Alps of Switzerland have a sub-horizontal structure except along the front, where they are crumpled into narrow folds. Where the reaction between moving force and resistance is greatest is where crumpling would be expected: this is at the front with gravity sliding, at the back with push. The crumpled zone often has cascade folds with axial planes dipping at 45° or even less which can hardly be explained except as a series of beds sliding into a depression under the influence of gravity.

   The nappe-front breakers are not restricted to the frontal part of external nappes, but normally occur in the frontal part of every nappe or nappe lobe.

2. The youngest beds travel farthest. In the Helvetide nappes, for example, the farthest-travelled nappes in the north are mainly formed by Cretaceous rocks; then comes the Jura nappe formed mainly by Jurassic rocks; and lastly the Verucano nappe which is mainly Permian.

   This follows from gravity tectonics, where the highest units have greatest potential energy. It is hard to see on a tangential-pressure model how a differential push could move the higher parts of a series several tens of kilometres further than the lower parts. How could such a force continue to be applied once the nappe had moved away if the pushing part is itself rooted in the crust?

3. 'Chevauchement intercutanes'. This refers to a structure where a series of overthrusts are found to exist in a seemingly tranquil, apparently unmoved strata. This is common in the Alps, and also many other places. The intercutaneous nappes mark an internal zone of great disturbance that took place inside a thick series of sediments without disturbing either underlying or overlying strata. It is hard to see how this could be obtained by lateral compression, but could, easily result from a gravity slide during deposition of the sedimentary pile.

4. The Pelvoux massif in the south of France is surrounded by nappes. This massif had emerged from the sea by the end of the Oligocene, and so presented an obstacle against which nappes were brought to a standstill when they collided, buckling and imbricating the strata of the toe. Nappes from the main axial region of the French Alps moved westwards, and piled up against the eastern side of Pelvoux massif: nappes from the north piled up on the northern side; nappes from the south piled up on the southern slopes. Thus there is a centripetal pattern of structures and a centripetal pattern of forces. No scheme of lateral compression can account for this pattern, but

Fig. 9.10 Section through the Naukluft Mountains from NW to SE showing the undeformed Unconformity Dolomite between older and younger strata that have been intensely folded and faulted. The basement rocks are disturbed only by block faulting (after Korn and Martin, 1959).

gravity sliding presents an easy explanation. Pelvoux was an island-like obstacle surrounded on three sides by slopes on which gravity sliding took place. The nappes of Pelvoux now reach heights of more than 3 000 m, higher than the source area of the nappes.

5. Perhaps the finest example to demonstrate these gravity-slide ideas is provided by the Naukluft Mountains in southwest Africa described in masterly fashion by Korn and Martin (1959). The situation is shown in Fig. 9.10. The bedrock of Precambrian rocks is

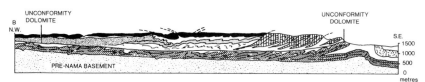

unfolded, but is overlain by a series of intensely-folded rocks which evidently moved from northwest to southeast. The intensity of deformation increases to the southeast, where the repeated thrust faults known as imbricate structure are found. This is the effect of breakers-at-the-nappe-front already described. A lateral push in the east would cause greatest deformation where applied, and its effects would die out to the west, just the reverse of what is found. A sliding mass of rock, however, would be most deformed at the front end, where it crashed into the obstacle that brought it to rest. This is what is found. But there is an even more remarkable feature in the Nauklift. After the faulted and folded rocks were emplaced they were eroded to a plain, and then a new series of sedimentary rocks were deposited unconformably on the lower, folded and faulted series. Originally the upper series would have been nearly horizontal. But this upper series later underwent another phase of folding and thrusting. It slid down the unconformity and was intensely deformed without any effect on the underlying rocks.

The unconformity between the two sets of folded rocks is marked by a distinctive yellow dolomite only 5–10 m thick. This evidently provided the lubricated layer allowing décollement of the overlying mass. Even so the lower part of the dolomite is virtually undisturbed, and the upper part becomes gradually deformed towards the upper adjacent rocks. A slab of rocks several kilometres thick slid on a lubricated plane over a dolomite layer only 10 m thick, and without disturbing any of the underlying rocks at all. There can be no question of crustal shortening here. Everything points to gravity sliding on a huge scale. Examples of gravity sliding can be drawn from all over the world: Fig. 9.11 shows examples from Papua New Guinea, Fig. 9.12 examples from Montana.

Lateral compression versus gravity tectonics

The idea of lateral compression to form fold mountains arises from early ideas of a shrinking earth. A cooling, shrinking earth would cause folds and mountains, like the wrinkles on the skin of a shrivelled apple. Although few now believe the earth to be shrinking, and despite the certain knowledge that the strength of rocks is insufficient to permit folds to be created by lateral compression, it is proving very difficult for geologists to rid themselves of these outmoded concepts. The notion of vice-like compression of sediments has a vice-like grip on the minds of geologists.

Fig. 9.11 Gravity slides in Papua
New Guinea. (*a*) Geological map.
(*b*) Cross-sections south of the
Kubor Range (after Findlay, 1974).

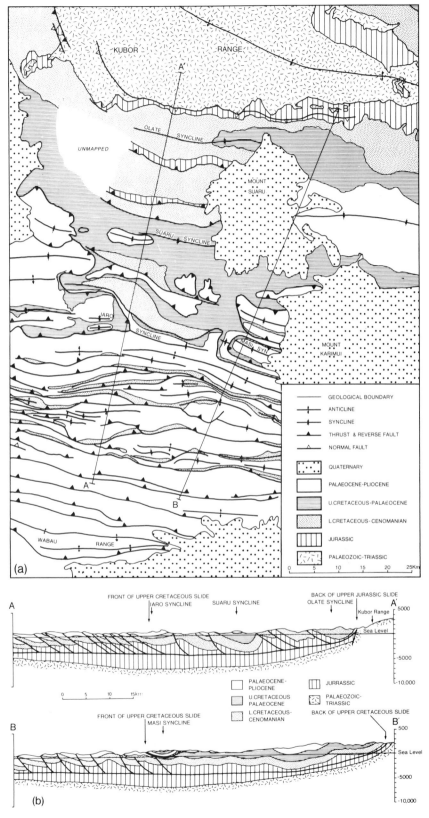

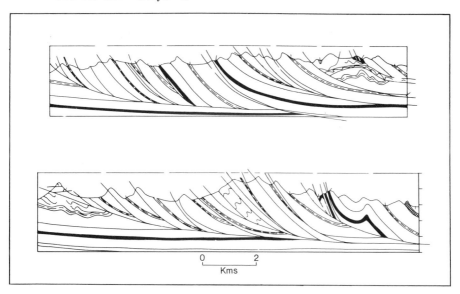

Fig. 9.12   Structure sections across central Sawtooth Range, Montana (after Deiss, 1943).

The Jura folds are often likened to the folds that can be generated in a table-cloth when pushed across a table. The table doesn't fold, but the table-cloth does. This homely analogy is very misleading. It is simply not possible to push a sheet of rocks sideways like a table-cloth. If the table were tilted and the cloth slid down into folds the analogy is better.

This fact must be stressed. It is not possible to fold rocks by a lateral push. Ideas gained by looking at folded table-cloths or at squashed clay in a squeezebox or vice do not relate to the folds we see in mountains. The compressive strength of rock is very much lower than the force that would be needed to push a slab of strata a few kilometres thick and hundreds of kilometres wide. If a force capable of moving such a slab were somehow applied the slab would not move, but would simply be crushed at the point where the stress was applied.

Gravity applies a body force that acts on every grain, every molecule, in the rock mass, and affects the whole mass in a way which an applied force cannot do. If this is so then folds are not brought about by lateral shortening of the earth's crust: they are superficial features and the folds relate only to the few kilometres of folded rock. The rock mass slid under gravity and was folded as a result of its sliding motion. Compression and shortening is confined to the folded beds. We can theoretically unravel the folds and see how big the sheet of rocks was before folding, but this is not a measure of crustal shortening.

The mechanism of folding by gravity sliding is so important that it is given the name of gravity-slide tectonics, or simply gravity tectonics (see DeJong and Scholten, 1973).

Of course rocks move downwards by gravity, so gravity folding will not make mountains. If folds are found in mountainous areas then there has been uplift subsequent to the folding. Folding does not make mountains. Some lowlands and some flat areas are underlain by folded rocks, and some mountains consist of folded rocks. There is no simple correlation between folding and the existence of mountains, or of folding and the former existence of mountains. Much folding probably occurs during sedimentation and is not related to mountain building and uplift in any direct way. There are no fold mountains caused by compression of sedimentary rocks as popularly conceived.

The search for a mechanism

Even though field evidence for gravity tectonics is very powerful, the description of the mechanism was for a long time difficult (see Voight, 1976). How could such enormous slabs of rocks be brought into motion on such gentle gradients? What sort of force could trigger the process? How could the internal frictional resistance of the rocks be overcome?

If sliding is brought about under gravity the strength of the rock is irrelevant, but nevertheless with normal friction there are also limits that prevent sliding. Adhesion at the fault plane cannot be reduced to zero because of irregularities. Certain materials seem to be particularly suitable for promoting slides, by acting as a lubricating layer, and others seem to be produced by the sliding and then act to increase lubrication. Sands flow very well if the pore pressure is high enough, a solution considered in more detail later. Salt flows very well, and is responsible for much sliding as well as salt-dome tectonics (p. 143). Clay often works as a lubricant in very surficial slides; in deep stratigraphic columns it tends to be de-watered and often serves more to seal water into underlying sand layers than as a mobile layer itself. Dolomite seems to be capable of solid-state flow and is found at the base of nappes in the Naukluft, and also at the base of the Katmandu Nappe in the Himalayas. Dolomite flow may be assisted by gliding along mineral cleavage planes, and it has also been found that as limestones become granulated the strain rate changes rapidly, and there is possibly a state of super-plasticity when the rock, if granulated very finely, flows like treacle. This is assisted by heat, which may be frictional. In very deeply-buried rock it is possible that other materials act in this way to produce mylonite and pseudo-tachylite, materials found associated with very deep thrust faults.

In deeper zones another mechanism is available for aiding the flow. Enhanced ductility is expected if deformation occurs during phase changes in minerals or during metamorphic reactions (White and Knipe, 1978). Transformation-enhanced ductility arises from stresses generated by mineral transformation, and reaction-enhanced ductility is due to fluid release and more complex factors. Thus the very formation of new minerals during metamorphism may assist the flow of nappes or diapirs.

The most probable solution to reduction of friction was provided by Hubbert and Rubey in 1959.

Imagine a layer of sand laid on the sea floor, which is then covered by a thick stratum of clay, and eventually many other sediments. The buried sediments are compacted by the pressure of the overlying load, which generally involves some reorientation of grains, the formation of secondary (diagenetic) minerals, and most importantly, the squeezing out of the water filling the space between grains. But in the situation just described water may be sealed in the basal sand layer by the clay above, and this has serious effects on its behaviour under pressure for water is essentially incompressible. The pressure of the overlying rock (known as the lithostatic pressure) results simply from its weight. The hydrostatic pressure on the interstitial fluid also results from the load, and if water cannot be expelled the pressure can eventually rise as the weight of sediment increases until the hydrostatic pressure is almost equal to the lithostatic pressure. At this stage the mass of sediments is virtually floating on the water-bearing stratum. Just as it is possible to push a huge ship away from a quayside, if it is floating, so a very slight force may set our postulated mass of sediments in motion. The mass of the material hardly matters so long as it is highly buoyant. Thus if the pore pressure in deep sedimentary layers can be increased sufficiently a very slight force may be sufficient to induce gravity sliding on slopes of very gentle inclination.

A homely experiment can be used to demonstrate the principle of buoyancy and sliding. On a thoroughly-wetted sheet of glass, place a can of beer. To make the beer can slide over the glass, the sheet will have to be tilted to about 25°, and the can may tip over rather than slide.

Now on the same sheet of thoroughly-wetted glass place an empty but cold beer can upside down. The air inside the can will expand on warming, increasing the buoyancy of the can to the stage where it will slide down the glass under its own weight on slopes of less than one degree. Furthermore as soon as it reaches the edge some air escapes, buoyancy is lost, and the can becomes instantly stationary.

Buoyancy has the effect of almost total reduction of friction and allows huge masses of rocks to slide. Such masses have great momentum and when brought to rest are liable to become very deformed, especially in the collision zone at the front of the slide.

The buoyancy mechanism of Hubbert and Rubey is not restricted to water imprisoned in thick sediments. In sediments undergoing metamorphism the hydrous (OH-bearing) minerals such as clays, chlorites and micas may lose water which will remain as high-pressure interstitial vapour until it can escape. Granitization may produce further 'connate' water. Zones of metamorphism and granitization are precisely those where uplifts and complementary downwarps tend to be concentrated, providing the slopes for nappe movement. The radioactive dating of many massifs shows mobilization of the core at the same time that nappes were mobile, as shown by stratigraphic evidence.

There is an easy transition from landslides to nappes of the Naukluft type, but even the giant nappes of the Alps, with cores of migmatite and granitic basement rocks, can be explained in the same way. Both kinds depend on the accumulation of potential energy due to gravitational stress fields developed between two regions of differing elevations and both aim to lessen that potential and restore a state of equilibrium. The two types are Van Bemmelen's Epidermal and Mesodermal sub-divisions of gravity tectonics.

Slumps or nappes are deformed to varying degrees. A very slow moving slump may retain the strata in the rotating block in correct relative position, but a rapid slump may merge into a debris flow, and the mobile material breaks up into countless fragments which become thoroughly mixed. The same is true of nappes. The extreme type of chaotic breakdown is the so-called argille scagliose (scaly clay) of the northern Appenines which consists of a matrix of heterogenous clayey sediments of various ages, with blocks and slabs of older rocks, some several kilometres wide, floating at random within it. Many other chaotic melanges are known, some associated with ophiolites and serpentinites.

To summarize, it seems there is a continuous series from small landslides to vast gravity slides. These are responsible for moving material downwards, and creating much of the folding that is seen in rocks. Most folding is a superficial effect and the rocks underlying a folded sequence are commonly found to be unfolded. Folding results from superficial forces and not from crustal compression or shrinkage of the earth, and folding does not build mountains.

10 The flow of ice and rock

When alpine glaciers were first studied they were thought of as rivers of ice, for they flow downhill, though very much slower than rivers, rates of a few metres or tens of metres per year being normal.

The mechanism of glacier movement was fairly obvious, it was thought. Snow falls on high ground, and by compaction turns into ice and thus into a glacier. The glacier slides downhill, lubricated by meltwater between the ice and the bedrock, until it reaches a sufficiently warm altitude to melt.

The glacier thus had a budget, with a balance between growth from snowfall, flow, and losses from melting and evaporation (ablation). A number of years of high precipitation should cause glaciers to advance if other conditions did not change; a rise in temperature should cause a retreat of glaciers, other conditions being unchanged. Historical records show that the position of glacier snouts has indeed changed considerably, and geological evidence shows even greater changes. The flow rate was also seen to be variable, and at times glaciers undergo surges when the velocity increases severalfold.

Cracks in glaciers, known as crevasses, may be hundreds of metres deep, but they do not penetrate right through the ice to bedrock. They have distinct patterns in plan which may sometimes be related to shear patterns along the valley side, or to mounds or steps in the underlying bedrock.

It was found that if a series of rocks or stakes are placed in a straight line across a glacier and observed over some years, the middle ones move down-valley fastest and those near the edge move very little. This showed that the glacier does not merely slide down as a whole, but there is differential movement within the ice – the ice itself is flowing. How can a material brittle enough to have crevasses also flow like a viscous liquid?

Faraday's discovery of regelation offered some sort of solution to this problem. Pressure on ice can cause local melting; reduction of pressure causes the water to freeze again (regelation). Some sort of mechanism could be devised whereby ice in a glacier could be cracked by applied forces, the opposite sides undergo some relative movement, and then the crack healed by renewed pressure and regelation. The details of how an applied force could sometimes break ice and sometimes heal the cracks allowed great scope for nineteenth-century controversy.

One clue to glacier flow came with the discovery that lake ice could flow at a stress much below the shear strength of 'regular' ice if the stress was applied parallel to the lake surface, but the significance of this could not be realized until the discovery of the crystal structure of ice by X-ray diffraction.

It turns out that ice crystallizes into almost hexagonal crystals with glide planes parallel to the base. Lake ice can be envisaged as an array of hexagonal crystals with parallel crystal axes, rather like a horizontal sheet of columnar basalt with vertical columnar jointing. In this position all the glide planes are parallel, and it is easy to see how the ice may be deformed by quite low stress parallel to the glide planes. Much greater stress will be needed to deform it perpendicular to the glide planes (Fig. 10.1).

Fig 10.1 (*a*) Hexagonal ice crystal, with glide planes parallel to the base. In lake ice the c-axis of the crystals are vertical and the glide planes parallel to the water surface. (*b*) Crystal deformed plastically by shear stress parallel to the glide planes. (*c*) Elastic deformation of crystal by shear strain normal to the glide planes.

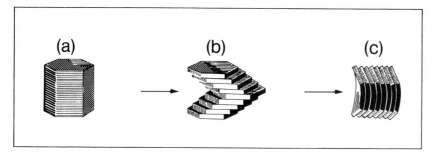

Another method of flow is also relevant to movement of crystalline materials of this kind. There is a constant gain-and-loss of atoms between different crystals in a mass of ice, and in the absence of any stress any individual grain of ice will lose about the same number of atoms that it gains, and so remain unchanged. But if a crystal is stressed it will lose more atoms than it gains and so shrink while a nearby unstressed grain will gain more than it loses and so grow. In this way there will be preferential growth of those ice crystals which are oriented in such a way that their glide planes are parallel to the stress, and grains in other orientations will tend to disappear. This is observed in glaciers, where it is found that not only does a marked crystal orientation appear with distance down-valley, but the ice crystals at a glacier snout may have a diameter about a thousand times greater than that of the first-formed ice crystals at the source of the glacier. This observation cannot be explained by mechanisms that ignore the crystal structure of ice.

The flow of material in a solid crystalline state is known as creep. There are several laws of creep, which are relevant to the behaviour of glaciers.

1. Creep is proportional to temperature. The closer the temperature comes to the melting point the greater the creep rate. In experiments at a fixed stress it was found that the creep rate at −1 °C is 1 000 times greater than at −20 °C. Ice becomes increasingly rigid at low temperatures. Actually all alpine glaciers are at a temperature very close to melting point, because the latent heat of ice is very much greater than its specific heat: very little heat is needed to raise the temperature of an ice block from −1 °C   to

0 °C compared with the heat needed to turn the same ice block at 0 °C into water at 0 °C. Since the temperature does not vary this first law of creep does not affect valley glaciers.

But ice caps are very different. They can be cooled to temperatures well below the freezing point, which reduces their capacity to flow very greatly. Ice caps may be thousands of metres thick, and their warmest part is actually the base where geothermal heat causes an increase in temperature and so an increase in flow rate. It seems probable that most of the flow beneath a thick ice cap occurs in the bottom few metres.

2. Creep is proportional to applied stress. The greater the stress the faster the ice will flow, and since the function is exponential this means that if stress is doubled the flow rate will be more than doubled. This factor accounts for the fact that big glaciers (with greater weight and therefore greater force) flow much faster than small glaciers. It also accounts for surges in valley glaciers, for a fairly slight increase in thickness of the glacier may lead to a considerable increase in the rate of flow.

3. Below a minimum or 'threshold' stress creep does not operate. This is known as the yield stress.

On the surface of a glacier there is no superimcumbent load, no stress, and so the surface of a glacier is below the yield stress. At the base of a glacier the full weight of the glacier is applied, so this will be well above the yield stress and so will flow by creep. Between these two points the stress varies from zero at the surface to some high value at the base, and at a point somewhere in the middle the stress must equal the yield stress. Below this point the ice will flow; above this point it cannot flow and is therefore rigid and can therefore be fractured. The yield stress point determines the depth of crevasses. A surface within a glacier divides the lower, flowing part from an upper, crevassed part. The upper part might be a jumble of crevassed and shattered ice; the lower part has flow structures (Fig. 10.2).

The discussion of glacier flow here has been mainly to introduce the idea of flow in a solid state and concepts of creep. Real glaciers apparently depend much more on basal sliding than suggested here. Raymond (1978) has presented an up-to-date detailed account of the mechanism of glacier flow.

Fig. 10.2 Diagrammatic long section of a glacier, showing the effect of yield stress. Ice stressed beyond yield stress will flow plastically; ice with stress less than yield stress will remain brittle.

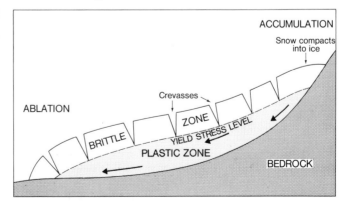

The flow of rock

The rocks that really show flow structures are the metamorphic rocks, schist, gneiss and marble. Actually ice is a metamorphic rock consisting of one mineral, ice, but we will now move on to see what happens in materials more commonly regarded as 'rock'. The similarity between structures in ice and other rocks has been frequently noted, as by Carey (1954) and Huddleston (1977). Schist and gneiss have a host of flow structures and a very marked preferred orientation of minerals. How do these relate to flow and the laws of creep?

Geothermal heat increases downwards, to we should have greatest flow at depth. Stress of course is greatest at depth, so flow should occur more in deeply-buried rocks, although it is well known that some rocks can flow, even at very low stresses. Marble mantlepieces and tombstones sometimes bend under their own weight over a few years.

Perhaps the most interesting comparison between glaciers and continental rocks relates to the yield stress concepts. If we apply the same reasoning used in glaciers to continents, we should find a plane at a certain depth below which rocks flow and above which they crack. We thus have a division between the upper crust with its faults, grabens, horst, and other features that result from the fracture of rigid blocks, and a lower zone where faulting is impossible and rocks flow. Of course the situation is enormously more complicated because the rocks consist of many minerals, each with its own thermal and crystal properties, but the basic idea seems reasonable.

If this is accepted we can see how the deep zone can flow beneath a more stationary upper crust. This is shown at its simplest in Fig. 10.3, where a stress is applied by sedimentation onto a limited area of crust.

Fig. 10.3 The effect of erosion and deposition at a continental margin, causing lateral flow and uplift in the plastic zone, with accompanying faulting in the brittle zone.

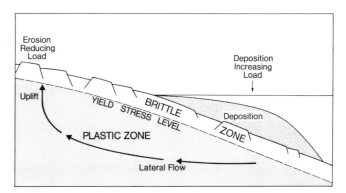

Below the yield stress level the rock flows away from the applied stress in a dominantly horizontal direction. The crust has been thinned in the erosion zone providing the sediment of the basin, so this will be an area of reduced vertical stress. Here the flowing rock might thicken, producing vertical uplift which might even push up the super-incumbent rigid zone to form a dome. This is where vertical foliation might be found in the metamorphic rocks.

Of course this story is greatly simplified, even for ice, but the basic idea of flow in the solid state in the lower zone where metamorphic rocks are created is very important for understanding tectonic relationships.

The geological role of creep in rocks and minerals was reviewed in a symposium (McClay, 1977), with particular emphasis on Coble creep, distinguished as follows: If a grain boundary is subjected to compressive and tensile stresses, then a chemical potential gradient for vacancy flow is set up. The flux of matter leads to depletion of material at relatively compressive boundaries and deposition at relatively tensile boundaries. If the diffusion of matter is essentially through the grains the process is termed Nabarro–Herring creep; if it is predominantly around the grain boundaries the flow is called Coble creep. Coble creep seems to be the most important kind in rocks. For tectonic purposes it is important to know that creep seems very feasible for the olivine rocks of the mantle. A review of creep in olivine is given by Goetze (1978).

Salt dome and diapirs

In many regions where salt deposits are found the salt was undoubtedly laid down in the first place as horizontal sheets, like the accompanying strata, but it is now found in the form of intrusions, varying in shape from circular to irregular, cutting across the overlying strata and pushing them aside. These structures are called salt domes or diapirs. A diapir pierces the overlying strata. Salt domes result from the flow of salt in response to variations in pressure and specific gravity. Irregularities in pressure may result from variations in the thickness of salt, or thickness of overlying strata, or from other causes. The salt will flow to areas of low pressure, and then, being lighter than the surrounding rocks, will start to rise through the overlying sediments.

Halokinesis is a term used by Trusheim (1960) meaning salt movement and salt tectonics. As a salt plug rises a basin develops around its margin because of the movement of the underlying salt towards the plug. As long as the movement continues the marginal basins accumulate extra sediments, which incidentally date the phase of salt movement. From such deposits the rate of salt movement has been calculated, at about 300 B (0.3 mm/y). Trusheim depicts a progression from salt pillows to salt stocks to a salt wall (Fig. 10.4).

$.03 \times 10^4 = 300$

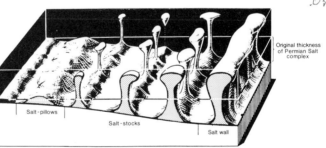

Fig. 10.4 Salt-pillow, salt-stock and salt-wall structures in relationship to depth and original thickness of Permian salt complex of northwest Germany (after Trusheim, 1960).

It seems that tectonic movement is not necessary to initiate salt domes, and in some places such as the Gulf Coast of North America they are found in beds that are otherwise very little disturbed. However, if there is tectonic movement, salt beds will be very susceptible to flow leading to décollement, location of nappes, and so on.

Diapirism has also been invoked to produce a trench-slope break bordering deep-sea trenches (Fig. 10.5), by outward and upward migration of mudstone derived from the continent, while subduction simply removed ocean-floor deposits (Scholl, Marlow and Cooper, 1977).

Dome intrusion is not restricted to salt and sediment, but occurs also in granites. Some granites – the 'granitization' granites – are simply altered earlier rocks that have been converted to granite, often

Fig. 10.5   A model in which the trench-slope break forms when a mass of rapidly deposited and underconsolidated mudstone migrates outward and upward to form a diapiric front. This model calls for in place deformation of continental or island-arc slope deposits and the subduction of oceanic beds. A = early stage; B = late stage (after Scholl, Marlow and Cooper, 1977).

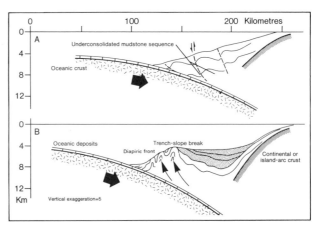

retaining earlier structures and 'ghost stratigraphy'. Once formed, the granite may be sufficently mobile to flow, and to rise.

It seems probable that some granite plutons rise through the surrounding rocks pushing aside and doming up the neighbouring rocks, but differing from salt domes because they are hot, and can to some extent incorporate bits of the neighbouring rock as xenoliths (foreign stones – bits of bedrock that may have been partially converted to granite).

Bathydermal tectogenesis is Van Bemmelen's term for the diapiric injection of granitized material from the core of a fold belt. Rock materials mobilized by granitization behave in the deeper part of the crust like evaporites. Instead of salt domes there are batholiths of granitic rocks, generally on a much larger scale than the salt diapirs.

The basic types are simple symmetrical domes (20–40 km across). These often spread after rising, sometimes symmetrically to give a mushroom structure, sometimes asymmetrically to give a tongue-like structure. The spreading into mushroom shapes, recumbent folds and the like occur where lateral flow met with less resistance than continued upward migration.

Fig. 10.6   Domes of granite intruded into schists. Zimbabwe (after MacGregor, 1951).

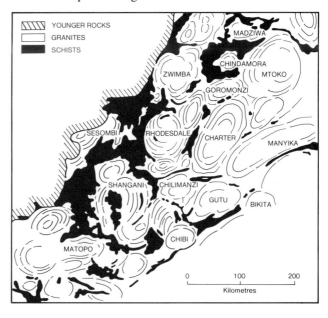

Flow lines preserved in the granites will generally form roughly circular patterns where the ground surface cuts across the dome or the 'stem' of a mushroom. A great mass of coalescing domes may give the picture of the Zimbabwe orogenic belt (Fig. 10.6).

Movement of a granite diapir folds and faults the surrounding rocks as a result of at least two processes:

1. Uplift and extension over the top of the batholith, possibly leading to nappes and cascade folds
2. Differential drag upwards on the flanks of the rising batholith and differential subsidence around the margins.

It must be stressed that the granite batholiths are emplaced by rheid flow, not liquid flow. They flowed like salt domes or glaciers, not like water. Oldfashioned ideas of 'magma' may give the wrong impression: granite magma is only fluid in acid volcanic rocks at high levels in the earth's crust. Gneiss domes, discussed in Chapter 7, are a related diapiric phenomenon.

Ramberg (1967), who studied the diapiric process experimentally, has outlined the patterns produced during dome formation and growth. The first stage after development of an unstable situation, where density of overburden exceeds that of the source material, is the formation of a series of waves. The amplitude of the waves increases until a series of evenly-spaced culminations develop on anticlinal ridges. These soon change into circular domes which rise until their progress is impeded by a feature such as an interface or the land surface. Granite plutons often have a distribution and spacing consistent with the early formation of a series of waves, as shown in Fig. 7.2, p. 96.

To make models realistic the effect of gravity must be taken into consideration, and in laboratory models a larger force must be substituted. By using a centrifuge Ramberg could produce up to 3 000 g, which enabled him to use material with considerable strength and high viscosity, necessary to study the flow of magma in dynamic models which meet reasonable scale requirements. He found that quite different types of structures developed depending on the difference in viscosity between the moving materials in the models and the surrounding 'crust'. With a strong contrast the flow pattern is controlled by structure in the surrounding crust. With a small contrast in viscosity the structure of the surroundings has little effect. On the contrary, the latter structures are more or less strongly deformed in contact with the moving bodies, secondary conformity often being developed. The experimental results strongly suggest that large domes and batholiths did not rise as magmas chiefly consisting of a liquid portion. The shape of such plutons indicated a moderate viscosity contrast between the rising body and the surroundings such as can be expected between crystalline rocks of unlike composition.

Rheidity

A valuable concept in discussion of the flow of rocks is rheidity, a term introduced by Carey (1954) to describe the capacity of a rock to flow. This is arbitrarily measured by the amount of time necessary for viscous flow to exceed by 1 000 times the elastic (or recoverable) deformation under specified conditions of temperature, pressure and shear stress. The rheidity determines whether a rock will behave as a fluid or solid in a particular situation.

Ice, for example, has a rheidity measured in weeks, say a million seconds, so has a rheidity of 10^6s.

Salt has a rheidity of about a year, 10^7s.

Gypsum has a rheidity of about ten years, 10^8s.

Igneous rocks have rheidities of tens of thousands of years (10^{11}s), and are properly considered rigid solids at the surface of the earth. Deep in the crust, however, rheidity must be low for isostatic compensation to occur and is perhaps 10^9s, hundreds of years.

A rheid deforms as a fluid but it is not a liquid. The states of matter are generally defined without reference to the time scale of deformation, and the concept is perhaps best explained diagrammatically (Fig.10.7).

Fig. 10.7 The states of matter (after Carey, 1954).

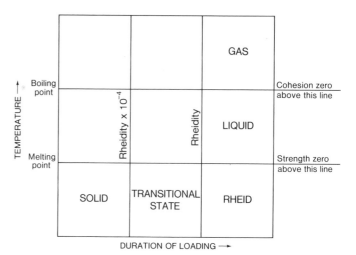

A rheid may now be more accurately defined as a substance whose temperature is below its melting point, and whose deformation by viscous flow during the time of the experiment is at least three orders of magnitude greater than the elastic deformation under the given conditions.

When loads are maintained for longer than the rheidity, the rock deforms as a fluid. Since tectonic loads are maintained for long periods – say ten thousand to ten million years (10^{11} to 10^{14}s) – and the asthenosphere has a rheidity of perhaps a hundred years (10^9s), the asthenosphere behaves as a fluid for tectonic phenomena.

The threshold value between brittle fracture and flow, in either a glacier or a continent, reflects changing rheidity. It is rheid flow in the deeper ice that enables a glacier to flow, carrying the brittle crust of the glacier, 50 m thick, on its back. On the continents the rheid flow in the deeper part causes the buckling, uplift, sagging and fracture in the upper, brittle crust, thus creating tectonic landforms.

The very great difference between the rheidity of evaporites and dolomite, compared with other rocks, explains why they are so important in processes such as décollement and diapirism. But when it is realized that only a change of rheidity is involved it is not surprising to find that granite may reproduce some of the diapiric features of more mobile salt domes, or that siliceous rocks may develop the flow (rheid) structures that we see in schists and gneisses.

11 Planation surfaces

Much geomorphic evidence for tectonic history comes from the study of plains cut across varied rocks and structures, called planation surfaces. An aside is needed here because for some years a swing in the emphasis of geomorphic work moved to process studies and systems thinking in an apparent reaction to 'Davisian geomorphology'. Since W. M. Davis pioneered work on planation surfaces this approach was strongly attacked, to the extent that even the existence of such surfaces was doubted. In fact they exist as directly observable landforms, and in many of the southern continents are so extensive that they dominate the landscape.

Further evidence for planation comes from stratigraphic and palaeogeographic studies. The Palaeozoic geosyncline that occupied much of eastern North America, western Europe and northwest Africa was a trench bounded by mountains for much of the Lower Palaeozoic, but by Triassic times the trench had been filled and the surrounding land worn down. The stratigraphic record of the period leaves no doubt that this part of the world had been largely planated. No one would claim it was plane like a billiard table, but it was flatter than it had been for millions of years previously.

The stratigraphic evidence reveals occasional great spreads of the sea over the continents, of which perhaps the greatest is the Cretaceous flood. The flooding is brought about by a relative rise of sea level, but extensive spread of the sea can only be brought about if the country is flat enough for the sea to spread widely. Much of the continents had been planated by the end of Cretaceous times.

Accepting the fact that plains exist, and that some process of planation cuts them, we can move on to examine briefly some of the controversies that have exercised, confused and wasted the time of geomorphologists for the past 80 years. There have been long debates about the names for the plains of different types, arguments about how they were formed, their relevance to geomorphic history, and their relationship to uplift.

Peneplains The writers early in the century used the term 'peneplain' to describe planation surfaces that had low relief, were of subaerial origin, and cut across geological structures. Residuals of higher country standing above the peneplains were called monadnocks.

Remnants of one or more earlier peneplains were identified in some areas. The problems that confuse studies of planation surfaces soon set in. People identified different surfaces, different numbers of surfaces, and correlated surfaces in different ways. Surfaces that some thought to be peneplains were explained by others in different ways, such as plains of marine erosion.

Pediplains   A challenge to the peneplain concept came from King (1953) who propounded his concept of the pediplain.

Pediplains have many of the same properties as peneplains – they have low relief, subaerial origin, and cut across geological structures. The big difference between peneplains and pediplains is not their descriptive features but their mode of origin. Peneplains are said to be formed by downcutting of rivers followed by valley widening and slope decline; pediplains are produced by slope retreat after a period of downcutting producing a new plain near the new base level (Fig. 11.1). 'The secret of landscape evolution lies, evidently, in the mode of development of hillslopes' (King, 1953).

Fig. 11.1   (*a*) Valley widening by slope retreat. (*b*) Valley widening by slope decline. Erosion by parallel slope retreat is said to create a pediment, and eventually a pediplain, while slope decline leads to the formation of a peneplain.

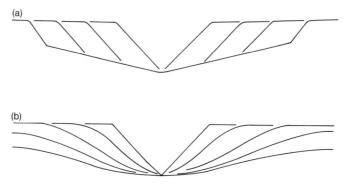

An important feature of pediplains is that as the scarp retreats a new bit of pediplain is created, and so the pediplain is a diachronic erosion surface getting younger towards the scarp. Landscapes may consist of pediplain remnants of various ages, giving a stepped landscape.

Panplains   Crickmay (1933) proposed the term panplain for those plains created dominantly by lateral erosion of rivers. Although doubtless an effective process in some areas, it is seldom possible to demonstrate that lateral erosion was responsible for older plains and plain remnants and in some plains it can be clearly shown that lateral erosion was not the process responsible. The concept has not found general acceptance.

Crickmay realized that his arguments for panplains were somewhat unconvincing because of the absence of present-day examples in many places. He attempted to explain this as a result of Quaternary climatic change, but he was probably unaware that practically all present-day valleys are aggraded by the post-glacial rise in sea level and do not show 'normal' erosional processes, at least in their lower reaches.

Structural plains   Sometimes plains or plateaus are on horizontally-bedded rock. It is then hard to prove that the plains are truly erosional plains cut at some particular base level, for they merely represent the top of an old bedding plain from which softer rocks that once were on top have been removed. In reality sedimentary rocks are rarely horizontal over the areas occupied by broad plains, but small remnants of old plains are very hard to separate from structural surfaces. Regions of horizontal rock are not good for the study of erosion surfaces.

Exhumed plains A variant on a structural plain is the exhumed plain. This is a plain which has been formed, buried under younger sediment, and then re-exposed by removal of the younger sediment. With conformable sedimentary rocks this is the same as a structural plain, but if there is a marked unconformity, and especially if the upper, stripped sediments are of terrestrial origin, the term exhumed plain is preferred.

Palaeoplains Hills (1975) describes the oldest surface of which we have evidence as a palaeoplain. Because we don't know the details of formation of old plains Hills suggests the non-committal term 'oldland' may be preferable to any one of peneplain, panplain or pediplain for the final product of planation. 'Palaeoplain is the term for oldlands that have been affected by later events of major kind and have been dissected so that only relics now remain.'

Plains of marine erosion When the term 'peneplain' was first proposed, debate centred not on the details of whether plains were eroded down or back, but on whether they were really subaerial erosion surfaces or plains of marine erosion. The sea is very effective at carving plains at the coast, and with sufficient time could it not carve plains of continental dimensions? Stratigraphers were aware that most unconformities have marine sediments on top, and the plains of unconformity could most probably have been carved by the same advancing sea that deposited the marine sediments. Were the plains on land carved in the same way?

Nowadays there is no doubt that many continental plains are carved by some form of subaerial, not submarine erosion. Furthermore marine erosion can only carve a plain of very limited width, for the very existence of a plain of marine erosion causes waves to lose energy and erosive power in crossing it. The only way to get a really broad plain of marine erosion is to have a slow rise of sea level accompanying coastal erosion. This is indeed what often happens in a marine transgression, and many unconformities are created in this way.

Some surfaces that are exposed plains of marine unconformity have been called peneplains. This is incorrect for a peneplain is, by definition, a plain formed by subaerial erosion. The so-called Schooley peneplain of the Appalachians appears to be a stripped sub-Cretaceous unconformity, in which case it should be called the Schooley surface, or Schooley unconformity, but not Schooley peneplain.

Planation surfaces It is certainly very desirable to have a non-commital term for describing planation surfaces whether it be erosion surface, peneplain, pediplain or some other term. The differences depend not on what we can observe of the plain, but on the deduced or observed mode of formation. In many places, especially

when we are dealing with small remnants of erosion surfaces, there
no way of determining the details of formation, and Mulcahy (196(
suggested that 'peneplain' be used in the sense of 'nearly a plain' fc
descriptive purposes, without connotation of the mode of origin.

When peneplain is used in this book it will be used in this way bu
usually the even more non-committal term 'planation surface' will b
used.

Multiple planation surfaces

In many parts of the earth the landscape is made of plains at severa
levels. Figure 11.2 from central Australia clearly shows two plains: th
lower one can be shown to be an erosion surface, the upper plain
presumed to be one also. High and low surfaces can be found on rock
of complex structure where erosional origin is quite definite, an
sometimes a succession of several surfaces can be found.

From India, in the region behind Madras, Büdel (1965) describe
multiple erosion surfaces in an area entirely underlain by granite. Th
lower plain, the Tamil Nadu Plain, broadly follows the coastline an
reaches inland an average of 100 km and in places 200 km. It rises t
an altitude of 200–500 m (with a slope of no more than 2 per cent) an
is succeeded by a steep step some 100 m in height which leads to th
higher (Miocene) Bangalore Plain (750–900 m), surmounted b
inselbergs up to 1 500 m.

Fig. 11.2 (*a*) A planation surface cut
across varied and complex Palaeozoic
rocks, dissected by Bungonia Gorge, New
South Wales, Australia (photo C. D.
Ollier).
(*b*) Planation surface with remnants of an
older planation surface preserved as
flat-topped hills in the distance. Coober
Pedy, South Australia (photo C. D.
Ollier).

Much debate has been spent on accounting for stepped landscape
a lot of it rather theoretical. Davis had cycles of erosion, suggestin
briefly that peneplains were formed during periods of tectonic quie
which were followed by perhaps rapid periods of uplift after whic
erosion cut down to sea level again and then a new peneplain wa
formed. Penck described treppen (stepped) landscapes bu
propounded periodic scarp retreat to form the surfaces. Lester Kin

has simplified the idea and postulated a more probable process in pedimentation to produce what he would regard as landscapes of stepped pediplains. As with the origin of simple plains, many arguments concern processes and mechanisms, but two other facts are now brought in. One is the relationship between erosion and uplift. Does uplift really come in jerks as Davis suggested, or is it continuous? Can erosion keep pace with uplift to create plains? Is it possible to find a process that will create several distinct plains even if uplift is continuous? Some of these questions can be answered with modern knowledge on the rates of the various processes concerned, a topic that will be discussed in Chapter 17.

In the southern Atacama Desert, Chile, Mortimer (1973) recognizes four episodes of landscape evolution, each of which led to the development of a regional erosion surface. The first began with the elevation of northern Chile above sea level in the late Mesozoic. Each subsequent phase started with incision of drainage into the preceding landscape. Radiometric dating of volcanic rocks show that the earliest elements of the topography existed in the Lower Eocene, and the landscape had formed by Upper Miocene, since when the present phase of canyon cutting commenced.

The fundamental reason for just four main erosion surfaces is not clear, for the complex geomorphic history might have produced a much more complicated landscape. Tectonic elevation of the Cordillera started early in landscape history and remobilization along major structures has since taken place. The eruptive centres have migrated eastwards since Miocene times, and there have been many relative sea-level changes, some of which are thought to be induced by relatively local tectonism. Incidentally, this is an area where geomorphology has some commercial application: supergene enrichment of copper deposits took place under the older erosion surfaces and the economic viability of a mine prospect can be largely predicted on geomorphic grounds.

The morphological development of the northwest Dolomites shows the close interaction between tectonic processes and erosion, and according to Engelen (1963) shows the following stages:

1. After the first phase of Alpine uplift in the Lower Tertiary the area was planated by an extensive erosion level of Oligo–Miocene age. The rather widespread preservation of this oldest erosion surface shows the slow rate of erosion of the surface, explained by the fact that the truncated rocks are mainly limestones and dolomites with no surface drainage.

2. A second phase of uplift during the Miocene and Lower Pliocene led to downcutting of valleys by about 1 000 m.

3. A new planation surface formed in the Upper Pliocene, now ranging in altitude from 1 900 to 2 300 m, especially well developed on the less resistant Middle Triassic strata around the reef masses.

4. A third phase of uplift in the Upper Pliocene and Quaternary was accompanied by renewed vertical river erosion of 1 000–1 500 m.

A second set of arguments centres on the reality of older erosion surfaces. Are they real or imaginary?

A specific challenge was presented by Hack (1960), who suggested that the concept of dynamic equilibrium provides a better basis for the interpretation of landforms than a theory of cyclic planation. He suggests that when the topography is in equilibrium all elements of the topography are downwasting at the same rate, and differences in relief are explained as structural complexities rather than in terms of evolutionary development. He envisaged large areas of erosionally-graded topography, maturely-dissected peneplains, as equilibrium landforms, but went so far as to suggest that real planation surfaces did not exist, even when weathering profiles and such things as bauxite deposits were associated with particular surfaces. This idea will be discussed further in Chapter 21. What sort of evidence is required to show that a plateau of today was once a low-level plain, and what sort of evidence is required to show that the existence of a former plain can be inferred from scattered ridges and hilltops? Since scepticism has outpaced understanding of old erosion surfaces for the past few decades this evidence will now be reviewed.

Evidence for planation surfaces

Most people who are not blind or stupid can tell when they are in an area of relatively flat country: they can recognize a plain when they see one. After this it is relatively easy to determine whether the plain is a depositional plain built up of alluvium, marine deposits, lake sediments or whatever, or whether it is an erosional plain cut across rock. If it is erosional it may be given one of the names offered in the previous section, depending on the detailed evidence and the taste of the investigator. In brief it is a planation surface.

An elevated plain can also be recognized without too much sophistication, and 'high plains' is the term given by farmers and foresters to various areas where common sense is enough to show that the country is both high and flat. Plateau is another term used for the same thing, usually when large. 'Tableland' and 'mesa' are other popular terms. High plains may cover areas of thousands of square kilometres or be tiny remnants. They are separated by a zone of steep ground from lower plains if such exist.

A number of small plain remnants at the same level suggest that they are remnants of a formerly more extensive plain, dissected by subsequent erosion until only fragments remain. This argument is particularly convincing if the surfaces are cut across varied bedrock.

In areas of folded or tilted strata the hard rocks form ridges and the soft rocks form valleys as described on p. 162. If the ridges tend to reach the same elevation over a wide area it might suggest that the fold had been planated before differential erosion brought out the hills and valleys. On a simply-dissected fold a wider range of elevations might be expected. However there is no agreement on how much variation in elevation might be attained by either processes, so simple summit accordance is never more than suggestive.

Better proof comes from the presence of bevelled cuestas. If a dipping bed of hard rock is simply eroded it develops a dip slope and a scarp slope (Fig. 11.3), with a rather sharp ridge (a cuesta) where these two slopes intersect. If the crest of this ridge is bevelled (has a level top)

Fig. 11.3   (*a*) A cuesta is a ridge formed by differential erosion of a dipping hard stratum. (*b*) A bevelled cuesta has a flat bevel that is a sure sign that an upper planation surface at about bevel level existed before differential erosion created the cuesta.

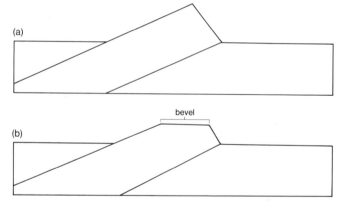

there is no doubt it is the remnant of a former erosion surface, for the rock structure would never develop a flat top unless a lateral erosion process was working at a particular base level. Figure 11.4 shows excellent bevelled cuestas in an area with plunging folds and well-developed joints.

Of course the planation surface on a bevelled cuesta may have been a very small landform such as a river terrace. But if there are lots of bevelled cuestas, all at about the same elevation, then a certain confidence may be placed in the former existence of a broad planation surface.

Bevels may sometimes fail to remove the entire ridge top but merely put a shelf onto it (Fig. 11.5). This happened on the Chalk escarpment of the North Downs of England and on the Chiltern Hills. Such a shelf is a sign of partial planation, followed by uplift and further erosion.

In more or less homogenous rocks the presence of bevels and accordant summit levels is of less significance. Some argue that as rivers cut down the interfluves are also lowered, and numerous 'phantom' planation surfaces may be revealed by accordant summits.

Fig. 11.4   Stereo air photo showing bevelled cuestas in an area of plunging folds and joints, Abner Range, Northern Territory, Australia (Crown copyright, reproduced by permission of the Director, Division of National Mapping, Department of National Development and Energy, Australia).

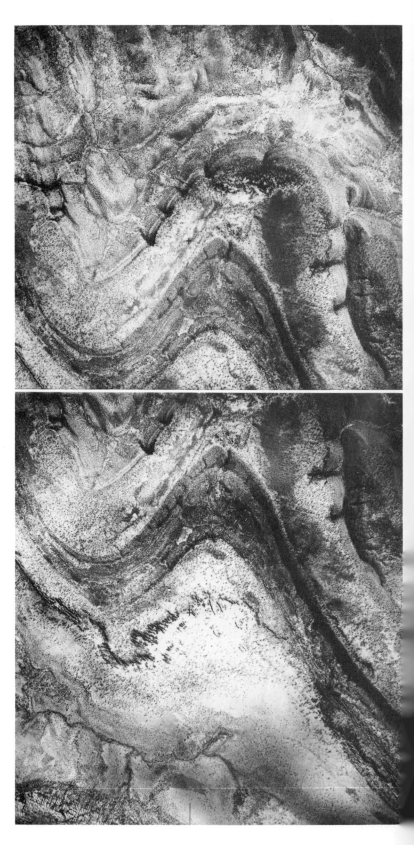

Fig. 11.5 Forms of Chalk dipslope on the North Downs. The Early Pleistocene bench. (a) is not related to structure but is a partial planation surface, which may be traced to other localities where it forms the summit surface of a bevelled cuesta (b) (after Wooldridge and Linton, 1955).

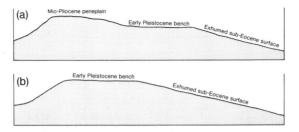

This idea is shown in Fig. 11.6. But in three dimensions the situation is not so bad, because normal interfluves will have a slope downstream, but those approximating to a true planation surface will have level profiles in a direction roughly parallel to the stream courses as well as across the stream profiles. The evidence for former surfaces must be considered in three dimensions. The same is true for reconstruction of old surfaces in folded sedimentary rocks, for if the strike sections of ridges show accordant summits then there is good reason to think there was a planation surface not much higher, even if bevelled cuestas are absent.

Fig. 11.6 A dissected planation surface may be detected at stage A by the accordant summit levels. Without other evidence this is a risky method because at later stages of erosion such as B, the present ground surface, the approximate summit accordance of the dissected landscape may suggest the existence of planation surfaces such as X and Y, but such 'phantom' surfaces never existed in reality.

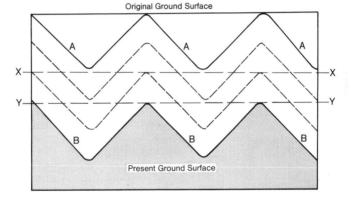

The Ozark Plateau brings together many of the lines of evidence discussed so far, and was described in detail by Bretz (1962) with the specific purpose of refuting the dynamic equilibrium hypothesis of Hack (p. 302) and his denial that peneplanation ever occurs under fluvial attack. It is worth describing in some detail, together with Bretz's list of criteria that support a peneplain interpretation.

The Ozark Plateau is structurally and topographically an imperfect dome about 400 km in maximum diameter, surrounded by lowlands. Drainage is irregularly radial, with long radial interfluves that cut across the strata because structural dip is slightly greater than topographic slope. A typical summit flat near Raymondville is over 1.5 km wide for a length of about 16 km and ranges in altitude only 25 m. The flat on which the town of Licking stands has a slope of only 4 m in a kilometre, but the gradient of the radial valleys is about 200 m/k.

As support for a cyclic erosion surface in the Ozarks, Bretz lists the following:

1. Relict shallow valleys on the summit flats.
2. Dissection of the main divide to leave wide, isolated summit flats interspersed with much-dissected stretches carrying the same summit accordances and composed of the same rocks.

3. Non-parallelism of divide and stream valley longitudinal profiles
4. Relict hills that rise above the even crest lines of the divides.
5. Continuity of profile of benchlands in valley heads with interfluve summits downstream.
6. Failure of rock and structure contrasts to appear in the upland topography.
7. Failure of escarpments to influence river course.
8. The bevelling of escarpment crests to conform to neighbouring radial summit flats.
9. Entrenched meanders that necessitate changes of base-level.
10. Clay fill in many Ozark caves.
11. Bedded stream gravels on uplands.

Various geometrical methods have been devised to detect former planation surfaces from contour maps, but they have not been very successful. This is partly because most maps are not accurate enough for the purpose, and also because the methods are tedious. The use of computers may make such methods attractive once more, especially in areas where altitude data are already stored on computer.

Perhaps the weakest evidence for the existence of a former plain is the 'eyeball' method, in which the eye sees a landscape of hills, mountains or ranges, and the mind casts a planation surface over it like a great blanket. However, it might be asked why hills do tend to reach accordant heights, and sometimes why the heights fall into particular pattern. The Gipfelflur of the European Alps is an imaginary surface that can be seen riding over the summits of the Alps in a huge dome. Cynics may disbelieve, but many experienced geologists feel they are seeing something real.

'From afar one always sees how the individual tumble of jagged peaks, so impressively irregular at close sight, approaches strikingly even upper limit. This summit level passes over the tops which are in most cases residual mountains of some former erosion surface. Many individual peaks are eroded below the surface, and few stick above it, and there are wide undulations of the summit plain itself. But as an overall feature this summit plain is a very real thing indeed' (Rutten, 1969).

A gipfelflur does not prove a former single peneplain, but it reflects an integration of former base levels which is today broadly unwarped. In the case of the Alps the gipfelflur was upwarped during the Late Pliocene and Pleistocene times.

The dating of planation surfaces
Relative dating

If there are a series of planation surfaces the normal age sequence is from oldest at the top and youngest at the bottom. This can be thought of, at the simplest, as resulting from a sequence of uplifts leading to rejuvenation of rivers and the cutting of a new plain at sea level, with accompanying reduction of the area occupied by the uplifted planation surface. The process can be repeated many times to give a stepped topography, and complications can arise in various ways depending on local structure and history.

Dating by overlying rock

If fossiliferous sediments lie on an erosion surface then the erosion surface must be contemporaneous with, or older than the sediment. The Pliocene Lenham Beds, for example, provide a date for the bevel on the Chalk of England described earlier.

Basalt, erupted from volcanoes, is a suitable rock for potassium–argon dating and where it lies on a planation surface the age of the basalt gives a minimum age for the erosion surface. If the basalt has been eroded in the creation of a younger planation surface it provides a maximum age for the younger surface.

Some of the East African volcanoes can be related to local erosion surfaces, and it is possible to work out the relationship between volcanicity and landscape development, and also to use the radioactive age of the volcanoes to put limits on the age of the erosion surfaces. Some examples from East Africa are shown in Fig. 11.7.

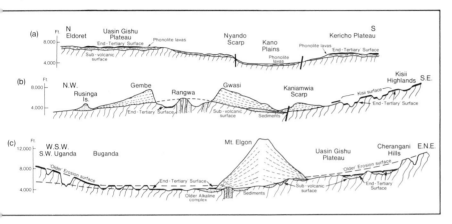

Fig. 11.7 Some examples of the relationship between volcanic rocks and erosion surfaces from East Africa. (*a*) Kericho to the Uasin Gishu Plateau. (*b*) Kisii Highlands to Rusinga Island. (*c*) Southwest Uganda to the Kenya Rift (after King, Le Bas and Sutherland, 1972).

Dating by soils and weathering profiles

Field evidence sometimes shows that particular surfaces have distinctive weathering profiles, soils, laterites, silcretes or similar features. These may be used to correlate surfaces and may also give a relative chronology, as in northern Australia where the oldest surface has a deep weathering profile, the next planation surface is produced by partial stripping of the weathering profile down to a silicified horizon, and the youngest surface is created by stripping the weathered profile down to bedrock. Palaeomagnetic studies of laterite horizons may, in favourable circumstances, provide an absolute date for the time of laterite formation and this gives a minimum date for the planation itself.

Dating relative to drainage patterns

Some indications of the age of drainage patterns will be given in Chapter 12, and the ages of erosion surfaces and drainage patterns are closely related and can sometimes be related to stratified deposits.

For example in the Western Rift Valley of Africa, the rifting that dismembered the drainage pattern created a trough which accumulated sediments. Since the oldest sediments are Miocene the rifting must be at least Miocene in age, and the drainage pattern older still. The plain on which the drainage pattern was formed must therefore be at least Miocene in age too, though in all probability it is much older (Gautier, 1965).

The ancient salt-lake drainage pattern of southwest Australia dates from before the severance of Australia and Antarctica, which is Eocene (Ollier, 1977c). The river system cannot be older than the planation surface it flows across, so this planation surface must be Eocene or older.

Furthermore the Western Australian palaeodrainage pattern has been incised into Lower Cretaceous marine sediments in the desert

areas, and is therefore no older than Cretaceous. In the Gibson Dese
some valleys containing Eocene deposits have been found, so th
drainage was in existence by Eocene times. Dating is more difficult i
shield areas which have been land since the Permian, but in a fe
places some present-day drainage lines follow Permian glacial valley
which do not appear to have been covered at any time by later deposit
so these drainage lines may have been active since the Permian. It i
improbable that any palaeodrainage patterns pre-date the Permian a
the Early Permian glaciation affected nearly all Western Australia an
would have probably obliterated any older drainage patterns (Van d
Graffe *et al.*, 1977).

Some problems of dating planation surfaces

The methods of dating erosion surfaces are somewhat loose and vagu
but even greater problems are presented by the lack of agreement o
how surfaces are created.

If something like Davisian peneplanation is true, each peneplai
has a fairly definite date of origin, which is the time of the uplift tha
caused rejuvenation and the start of renewed peneplanation. Th
planation period would last for a finite length of time, and stop when
new uplift led to the creation of yet another, younger peneplain.

With pediplanation the situation is different. Suppose a pedime
starts eroding to sea level, then the pediplain extends inland by scar
retreat. Scarp retreat takes time, and a plain that started at the coast i
say Eocene times, may have moved inland 500 km by Miocene time
and still be retreating today. The surface of a pediplain is diachroni
Suppose now that after the pediplain described above was we
established a change of base level was brought about by eart
movement. A new pediplain starts at the coast and moves inland. Bu
while the younger pediplain is retreating, the scarp backing the ol
pediplain is also retreating. Pediments are being created in tw
different situations at the same time, one on the high level and one a
sea level. With pediplains therefore we cannot really date the pediplai
as a whole, and the dating methods described earlier only refer to th
area of pediplain where the method is used – the date cannot b
extended to the whole plain. When the age of a pediplain is given (a
for instance asserting that the Moorland Pediplain of South Africa
Miocene) the phrase should be taken to mean that the pediplain wa
initiated in Miocene, but will undoubtedly be diachronic and is eve
extending at the present day.

Weathering and planation surfaces

Weathering may relate to planation surfaces and their suppose
cyclicity. One must assume that weathering goes on all the time, bu
there does seem to be one major event when deep-weathering profile
were formed and preserved for a very long time. The most probab
time for production of deep-weathering profiles was before th
break-up of Pangaea. Then the land masses had been exposed t
weathering for a very long time and were probably rather flat. Couple
with this longevity and low relief was a reduction in erosion, becaus
on a supercontinent it is much further to the sea, on average, than it i
with many smaller continents, long weathering and little erosio
produced a major planation surface, the Gondwana surface of King
which is very widespread and very deeply weathered. When Pange
broke up – an event which ranged from Triassic to Eocene – renewe
erosion would set in and create new planation surfaces graded to se

levels on the new continental margins. Stripping of the deeply-weathered rock would proceed rapidly, and the formation of the next surface would be relatively easy. This surface, largely created by stripping of weathered material, is the African surface of King (1962), and probably the Moorland surface of his later classification (1976). Once the deeply-weathered rock is removed, formation of subsequent erosion surfaces is much more difficult, for they have to be cut across fresh, hard rock. Younger surfaces are therefore likely to be smaller in area and less well planated than the older surfaces.

In Uganda the three surfaces can be mapped (Fig. 11.8), and have been recognized from soil surveys. The ancient Gondwana surface is

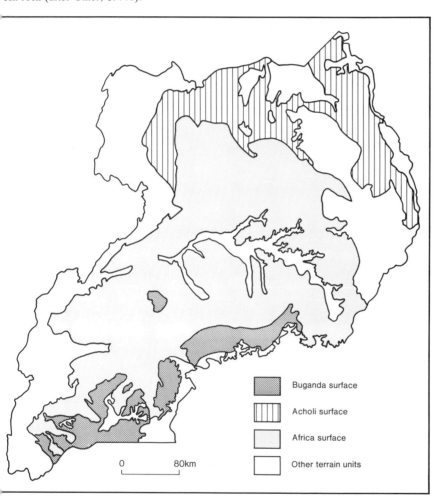

Fig. 11.8 Planation surfaces of Uganda. The Buganda surface is the highest, and in the area shown there are many plateau remnants. This is equivalent to the Gondwana surface and probably pre-dates the break-up of Gondwanaland. Rocks are very deeply-weathered beneath this surface. The African surface is essentially cut across pre-weathered rock. The Acholi surface is the lowest, deeply-weathered rock has been largely stripped off, and this surface is cut across fresh rock (after Ollier, 1977b).

Buganda surface

Acholi surface

Africa surface

Other terrain units

0   80km

preserved as planated summits on resistant rocks such as quartzites on watersheds: it is sometimes called the Buganda surface. The African surface is widespread and generally cut across deep regolith, though occasional inselbergs poke through indicating a very irregular weathering front. Only in northern Uganda is the old saprolite completely stripped, and the Acholi surface is cut across fresh rock. It has thin one-cycle soils in contrast to the two-cycle soils of the African surface (Ollier, 1959, 1960).

Planation surfaces and tectonics

Planation surfaces are related to tectonics in two ways. Firstly the initiation of new erosion surfaces requires uplift and rejuvenation of valleys before the new surface can be graded to the lower level. A sequence of planation surfaces, carefully interpreted, may provide a history of uplift movements.

Secondly, planation surfaces may show directly what tectonic movements have occurred since the planation surface was formed. Planation surfaces are approximately flat when first formed and if they are warped or arched then tectonic movement post-dates the planation. In eastern Australia, for example, the Eastern Highlands reveal a gentle rise of planation surfaces from central Australia, and arching over the Highlands, and a steeper slope towards the east coast: in Southern Africa the planation surfaces rise from the centre of the continent towards the rim, and (like the Australian surfaces) arch over and descend more steeply to the coast: in East Africa the erosion surfaces show a rise towards the rift valleys, with the highest part of the arch about 50 km from the rift valley faults. Examination of such erosion surfaces enables the geomorphologist to locate axes of uplift and work out the amount of uplift and perhaps even the date. The movements that tilt and warp planation surfaces will also affect the direction of drainage, and the study of planation surfaces is usually supplemented by drainage-pattern analysis. This is the subject of the next chapter.

Summary of planation surfaces

Adams (1975), in a summary to conclude a collection of papers on planation surfaces, suggests that the following appear to be areas of general agreement concerning planation surfaces:

1. Planation surfaces do exist, and are not merely steady-state topographic products of random dissection.
2. The most distinctive unit surface is the pediment.
3. In attempting to estimate the emergence of a land mass from present altitudes of planation surfaces, their origins, base levels and probable regional slopes must be considered.
4. Most planation surfaces are or have been graded with respect to some base elevation, usually sea level.
5. The most important time aspects of a planation surface is from the latest possible time of initiation of the cycle that produced it to the earliest possible time that it ceased being shaped because of burial or uplift.
6. A planation surface need not be presently under the same climatic regime as when it was initially developed.
7. Planation surfaces should be considered in the general framework of plate tectonics, sea-floor spreading, and the resulting change of continental areas with respect to the position of climatic zones.

12 Drainage patterns, rivers and tectonics

Since the study of river patterns, once a main theme of geomorphology, is now unfashionable, it may be worth recalling some features of valley morphology. Simple river systems have a dendritic pattern with tributaries joining a main stream at acute angles that point downstream. Tributaries have steeper gradients than the main stream, but their junctions are at precisely the same elevation (Playfair's Law). Simple valleys increase in width and become flatter in the downstream direction. If these conditions are not met we may assume that a valley has been affected by some complicating factor of structure or geomorphic history or both.

Let us begin with the drainage in a non-tectonic setting – plains. If we consider alluvial plains that result from aggradation we have no direct tectonic features (though even here it is probable that some tectonically-uplifted mountains provided the alluvium, and some sort of tectonically-controlled basin has been filled in). Alluvial plains have remarkably gentle gradients, frequently less than 1/1 000, and are covered by primary alluvial landforms. These can be quite varied and alluvial plains have many differences in detail: the Canterbury Plains of New Zealand have braided river channels often several kilometres wide: the plains of southeast Australia consist largely of ancient alluvium, traversed by hundreds of tiny, sinuous old stream courses only a few metres wide (prior streams) and meandering old river courses (ancestral rivers) as well as present-day meandering rivers: the Fly River floodplain in Papua New Guinea consists mainly of levees and backswamp deposits with higher remnants of old alluvium. Whatever the local details, overall drainage patterns are dendritic, the dominant drainage pattern on any simple sloping surface. The dendritic, but here variations from dendritic pattern are most likely to features such as meander patterns, crevasses, deferred junctions, autopiracy and other purely fluvial features.

The overall pattern on a plain cut across bedrock should also be dendritic but here variations from dendritic pattern are most likely to come from structural control, not from primary alluvial features as on alluvial plains.

Structural control of drainage patterns

Drainage patterns on horizontal sedimentary rock

Approximately horizontal sedimentary rocks tend to form plateaus plains and stepped topography, with harder rocks making the flats and softer rock being stripped away. The simple dendritic pattern can be well maintained in most instances, but is likely to be modified by joints in the harder rocks. These may form an angular pattern, commonly with two dominant joint directions almost at right angles. The statistical analysis of drainage patterns in such areas by determination of length of straight streams in various directions leads to a quantitative measure of the effect of joints. Some rocks, especially sandstones, have only one dominant joint pattern (Fig. 11.4, p. 154).

Drainage pattern on tilted sedimentary rocks

In a series of tilted sedimentary rocks, the soft rocks such as clays and shales, are eroded faster than the harder rocks, so a pattern with rivers running along the strike is very common. Strike valleys run in the strike direction, and are separated from each other by strike ridges on harder rock.

Tributaries to strike valleys commonly enter almost at right angles. Streams that flow down the dip slopes of the hard beds are called dip streams: those that flow in the opposite direction are called anti-dip streams (Fig. 12.1). When the dip is not very high the dip streams will

Fig. 12.1   Block diagram and map showing the structural relationships of streams on dipping strata.

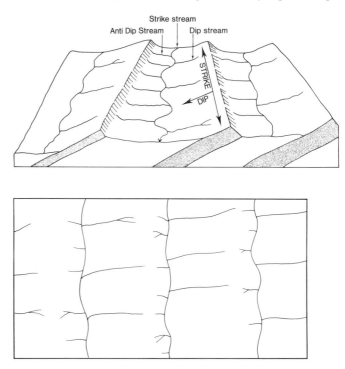

be longer than the anti-dip streams, as they flow on longer slopes. Also the anti-dip streams are located on the softer rocks, which if clays and shales are likely to have lower infiltration and higher run-off than the rocks of the ridges, and so have a greater drainage density. This results in a drainage pattern consisting of parallel strike streams with relatively long and widely-spaced dip streams coming in on one side, and relatively short but closely-spaced streams on the other. Such a pattern clearly indicates not only the effect of tilted sedimentary rocks, but shows the direction of dip, the location of more and less permeable strata, and even provides a rough idea of the amount of dip. Such a pattern (Fig. 12.2) is known as a trellis pattern.

Even if the dip of the rock is such that the ridges are roughly symmetrical, commonly known as hogback ridges, the direction of dip can generally be deduced from the drainage density being greater on the anti-dip slopes. Contour maps also show that the anti-dip slope is usually more scalloped and embayed, for the anti-dip streams take advantage of any joints in the hard stratum while dip streams simply flow over its surface; the particles of the rock are oriented parallel to bedding and resist erosion on the dip slopes more than the edge-on fabric exposed on the dip slope.

An older nomenclature for the streams of a trellis pattern is shown in Fig. 12.3. This is established in many textbooks but should be discontinued because it builds into the nomenclature a presumed (and erroneous) time-sequence for the origin of the various streams. It was thought that a main stream flowed down an initial erosion surface cut across the strata, and this was called a consequent stream, being a consequence of the initial dip of the presumed erosion surface. The strike streams were thought to be developed subsequently to the consequent streams, so were called subsequent stream. Once these had carved valleys, secondary streams would form on the new valley sides; those parallel to the consequent stream were called secondary consequent streams, so were called subsequent streams. Once these had streams). In reality the whole of any land area must be drained, and run-off and initiation of drainage affect all parts of the land surface right from the beginning. No part of the land area awaits the

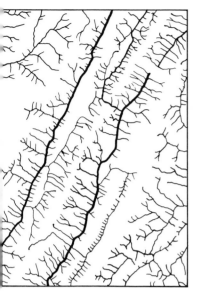

Fig. 12.2   Trellis pattern. Monterey, Virginia quadrangle (after Thornbury, 1969).

Fig. 12.3   Outdated nomenclature for structurally-controlled streams in regions of dipping strata.

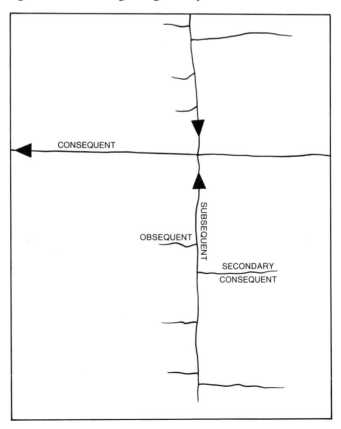

establishment of main drainage before becoming drained itself, and 'subsequent' is misleading, as well as being less descriptive than 'strike' valley. The early history of a drainage pattern may be difficult to work out, but the structural relationships are clear, so structural terms for drainage pattern description are to be preferred.

Drainage pattern of folded sedimentary rocks

A fold consists of two limbs, each of which has a simple tilt. On each limb there will be a trellis drainage pattern, but the directions of the dip and anti-dip streams will be different on opposite limbs. This results in a trellis drainage pattern that is symmetrical about the fold axis. An anticline has infacing escarpments, so the dip streams on opposite limbs flow away from the fold axis, and the anti-dip streams flow towards it. Thus the trellis pattern not only indicates details of hard and soft beds in each limb, as outlined earlier, but clearly indicates the position of the fold axis.

A succession of folds simply provides a succession of tilted beds alternately dipping in one direction and then the opposite direction.

If fold axes were horizontal the strike streams on opposite limbs of a fold would be parallel, but this is not common. Folds plunge, and the strata swing round the nose of the plunging fold to produce curves in the strike ridges and valleys. Each limb of a plunging fold has a simple trellis drainage pattern, but this trellis pattern is bent around the nose of the plunging fold. This distorted trellis pattern therefore not only indicates dips, strike and fold axis position, but also indicates the direction of plunge. The folded rocks of the Appalachian Mountain provide excellent examples of this drainage pattern, which is sometimes known as Appalachian-type drainage.

In areas of folded rocks many valleys are formed along softer sediments and the country has a distinct 'grain' parallel to the strike. On such a grainy basis we can expect overall drainage patterns to have some major stream directions resulting from the original slope of the general ground surface but the strike of the rock will show through in many tributaries. The drainage patterns of the Upper Lachlan and the Upper Murrumbidgee illustrate the effect of a general slope to the northwest in a region of uniform strike to the north (Fig. 12.4).

Modifications of river valley landscapes

From the moment they are initiated river valleys continue to be modified by the action of river erosion. There is downcutting (vertical erosion) and valley widening by slope processes and lateral erosion by the river. Downcutting leads in turn to extension of the river upstream, that is headward erosion. The river cannot cut below the level of its mouth, which is known as base level: the sea is the ultimate base level, but temporary base levels of erosion may be created at rock barriers or lakes. Lakes are rather ephemeral landscape features so the temporary base levels are indeed short lived, but even the ultimate base level of the sea is varied by eustatic changes in sea level or by tectonic movements of the land.

When vertical erosion is reduced (by nearness to base level) valley widening occurs, that is lateral erosion takes over. If vertical erosion increases again (because of uplift of the land or for any other reason) the river is said to be rejuvenated: rejuvenation simply means increase in vertical erosion. Tectonic uplift is likely to cause rejuvenation, but caution is needed in interpretation because such things as climatic change, chance exposure of softer sediments, marine erosion at a river

Fig. 12.4 Upper Lachlan and Upper Murrumbidgee drainage pattern showing the 'grain' produced by the regional northerly strike in an area sloping to the northwest (after Ollier, 1978).

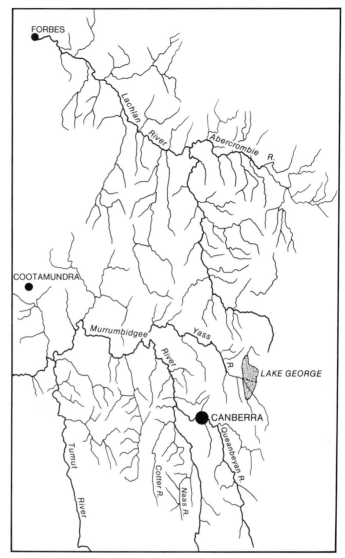

mouth, or eustatic changes in sea level can all have the same effect.

Alternating vertical and lateral erosion may lead to the formation of step-like landforms on valley sides known as river terraces. Some terraces are rock cut, others are formed of thick alluvium laid down when a river is aggrading, that is building up its bed. Terraces are first formed at river level, as flood plains, and only become terraces when vertical erosion cuts though them. They are valuable because they give indications of former river gradients, and successive terraces show successive river positions. In areas of tectonic activity river terraces may be warped in various ways, giving valuable clues for interpretation of tectonic movement.

River capture is the name given to change in river pattern when a stream which is cutting down rapidly extends headwards into the catchment of a stream which is cutting down only slowly. Thus in Fig. 12.5a stream CD is steep and is eroding downwards and headwards quite quickly. Stream AB has a gentle gradient and is cutting down only slowly. Figure 12.5b shows the result of capture. Stream CD has captured the waters of river AB at x. Stream xA is reduced in volume, and the waters of Bx are added to CD increasing its vertical erosion even more. The sharp change of direction at x is called the elbow of capture, and the sudden increase in gradient at x is called a knick point. The knick point may eventually migrate upstream.

Fig. 12.5   Diagrammatic representation of river capture. For explanation see text.

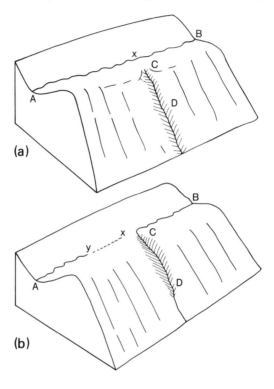

(a)

(b)

The old section of stream AB at y no longer carries water, and is known as a wind gap. Old river gravels and alluvium mark the former course of the river at Y, and if AB was a complex valley before capture the section y may even have old river terraces or other fluvial features.

A simple example of river capture or diversion appears to be provided by the Barron River near Cairns, Queensland. The upper course of this river is on a plateau, but it turns seaward and drops 300 m via a series of large waterfalls to the Pacific. It appears that the headwaters were originally tributary to the Mitchell, which even now rises within 15 km of the Pacific yet flows 550 km to the Gulf of Carpentaria (Fig. 12.6). There is a continuous belt of river gravel marking the old course of the river.

Another good example comes from Marulan, New South Wales. The Upper Shoalhaven flows north to this point but then takes an abrupt turn to the east and flows to the sea: a more spectacular elbow of capture would be hard to imagine. Alluvial gravels are found along the postulated link between the elbow of capture and the Wollondilly River which is presumed to be the former extension of the Shoalhaven.

Suppose a river is flowing across a plain that has been cut across varied rocks, and is then rejuvenated. The river will cut down to the new base level, and with its tributaries will excavate its drainage basin towards the new base level. Softer rocks such as clays and shales will be

Fig. 12.6 River capture of the headwaters of the Mitchell River, Queensland, by the Barron River.

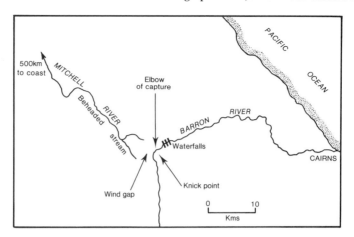

removed readily, but any harder rocks may resist reduction to the new level. However, the main river may retain its course where this accidentally crosses a band of hard rock. When much of the area has been reduced to this new plain level the river course may seem incongruous. For example, the Bristol Avon flows across a broad plain that continues out to sea, but the river leaves the lowland and takes a remarkable course through a ridge of Carboniferous Limestone to form the Clifton gorge. This river course is in fact inherited, and is an example of superimposed drainage.

Some superimposed drainage is initiated on a cover of rocks over a more varied assemblage of basement rocks, but this is not essential and some superimposed drainage originated on a plain cut directly across complex basement rocks.

Another way to get remarkable gorges is for tectonic uplift to raise a strip of highland across a river by warping or faulting, but if the uplift is sufficiently slow for the river erosion to keep pace it will maintain its course. This is called antecedent drainage.

The Fitzroy River of Queensland rises some considerable distance from the coast and crosses a range of high coastal mountains. How does the Fitzroy cross ranges that are higher than its source? Either there has been river capture of the inland drainage by a headward-eroding coastal river, or (more probably) the Fitzroy is antecedent where it crosses the ranges. Similarly in the Burdekin system long strike rivers unite and these flow across a gorge in the Leichhardt Range (said to be impassable to four-footed animals) to the coastal plains. Again antecedence seems to be the best way to account for a river that takes a course through an impassable gorge across a mountain range higher than the source area.

The Tennessee River has persisted in its present course through the Walden Ridge probably since the close of the Cretaceous Period at least (Johnson, 1905).

Two other important effects occur in the lower reaches of rivers. A relative rise of sea level may drown the lower course of the river, drowning the valleys. The rivers are said to be betrunked. Conversely a relative fall of sea level may expose broad coastal flats and the river is then extended over the flats. The old Rhine–Thames river system is an example of betrunked rivers. Extended rivers are found on the coastal plains of southeastern USA, especially around the Gulf of Mexico. Australian examples are found around the Gulf of Carpentaria. Very young examples are found along the south coast of Honshu, Japan where successive tectonic uplifts have caused two parallel strips of sea floor to emerge.

Effects of warping

Figure 12.7 shows some of the features brought about by warping affecting an area with initially simple dendritic drainage to the west. At (1) an unwarp has risen across a river; to the west of the warp the river continues to flow in the same direction with the same drainage pattern, only the amount of water carried will have changed, and such a river is said to be beheaded. Along the ridge axis of the warp (2) there may be

Fig. 12.7   Effects of warping on drainage systems. For explanation see text.

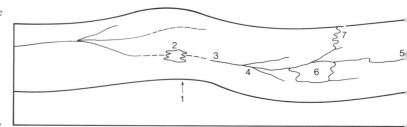

Fig. 12.8   Tilted lake shore, Mulurulu, New South Wales (reproduction by permission of the Department of Lands, NSW, Australia).

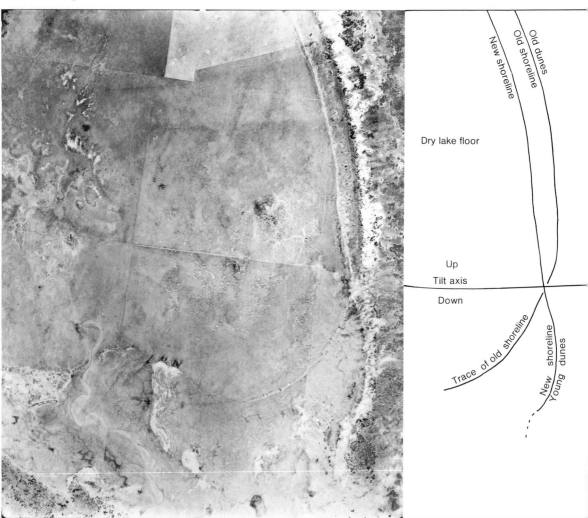

areas of indefinite drainage with negligible gradient, and there may be swamps or shallow lakes.

To the east of the warp the drainage is reversed and the main drainage flows to the east (3). However the tributary valleys (which have steeper initial gradient) may continue to flow in their old direction, so we get a barbed drainage pattern (4). In Fig. 12.7 the original drainage direction is maintained at (5). A lake is likely to form in the tectonic sag or depression between the normal and reversed drainage (6) which may find an outlet by flowing along the depression (7). In suitable circumstances features of lake shores may be tilted, revealing the nature of younger tectonic movements (Fig. 12.8).

Another kind of drainage modification is shown in Fig. 12.9, in which the westward drainage is blocked by an upwarp and flows along the foot of the warp (1). The drainage is said to be defeated. If the upwarp rises sufficiently slowly it is possible for the river to keep pace by vertical erosion and so maintain its course. Valley widening does not proceed very fast, so the river on such a warp carves a gorge (2). Since the river was in position before the warp, this is called antecedent drainage. An antecedent gorge may also form on the flank of a warp when a river has such a flank position before uplift (Fig. 12.9(b)).

Fig. 12.9   Further effects of warping on drainage systems. For explanation see text.

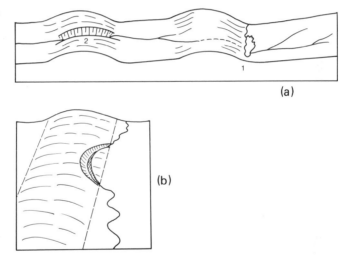

(a)

(b)

To consider some real examples, streams of the Sydney–Shoalhaven area rise close to the coast and start their courses with simple dendritic drainage to the west (Fig. 12.10). This westerly flow is not maintained, and the rivers are all diverted by tectonic warps and flow eventually into the Pacific.

Central Arizona stood higher than the Grand Canyon region when streams carried gravels to the north. Potassium-argon dates on basalt pebbles show this was 10 m.y. and older. The present south-flowing drainage of the Verde River and neighbouring streams resulted from uplift of northern Arizona, some time before 5 m.y., the age of some basalts that flow into the Verde River. From this it is possible to conclude that the major uplift of the Colorado Plateau was between 10 and 5 m.y. ago, and this was also the period of major canyon erosion (McKee and McKee, 1972). A similar early age limit is provided by Lucchitta (1972) who demonstrated that in the Basin and Range Province there was no Colorado until 10.6 m.y. ago.

Fig. 12.10   Drainage features of the Sydney–Shoalhaven area, New South Wales. For explanation see text (after Ollier, 1978).

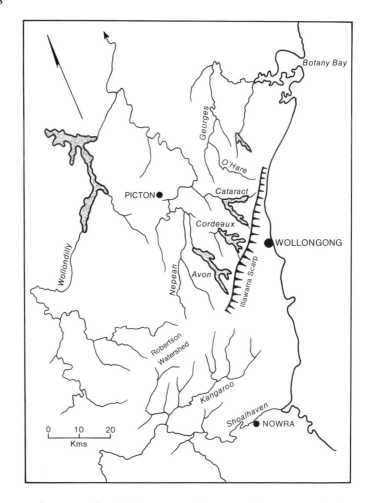

An example of primary warping on a river pattern is shown by the rivers on the south coast of New Guinea (Fig. 12.11). It seems that the drainage was originally simply dendritic to the coast, but an upwarp deflected the Fly River to the southeast while beheaded rivers continued to flow to the sea along old courses. A further uplift along the Morehead axis caused further diversions, but the Morehead River was able to continue its course as an antecedent stream. Of course the drainage pattern alone does not provide sufficient evidence but the interpretation is supported by topographic data and the nature of the river courses. At Moorhead the river flows in a gorge, just as would be expected of an antecedent river in this situation. The sedimentary history recorded in the Fly River delta also supports this interpretation.

A much greater example of an antecedent course is provided by the Arun River. This rises in Tibet and flows across the Himalayas. It could hardly be superimposed from higher ground still, for there is nothing to suggest that the Tibetan plateau was ever higher than the Himalayas, and it does not seem possible for a river on the southern flanks of the mountains to cut right through the range by headward erosion and capture a stream on the Tibetan plateau. The only remaining hypothesis, and one which fits other facts, is that the Arun is antecedent to the uplift of the Himalayas (Wager, 1937). The further

Fig. 12.11   Effect of warping on rivers of southern New Guinea. Before uplift of the Oriomo axis the Upper Fly flowed into the Baim (parallel to the Digoel) and the Strickland flowed into the Merauke. The Morehead axis of uplift created a ridge, but the Morehead river maintained its course and cut an antecedent gorge (after Blake and Ollier, 1970).

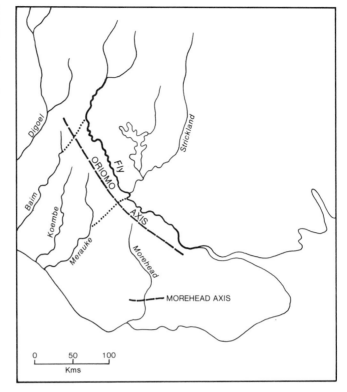

evidence is provided by river terraces. These mark old river-bed levels, and they indicate successive and differential warping of the river bed along the Arun. The zone of maximum warping is of course the position of maximum uplift, and this is indicated by the river terraces better than by any other geological feature. The axis of uplift appears to have remained stationary, rather than having migrated upstream or downstream, indicating pure vertical uplift.

The Kangaroo–Shoalhaven (Fig. 12.10) looks like a perfect example of barbed drainage (4, on Fig. 12.7), and the drainage pattern suggests reversal of the Lower Shoalhaven (3) while the Kangaroo (and several other tributaries) maintain their original direction. Some of the Ugandan rivers shown in Fig. 12.17 provide further good examples of barbed drainage patterns.

Effects of faulting

Figure 12.12 shows the effects of east-facing faults on west-flowing streams. At (1) we have beheaded streams that continue to flow in their original direction. East of the fault the drainage is blocked and may form a fault-angle lake (2) and flow out along the fault-angle depression (3). The former continuation of the river is now a beheaded river (4). A west-facing fault will give rise to a fault-angle lake only if the downthrown block is also backtilted, with reversal of drainage (5). It is possible for river erosion to keep pace with the uplift of a tilt block or horst if uplift is sufficiently slow, and as in the same situation with upwarps the result is antecedent drainage: the early-formed river maintains its course across the later tectonic topography. The antecedent river usually flows in a gorge (6).

An example of simple faulting across a river is demonstrated by the Cadell tilt block which lies across the Murray River near the town of

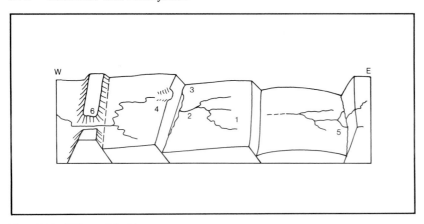

Fig. 12.12   Effects of faulting on drainage systems. For explanation see text.

Echuca (Fig. 12.13). Here a fault block diverted the old Murray, first in a route around the north of the tilt block and later round the south, the present course of the Murray. The ancient course of the river can be seen in Green Gully on top of the fault block. Again, though the drainage pattern alone is very revealing, the full story has been worked out in greater detail by the study of sediments in the stream courses and in the deposits of the old lake deposits in the fault angle.

Fig. 12.13   The effect of faulting across the River Murray (after Hills, 1975).

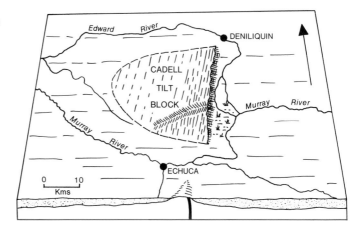

In New South Wales, Lake George and its associated features appear to provide a simple and elegant example of drainage strongly modified by faulting. The situation is shown in Fig. 12.14. Before faulting, Taylors Creek was a tributary of the Yass River, and to the south the Molonglo flowed along the line of its present route. Faulting later produced a horst athwart the Molonglo bounded by the Queanbeyan Fault on the west and the Lake George Fault on the east, the horst now being known as the Cullarin horst. The Molonglo River eroded its bed to keep pace with uplift, forming an antecedent course across the horst with gorges where the river enters and leaves the fault block. But Taylors Creek was defeated by the faulting; Lake George was formed on the downthrow side and Geary's Gap, marked by many patches of river gravels marks the former continuation of the valley.

If the basin were to fill, water would overflow through Geary's Gap and not over the watershed to the north and south, and old beach levels

Fig. 12.14   Effect of faulting at Lake George, New South Wales (after Ollier, 1978).

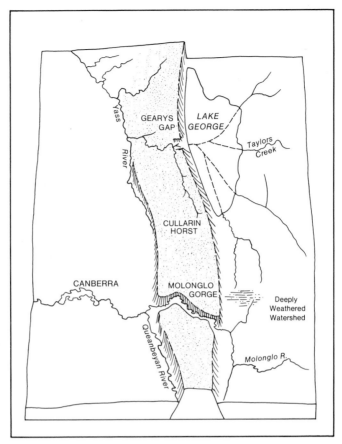

are found around Lake George at the elevation of Geary's Gap. The oldest sediments in the Lake George basin would give an age for the faulting, which is Miocene.

Another example of disruption of drainage by faulting is provided by the lower River Murray (Fig. 12.15). This has a westerly course as far as Morgan, where it turns abruptly south. Before the uplift of the Mt Lofty fault block the Murray continued in a northwesterly direction along what is now the Broughton River. In this case the Broughton River is the beheaded remnant of a former Murray. Certainly the Broughton has every indication of being beheaded, because it now fails to reach the sea, though formerly it carried enough water and sediment to build a large delta near Port Pirie. The sediments at the base of the delta are of Eocene age, so the Murray had this course in Eocene times and was subsequently diverted by uplift of the Mt Lofty Mountains (Williams and Goode, 1978).

Rivers like the O'Hare, Cordeaux, Cataract and Kangaroo (Fig. 12.10) rise within a few kilometres of the coast and flow to the west, with broad valleys and a perfectly simple dendritic drainage pattern. It would take a long time to erode these valleys, and when they were produced there could not have been coastal cliffs at their source as at present. It seems inescapable that there was an extensive catchment area to the east which has somehow disappeared. This is an example of (4) from Fig. 12.12, with the downfaulted block missing.

Similarly in the north Queensland area the steep scarps close to the

Fig. 12.15 Drainage diversion of the Lower River Murray, South Australia (after Williams and Goode, 1978).

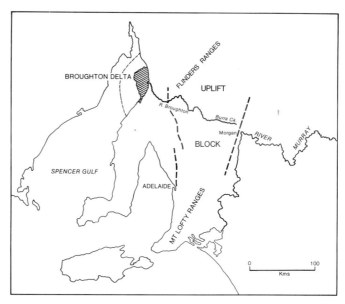

coast may be fault scarps, and as early as 1911 Griffith Taylor was suggesting,

> It is quite probable ... blocks of the earth's crust bounded by north-south faults have also determined the topography here. We may imagine (as Dr Danes suggested) a tertiary extension to the east for several hundred miles with a normal consequent drainage. The greater portion of this is faulted and subsides gradually, and is covered by the coral reefs of the Great Barrier.

This was the only model available to Taylor, in pre-Wegener days, but we can now think of another way of losing land – by continental drift and sea-floor spreading removing the land to the east. We have good evidence of sea-floor spreading in the Coral Sea and the Tasman Sea, which adds credibility to the deductions based on drainage patterns.

In Western Australia former drainage systems are marked by chains of salt lakes (Fig. 12.16). Some broad rivers have valleys 2 or 3 km wide, right on the watershed dividing north-flowing drainage from the narrow strip of coastal drainage. It appears that these rivers have been beheaded, like the New South Wales rivers described earlier, with loss of land to the south. These rivers were dismembered by continental drift and originally rose in Antarctica. Since drift started in the Eocene the drainage patterns must be at least Eocene in age, and probably Cretaceous or even older.

Combined warping and faulting: rift valleys

The drainage around Lake Victoria provides a classic example of both the initial dendritic pattern and later disruption by tectonic movements (Fig. 12.17). One great river system flowed from east to west across what is now Lake Victoria, the main route being the line of the present Kagera and Kotonga Rivers. North of this another drainage system, the proto-Kyoga, flowed in the same direction, and to the north of this the forerunner of the Murchison Nile had a similar pattern. These major rivers all arose far to the east, the southern ones rising in Kenya, and flowed west to the Zaire (Congo) River. These rivers were broad and the interfluves flat.

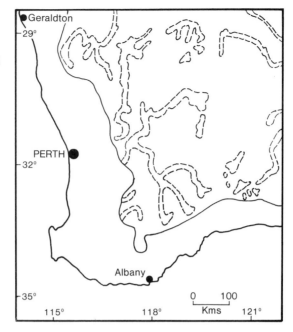

Fig. 12.16   Ancient drainage lines of Western Australia on an old uplifted erosion surface. The headwaters of the broad drainage lines appear to have originated to the south before continental drift brought about the separation of Australia and Antarctica (after Ollier, 1977c).

This pattern was broken up by the formation of the Western Rift Valley. Though the rift valley block went down there was a rise towards the rift on its eastern side, which made the old drainage basins saucer shaped. The southernmost 'saucer' filled with water to form Lake Victoria, which has an area of 70 000 km³ though at its deepest point it attains a depth of less than 100 m. Rivers to the east continue to flow into this lake in their normal way, though they have of course

Fig. 12.17   Drainage near Lake Victoria, East Africa. Original drainage was from east to west, with the Mara probably continuous with the Kagera. The drainage lines of the Katonga and Kagera are continuous between Lake Victoria and Lake Edward, as is the Kafu from Lake Kyoga to Lake Albert. Much of the drainage of these three rivers is now reversed as shown by arrows because of uplift in a zone parallel to the rift valley. The middle course of the Mara–Kagera river was drowned when downwarping created the Lake Victoria basin. Lake Victoria overflowed at the lowest point on its watershed at Jinja forming the stretch of the Nile between Victoria and Kyoga. Lake Kyoga was formed by similar back tilting of the Kafu River. It flowed up a northern tributary and overflowed in to the Albert rift valley to form the stretch of the Nile between Kyoga and Albert. From Lake Albert the Nile flows north along a continuation of the rift valley, and joins the Aswa which follows the line of a major mylonite band.

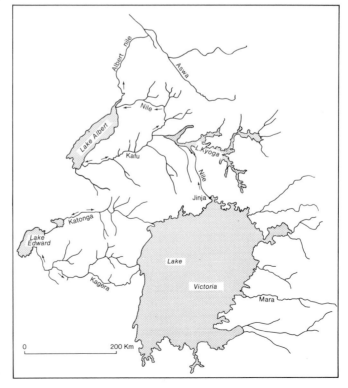

aggraded their valley floors. The rivers to the west were tilted back: the biggest rivers, having the gentlest gradients, were actually reversed and flow to Lake Victoria along the same river beds but in the opposite direction. Their tributaries flow in their original direction, but turn abruptly back when they meet the main stream, forming classicial barbed drainage patterns.

Water drained to Lake Victoria from both east and west, and the water deepened until it overflowed at the lowest point on its watershed, which happened to be at the location of the present town of Jinja. This overflow then drained into the Kyoga system, incidentally rejuvenating the Kyoga tributary that flows from Jinja and giving it quite a different character from the other tributaries of Kyoga. (Of all the southern tributaries to Kyoga only the Nile shows rejuvenation, so overflow is preferable to the suggestion of De Swart and Trendall that 'A tributary of the Kyoga system cut back to drain Lake Victoria.') The additional water no doubt contributed to the growth of Lake Kyoga. The proto-Kyoga had itself been defeated by the uplift towards the rift, and its easternmost tributaries flow back, with barbed tributaries, just like the proto-Victoria tributaries. Lake Kyoga in turn filled until it flowed over the lowest point in its watershed into the Murchison Nile. The rift valley dies out at this latitude, and the waters here flow down into the rift valley. The rift valley of course causes rejuvenation of the rivers, and a lower surface has been cut back as far as the Murchison Falls. The course of the overflow water from Lake Victoria at Jinja, through Lake Kyoga and to the rift valley is the upper course of the White Nile.

On the western side of the Albert rift the Zaire drainage pattern has been beheaded, and the Zaire lost a water volume equivalent to the upper Nile when the rift valley was formed. Broad valleys rise on the warped shoulder of the Zaire side of the rift, with no massive rivers available to carve them. The older river courses on the Uganda side simply continue over the divides as broad swamps with indistinct drainage, and eventually turn into rivers draining to Lake Albert. As many early African explorers discovered, the watersheds in this part of Africa are very gentle plains, not rugged mountain ranges. It is remarkable that the divide between two of the world's greatest rivers, the Nile and the Congo, should be so insignificant a topographic feature. It is in fact a tectonic warp in a region of low relief, as yet unaffected by post tectonic erosion. It is significant that, as clearly shown by the watersheds, the axis of the warp separating rift valley drainage from Nile drainage is not at the rift valley fault but about 50 km back from the fault.

The oldest sediments in the rift valley give the age of the faulting, which turns out to be Miocene. The old drainage pattern and the associated erosion surfaces must be older than this since they clearly pre-date the faulting and warping.

Summary of drainage pattern relationships

In this chapter we have seen that the simplest pattern of drainage is the dendritic pattern. This can be modified by differential erosion picking out rock structure, and can be modified by warping or faulting. Tectonic changes produce a whole range of minor landforms and associated changes so that there can be no doubt about many of the hypotheses of drainage modification. If drainage pattern were the only indication of tectonics any hypothesis of modification would be tentative, but when long profiles, terrace history, overflows, wind gaps, alluvial areas and so on all build up a consistent story we can use drainage pattern data with confidence to deduce tectonic movements.

It is also clear that many drainage modifications can be dated, even if approximately. Some of these dates are quite old. The East African rift valleys go back to the Miocene, and the major drainage patterns were initiated earlier. Lake George was formed in the Miocene, and the earlier drainage pre-dates the lake. Some drainage patterns in Western Australia pre-date continental drift, and are therefore Mesozoic.

Of course there are many younger drainage-pattern modifications, but for much of the world it seems that major drainage patterns were well established in the Tertiary, and many tectonic modifications to drainage patterns also have an old history.

The very largest rivers have features of a different order of magnitude from those discussed so far, relating to major features of global tectonics, and these need to be discussed separately.

Big rivers and their deltas

A river may be considered big because of the length of its course, the size of its drainage basin, the amount of water it discharges, or the amount of sediment it transports. It is hard to determine discharge and bed load transport, so a better appreciation is to be had from length and drainage basin area. Of the 50 biggest rivers in the world all but one are over 1 000 km long. These 50 rivers drain 47 per cent of the continents excluding Greenland and Antarctica; the five biggest, the Amazon, Zaire (Congo), Mississippi, Nile and Yenisei account for 10 per cent and the Amazon alone accounts for 5 per cent. Eleven of the largest rivers supply about 35 per cent of the suspended sediment load carried to the ocean (Drake, D. E., 1976).

Potter (1978) concluded that big rivers are an overlooked, under-exploited theme in earth history, and posed the following provocative questions:

1. What are the geological controls of big rivers?
2. Can big rivers be related to plate tectonics?
3. What is the relationship between depositional basins and their river systems?
4. What was the occurrence of big rivers in past earth's history? Is the Amazon the largest river ever? Were there bigger ones on Gondwanaland or Pangaea?
5. What conditions are required for long-lasting rather than short-lived big river systems?
6. What happens to ancient river systems? Are their sediments preserved in ancient delta systems or consumed at subduction sites?

Before trying to answer these questions some examples and ideas to emphasize the time scale involved and the geomorphology – tectonic relationships will be reviewed.

The time of origin of a river system is defined as the earliest date at which a continuing persistent river drained a region in question. A river dates from the last marine regression, the last significant tectonic uplift or warping, the waning of an ice sheet, or the cessation of lava outpouring.

A river system is terminated by marine invasion, new and different tectonic movement, glaciation, or lava outpouring.

McMillian (1973) has reconstructed the palaeodrainage system

Fig. 12.18   Reconstruction of Tertiary river system of eastern and central Canada (after McMillian, 1973).

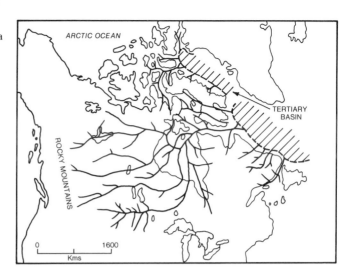

across much of Canada, with drainage towards a Tertiary basin between Canada and Greenland (Fig. 12.18).

The Colorado River has five broad stages in its development (McKee *et al.,* 1967) – withdrawal of a Cretaceous Sea, Early Tertiary deformation and erosion, renewed uplift, volcanism and faulting, and eastward headward erosion.

Martin (1975) described a palaeodrainage system in south-west Africa oriented towards the present African coast, suggesting the presence of a proto-South Atlantic in Late Palaeozoic time. Palaeozoic valleys with relief up to 1 000 m were infilled by Permo–Carboniferous glacial deposits, then by basalts (associated with tensional faulting when the Atlantic opened), and are now being exhumed.

Swann (1963) reconstructed the ancestral Mississippi, and showed that this river has been oriented south and southwestward across the North American craton since the Early Carboniferous. Furthermore the river system follows the axis of the Mississippi embayment which had its beginnings in the Late Precambrian.

Grabert (1971), quoted in Potter (1978), suggested that the Brazilian shield has been uplifted and exposed to weathering since Triassic or earlier times. Early drainage was radial so that much would flow to the Andean geosyncline. The mid-Tertiary uplift of the Andes caused the Amazon to flow east and the Parana to flow southward subparallel to the strike of the Andes. De Rezende (1972) has pointed out that there is also marked structural control for much of the lower course of the Amazon, which occupies a graben or linear depression containing lower Palaeozoic sediments in a long narrow belt extending across the Brazilian shield. The Parana in southwestern Brazil, Uruguay and Paraguay follows the axis of greatest thickness of a Cretaceous basalts, possibly because their weight depresses the crust.

Potter (1978) distinguished five types of big river in relation to tectonic setting, which may be simplified as follows:
1. Rivers that flow across a craton without any belts of fold mountains. The Nile, Niger and Zambezi are examples.
2. Rivers marginal to mountain belts that flow across a craton such as the Amazon and Mississippi.
3. Rivers that are marginal to a mountain chain and flow parallel to it, such as the Ganges, Mackenzie and Parana.

4. Rivers that flow within a mountain chain and parallel to the grain, such as the Mekong, Magdalena, Yukon, Brahmaputra and Irrawaddi.

5. Rivers that flow in superimposed courses across the strike of mountain chains, such as the Fraser, the Snake–Columbia and Danube.

This classification is very similar to one proposed by Audley-Charles, Curray and Evans (1977), who proposed a classification of the location of deltas as follows:

(a)   On the craton shelf and its submerged edge.

(b)   On the craton edge and extending on to ocean floor.

(c)   On the continental side of mountain chains.

(d)   On the ocean side of mountain chains.

(e)   Deltas on the western side of the Pacific associated with island arcs or marginal seas.

These different types of delta are expressed in the character of their basement and the lithology and stratigraphy of their deltaic pile. Recognition of these different delta types is useful for deciphering the palaeogeographic and structural history of ancient deltas, and Audley-Charles, Curray and Evans even claim that the structural history of deltas may be predicted with considerable confidence by the tectonic setting in which they accumulated.

Considering the relationship of big rivers, deltas and global tectonics, Potter (1978) pointed out that the 28 biggest rivers in the world all drain to 'trailing edges of continents' or to marginal seas. The twenty-ninth in magnitude, the Columbia, drains across a 'collision coast'. Twenty-five of the world's largest deltas are also found on 'trailing and marginal' coasts. Whether or not the plate-tectonics framework is correct, it does seem that the results reflect a continental asymmetry of watersheds with mountain belts on the 'collision side', and rivers draining to 'trailing edges' having the largest drainage area. Thus, in the broadest view, rivers respond to the megageomorphology of continents which in turn is related to global tectonics.

Some major rivers are rift controlled, and some of the rifts are thought to be 'failed' arms of triple junctions (aulocogens) formed during plate fragmentation (Burke and Dewey, 1973). The Niger is the classic example, but the Rio Grande (following a Miocene rift) and the Limpopo (Mesozoic) are others.

We can now try to answer the questions posed on p. 177. Some of the big rivers of today have been big for a very long time – at least throughout the Cenozoic. They seem to relate to plate tectonics, but this may be fortuitous to some extent: if a strip of highland is raised at a continent margin, rivers must drain from it and join some major river in a lowland. Some of the big rivers of today have courses that probably pre-date the break-up of supercontinents. If this is so it seems likely that the Amazon is not the largest river ever, but would have been exceeded by some of the rivers that drained a united Gondwanaland or Pangaea. The proto-Congo, for example, with headwaters in Kenya, could be one, and before the rifting of the Atlantic the proto-Congo may even have flowed across South America. The River Murray may have risen in lands to the east of Australia, no longer present, and its ancient course, traced so far to St Vincent's Gulf, might earlier have crossed much more of Australia or Antarctica.

Delta sediments of the present large rivers such as the Amazon and Niger deltas, are still with us, but this is because they are on passive sites, not subduction sites. Since major rivers do not flow to subduction sites we have no present-day check on the role of subduction of the sediments of major rivers. Why are some river systems long lasting while others are relatively ephemeral? Possibly there is a feedback mechanism that enables long rivers, once established, to be self sustaining. Their eroded headwaters tend to rise isostatically, while their depositional deltas sink isostatically. Isostasy thus provides some sort of mechanism for continuity, though admittedly the valley as a whole might be expected to show isostatic rise. Even if erosion is itself the cause of isostatic compensation and tectonic uplift, it cannot be the cause of sea-floor spreading, rifting and the formation of new continental edges and mountain ranges. But these processes, although related to the distribution of major rivers and deltas, actually break them up, and the long rivers of today have been disrupted rather than created by plate tectonics.

The degree to which rivers control and are controlled by tectonics remains to be worked out, and the elucidation of the interplay of geomorphology and tectonics will require a degree of geomorphic input that is not yet available.

13　Geosynclines

In folded rocks an arched or upfolded rock stratum is an anticline, a sag or downfold is a syncline. A geanticline is an uparched mass of rock of huge dimensions that may incorporate many minor folds. A geosyncline is a downfold of major dimensions.

But geosyncline is a term with many more connotations. Geosynclines are commonly thought of as a special kind of sedimentary basin which has a typical history going from sedimentary basin to fold belt, and finally to a mountain range. The simple idea is that sediments accumulate in a trough which is then somehow squashed, folding the sediments and forcing them up to form mountain chains. Geosynclines are relatively long and narrow sedimentary basins that exist for a length of time measured in geological periods. They appear to have bordered land masses for they contain rocks derived from continental areas. Classic examples of geosynclines include the Caledonian geosyncline of the British Isles, the Appalachian geosyncline of North America and the Tasman geosyncline of Australia.

The geosyncline concept has come to be in disrepute in recent years, because geosynclines have many differences (nothwithstanding what they have in common), and because early workers had various ideas associated with geosynclines that are now refuted. It is now fashionable to use the term 'fold belt' or 'mobile belt', but these terms seem to refer largely to the same thing. The geosyncline is a palaeogeographic feature, and not its altered equivalent, which is an orogen, orogenic belt, or fold belt.

Two types of geosyncline have been generally recognized: the miogeosyncline is filled mainly by continental debris, was originally shallower, and is not very deformed; eugeosynclines have more pronounced subsidence, thicker accumulation of sediment often including many volcanic products, considerable deformation and many igneous intrusions. Eugeosynclinal and miogeosynclinal fold belts occur in pairs in many parts of the world, including the western United States, British Columbia and Alberta, and in the Appalachians.

The Appalachian situation is shown in Figs. 13.1 and 13.2. Basically there are three elements in the geosyncline of the classical Appalachian type:

Fig. 13.1   Tectonic map of the southern Appalachian Mountains (after Cox and Cox, 1974).

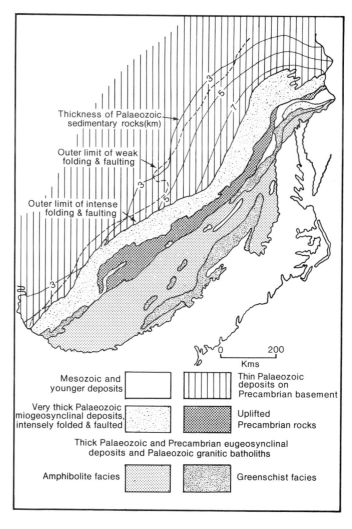

Fig. 13.2   Restored geosynclinal section of the Appalachian geosyncline in eastern North America (after Kay, 1951).

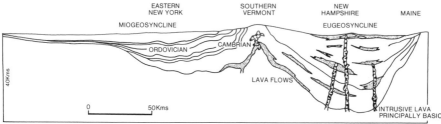

1. The mildly deformed marginal or miogeosyncline.
2. A middle zone of fairly continuous ultramafic rocks (ophiolites).
3. An outer very deformed eugeosyncline.

In the Alps on the other hand there is a separation in time between the two types of deposits. The first deposits are eugeosynclinal and after these have been deformed and mountains created, molasse basins become filled with debris derived from the mountains.

Aubouin (1965) attempted to represent geosynclines by specific examples, detailing their changes in space and time, emphasizing the presence of double geosynclines – eugeosynclines and miogeosynclines – and definitely relating them to tectonic development of mountains. This was perhaps the last major work on geosynclines with a traditional approach.

The latest fashion is to relate geosynclines to plate tectonics, each characterized by a topographic feature, a suite of sediments and tectonic setting at a plate boundary. There are several kinds:

Pacific type, at converging plate boundaries. This may be divided into

Andes type, where the trench is adjacent to a continent, and

Island-arc type, where a trench is parallel to an arc but separated from it by an arc-trench gap, and behind the arc is a shallow sea.

Atlantic type, with accumulation of sediment at the trailing edge of a continent, and

Mediterranean type, denoting intercontinental sedimentation.

Geosynclinal sediments

When Hall first described the Appalachian geosyncline in 1859 he noted that it contained many thousands of metres of sedimentary rocks that were laid down in shallow water. Clearly if the sedimentary basin were simply waiting to be filled it should first have deep water and collect deep-water sediments, and only have shallow-water sediments at the end. The only way to have shallow-water sediments throughout a thick succession is for the geosyncline to sink at a rate close to the rate of sediment fill. Perhaps the weight of sediment actually causes the sinking.

But not all geosynclines have shallow-water sediments throughout. The southern New England (Australia) geosyncline had deep-water sediments for most of the Palaeozoic era and only in the Late Carboniferous and Permian were shallow-water marine or terrestrial conditions prominent (Korsch, 1977). A 4.7 km bore through Cenozoic sediments off northwest Australia, which might be regarded as a modern analogue of geosyncline, showed a change from a deep-water situation in the Palaeocene through marine deposits of indeterminate depth in the Eocene, deep to inner-shelf sediments in the Oligocene to Middle Miocene, and reef conditions which have persisted from the Middle Miocene to the present day (Wright, 1977). Modern Californian basins all show a trend to deposition in progressively shallower water, culminating in mid-Pleistocene folding and erosion, and Upper Pleistocene units lie nearly flat across the upturned edges of older units.

The literature of geosynclines abounds in references to two sediment types, 'flysch' and 'molasse', which are both very difficult to define in a satisfactory way. The terms originated in Switzerland. Flysch in Switzerland consists mainly of dark shale and siltstone (generally calcareous) in thin layers, and layers of greywacke with graded bedding.

Greywacke is itself a term that has persistently defied all proposals at either precise definition or complete abandonment. Roughly, it is a dark-coloured sandstone containing abundant clay, rock fragments, and large detrital quartz and feldspar grains, and chlorite. It may have graded bedding, and it may be derived from a source area with limited weathering.

Molasse, in Switzerland, consists mainly of calcareous and arkosic sandstone and conglomerate. In some definitions the 'post-orogenic' nature of molasse is built into the definition.

It should be noted that the original flysch contains no volcanic rocks, so the term cannot be applied to eugeosynclines as defined by Kay (1947), and by the *Glossary of Geological Terms* prepared under the direction of the American Geological Institute which defines 'eugeosyncline' as an orthogeosyncline (a long narrow one) in which volcanic rocks are abundant. According to Cobbing (1978) the terms 'eugeosyncline' and 'miogeosyncline' are instantly recognizable by the majority of geologists as implying thick volcanic sequences on the one hand and sedimentary sequences on the other. Furthermore, the original molasse is quite different from the Permian and Triassic redbeds of North America to which the term is often applied.

Waterschoot van der Gracht (1931) introduced the concept of 'orogenic facies' into North America. Flysch was defined as 'sediment deposited previous to the major paroxysm' and molasse as sediment deposited 'during or immediately after the major diastrophism' emphasizing the early ideas of direct association of geosynclines, their sediments, and orogeny. It also seems that the prevalent view of orogeny was of 'one big bang', and the idea of orogenic facies was introduced accordingly.

Petrographic classifications of geosynclines came later, and certain rock types were optimistically related to parts of a geosyncline cycle which also relates to a geomorphic and a tectonic cycle. Krynine (1941) divided geosyncline cycles into three stages:

1. Peneplanation (or nearly geosynclinal stage), with cyclic deposition on a fluctuating flat surface after much weathering, characterized by first-cycle quartzites.
2. Geosynclinal stage, with basin deposition interrupted by local vertical movement, with greywacke as the typical sediment.
3. Post-geosynclinal stage, with uplift after folding and magmatic intrusion of the geosyncline, and arkoses as the typical sediment.

The model for this sequence was the Appalachians, but in the Alps arkose is found in the 'flysch' of the geosynclinal stage, and the post-orogenic molasse consists largely of rock fragments. There are too many factors affecting rock composition to allow its use as an indicator of geosynclinal or tectonic relationships. Later workers used sedimentary structures, especially cross bedding and graded bedding of turbidity currents as palaeogeographic indicators, and related orogeny to geosynclinal fill through the interrelationship of sediment supply and rate of subsidence. These are not always directly related and some deep basins accumulate little sediment.

Some other rocks found in geosynclines have assumed importance in recent years, possibly relating geosynclines to plate tectonics.

Ophiolites are more or less altered basic rocks, presumed to be volcanic rocks erupted on to the sea floor at the bottom of a geosynclinal trough or perhaps parts of the sea floor itself. The fully-developed ophiolite sequence contains deep-sea sediment, pillow lavas, dykes, gabbros, peridotite and other basic and ultrabasic rocks. Since deep-sea ophiolites are hard to study considerable effort has been put into the study of ophiolites where access is easier on land.

Ophiolites are thought to be important because they may mark, in some way, the suture zone marking the site of plate collision. Certainly many sutures bounding major tectonic units are marked by serpentine belts (metamorphosed ophiolite). The narrow zones of ophiolites in the Alpine–Himalayan mountain belt are interpreted by some as slices of oceanic crust and mantle. They may mark the lines along which continents collided. Similarly, the Ural and Appalachian–Caledonian mountain belts, which lie within ancient Pangaea, have narrow ophiolite zones, thought by some to mark the sites of vanished oceans. The Urals is thus interpreted first as an area of sea floor created by sea-floor spreading, consumed later when two continents collided, leaving only the ophiolite belt as a trace of the former sea floor.

Mélanges are tectonically chaotic rock units, megabreccias of two types: tectonic mélange, and sedimentary mélange or olistostrome. Both are found on the outer side of the trench-arc gap and both are thought to be related to underthrusting tectonics at convergent plate boundaries.

Olistostromes are sedimentary stratigraphic units containing numerous exotic and local, mainly sedimentary, blocks in a fine-grained matrix. They are formed by gravity sliding, on a topographic slope such as the side of a geosyncline, or trench. They are bedded mudslide deposits and have no direct indication of being subducted, though the Dunnage mélange of Newfoundland (Lower Palaeozoic) and the Gwna mélange of Anglesey, Wales (Late Precambrian) have both been attributed to subduction.

Tectonic mélanges are rock bodies composed of tectonically mixed blocks in a sheared shaly matrix. They may be very big, with individual blocks up to several hundred metres long. Rocks of very diverse origins may be included, such as cherts, glaucophane schist, greywacke and basalt.

A problem with these mélanges is that if a sedimentary olistostrome is sheared, it is difficult to distinguish it from a tectonic mélange. The definitions of melange and olistostrome just given both depend on knowledge of assumed mechanism of deformation, and often this is a mere assumption. Some prefer to use the term mélange in a purely descriptive manner, with no implication of the origin of the rock mass.

The tectonic significance of mélanges is that they might result from submarine slumping, or they might be produced tectonically by scraping sediment and ocean floor from the top of a subducting plate. Mélanges are often associated with pillow lavas, serpentinites and cherts, and look like trench-bottom deposits. Some mélanges are associated with blueschists (see p. 99) which indicate high-pressure and low-temperature metamorphism and could be formed in an environment such as that at the top of a subducted plate where pressure is great although the rock is so near the surface of the earth that temperature is still low. The association of blueschist with mélange supposedly indicates a former subduction zone.

Source of sedimentary fill

From the study of sedimentary rocks in geosynclines the direction from which the sediment was derived can be determined. In many geosynclines the direction was surprising for the sediments were derived from the side where no continent exists at present. The sediments could not have been derived from the ocean floor, for even if it were uplifted it would have quite the wrong composition. For example, many of the clastic sediments that are found in the Appalachian geosyncline thicken to the east. Such eastward coarsening is found in beds of Ordovician, Silurian and Devonian age in various parts of the eastern states. To provide these sediments a continent, Appalachia, was invented. Dana wrote in 1856 'The existence of an Amazon on any such Atlantic continent in Silurian, Devonian, or Carboniferous times is too wild an hypothesis for a moment's indulgence'. Now we know that the 'Atlantic continent' was present in the form of Europe and Africa before continental drift created the Atlantic Ocean. Similarly the Tasman geosyncline of eastern Australia was filled from the east, now part of the Pacific Ocean, but the source for the sediments is the former continental area which has since scattered to form the Lord Howe rise and the islands of New Zealand, New Guinea and New Caledonia.

Isaacson (1975) showed that during the Devonian the central Andes received their sediments from the west. 'The size of the western land source that yielded the volume of sedimentary detritus in the Bolivian portion of the basin alone suggests that more land area existed in the western South American continent than exists today. Abundances of detrital mica in the sediments indicates that there was provenance of highly micaceous rocks ...', not at all the sort of sediment that could be derived from an oceanic source. Carey (1976 p. 127) suggests that the missing land was northeastern Australia and Tasmantis – the now fragmented easterly source of Australian Palaeozoic sediment.

Folding in geosynclines

As seen at present the rocks of geosynclines are folded, and some of them are currently mountainous regions. The idea arose that geosynclines and mountain formation were closely related, and indeed in the pre-plate-tectonics day most ideas of mountain building were somehow related to geosynclines. At its simplest the idea was that a geosyncline existed as a sedimentary basin for several geological periods, but was finally compressed, thus folding the rocks and uplifting the mountains.

A number of features need further discussion. Firstly, the folding of the rocks. Geosynclines were not simply filled up and then folded, for there are many periods of folding, erosion and further deposition within geosynclinal sequences. Folding was not a feature that occurred only at the end of the geosyncline's life, but frequently accompanied deposition. A squeeze at the end is not enough to make the observed folds.

The long axes of folds are of course roughly parallel to the length of the geosyncline, a result that can be obtained either by crustal compression or by sliding down the sides of the geosyncline. Study of the folds often reveals the direction of movement, and it is found that some sediments moved towards the centre of the geosyncline. The term syntaphral tectonics was coined for this sort of folding, and a symposium on this topic brought together many pieces of evidence for the widespread nature of gravity sliding in geosynclines (Carey 1963b).

More often, however, the folds are away from the axis of the geosyncline. The impression is that the contents of the geosyncline were pushed towards a foreland, and the name foreland folding is given to this sort of structure. Early ideas suggested that the geosyncline is squashed, like the closing of a vice, and the sediments forced out, as seen in the foreland folds. There was some concern because it seemed reasonable to find folds symmetrically arranged about the centre of the geosyncline, yet in reality fold belts are usually very asymmetrical.

The side of a geosyncline adjacent to a continent is usually well preserved, and here we find that the sedimentary layers thin out over the continental foreland. The other side of the geosyncline is often so strongly deformed or covered by younger rocks that the nature of that margin at the time of deposition is obscure.

In plate-tectonics terms the deposits in trenches adjacent to island arcs of western America may be thrust towards the continent, plastered onto the inner wall of the trench and eventually be exposed as 'geosynclinal' deposits. It seems questionable whether the semi-consolidated deposits of the fill have sufficient strength to fold abruptly to form the steeply-sloping surface of the commonly 500–1 500 m high trench wall which in many areas is steeper than 15–20° and locally may exceed 40°.

Some geosynclines appear to have been inverted by vertical movement, which raised the altered and metamorphosed rocks of the centre, eventually exposed as a 'root zone', and the more superficial rocks slid away from the uplifted area to create foreland folds by gravity tectonics. This mechanism does not explain the lack of symmetry better than any other.

The folding of geosynclinal sediments, whether syntaphral or foreland folding, is not necessarily related to squeezing of the geosyncline. Nor is it necessarily related to uplift of mountain-building proportions. Folding often has nothing to do with any uplift, or with mountain building. Many geosynclines finish their life as belts of sedimentation and folding not as mountain chains but as plains. Subsequently mountains may be lifted on the same site, but that is another story.

Metamorphism and intrusion

As it fills the geosyncline gradually sinks, perhaps under the weight of its sediments, and the sedimentary pile thickens to several kilometres. Eventually a stage will be reached when the bottom of the geosyncline reaches a level of elevated temperatures and pressures, and at this stage metamorphism and granitization set in. Granite magmas may be created which, being lighter than the surrounding rocks, may rise diapirically as distinct plutons through the geosynclinal sequence. Clearly this process can only occur fairly late in the geosyncline's life, when great thicknesses of sediment have accumulated. By this time the nature of the geosyncline has changed with differences in tectonic movement, in rock density, rock strength and so on. This stage possibly marks the end of the active life of the geosyncline. There is no necessary connection between emplacement of granite and mountain building. The formation of granite is not necessarily associated with changes of volume but may be simply an isovolumetric change in the sediment pile, converting it to a material more suited to the prevailing pressure and temperature conditions. Although miogeosynclines, with their continent-derived sediments, would appear to be suitable for granitization, the volcanogenic sediments of eugeosynclines are most commonly converted to granite. The greater depth may be the reason for this, the eugeosyncline reaching hotter zones.

Burial of a sufficiently thick pile of sediment at a sufficiently fast rate necessarily leads to the creation of a metamorphic belt. Self heating and thermal transport from below necessarily lead to partial melting and regional metamorphism. In any geosynclinal position metamorphism is likely to happen, but whether it results from subduction or similar plate-tectonic compression, or simply from deep burial is hard to decide. However, the blueschist facies rocks,

representing high-pressure/low-temperature metamorphism, cann~
be produced by simple burial.

Modern analogues of geosynclines

Various analogies have been suggested from modern situations t~
account for geosynclines. The sedimentary deposits off the easte~
coast of North America provide one example (Fig. 13.3). Th~

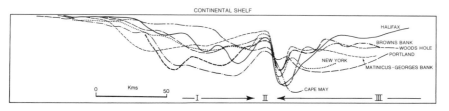

Fig. 13.3 The Atlantic continental margin of North America as an analogue of a geosyncline. Compare with the section of the Appalachian geosyncline in Fig. 13.2 (after Heezen and Drake, 1963).

nearshore deposits form one basin, equivalent to the miogeosynclin~
but a deeper trough has sediments containing many turbidity curre~
deposits which were thought to be analogous to a eugeosynclin~
although of course there is no volcanic component to the~
deposits. Another actualistic model for geosynclines was proposed b~
Dietz (1963) who suggested that the continental-terrace sedimenta~
wedges are modern miogeosynclines and the subjacent continenta~
rise sedimentary prisms are modern eugeosynclines, together formir~
a couplet. Thrusting by a spreading sea floor is supposed to convert th~
continental-rise sediments into folded eugeosynclinal deposits. Th~
mechanism is shown diagrammatically in Fig. 13.4. This model has n~
role for island arcs or trenches and would be applicable perhaps t~
sediments on trailing edges of continents where these features do n~
exist.

Fig. 13.4 Continental terrace and continental rise as analogues for miogeosynclines and eugeosynclines on a plate-tectonic model (after Dietz, 1963).

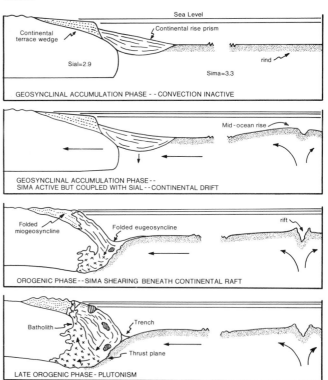

If the Atlantic margin provided a satisfactory analogy of a geosyncline, it might be expected that pre-Mesozoic Atlantic-type continental margins would be found. Several possibilities have been suggested, but none is satisfactory (Smith, 1976, p. 273).

Other possible modern examples of non-volcanic geosynclines are provided by the sedimentary basins of California. The Los Angeles basin has 12 000 m of marine deposits; the Ventura basin has 11 000 m of marine clays; the San Joaquin basin has 6 000 m of Miocene and Pliocene sediments and 300 m of non-marine Pleistocene deposits.

The modern Bengal Deep-Sea Fan is regarded by Curray and Moore (1974) as the uppermost 4 km of the geosynclinal pile of sediments filling the Bay of Bengal, in the northeast Indian Ocean (Fig. 13.5). The fan post-dates the first collision of India and Asia and the uplift of the ancestral Himalayas at the end of the Palaeocene.

Fig. 13.5   The Bengal Fan, a possible modern analogue of a geosyncline (after Curray and Moore, 1974).

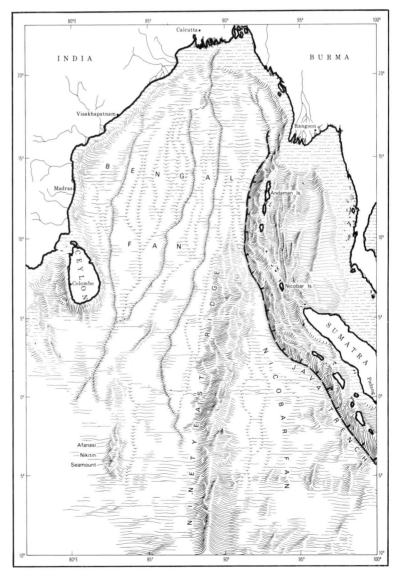

Underlying the fan are continental-rise sediments up to 12 km thick which extend into the Bengal and Assam valleys, deposited on the margin of India following its separation from Australia and Antarctica in the Cretaceous. Deformation is occurring simultaneously with deposition as the fan and geosyncline pass obliquely northwest into the subduction zone of the Sunda Arc and Indoburman Ranges.

Nowadays a more fashionable analogy is with island arcs. The basin between the volcanic arc and the continent is largely filled with continental debris, like a miogeosyncline; the trench between the arc and the ocean is a eugeosyncline and is filled mainly with sediments of volcanic origin, or with turbidite deposits from the continent.

Trenches adjacent to continents are another analogue for geosynclines, and the fill of older geosynclines in western America is thought to be represented in the Mesozoic rocks of the circum-Pacific eugeosyncline which are commonly assumed to be accreted to the continent by some sort of subduction process. The amount of underthrusting to produce these geosynclinal sedimentary bodies would be a few hundred kilometres, but plate reconstruction suggests that 5 000–7 000 km should have been consumed. Plate-tectonic considerations also suggest that oceanic sediments should be scraped off the downgoing slab, and by analogy with modern trenches and ocean flooors about 30 per cent of the rocks of the circum-Pacific eugeosyncline should be oceanic rocks. In fact the proportion is much lower (less than 1 per cent in the Franciscan Formation in California) and it seems that the bulk of the eugeosynclinal deposits in the deformed masses of the circum-Pacific belt are not trench deposits (Scholl and Marlow, 1974).

Aulacogenic geosynclines are another possible type of plate tectonic geosyncline. Aulacogens, the failed arms of triple junctions (p. 82) are linear structural troughs that cut across stable regions, and become filled with sediments and volcanic rocks. The Athapuscow aulacogen of the northwest Canadian shield has over 11 km of sediments in a fault-bounded trough ranging from 70 km wide where it joins the Coronation Gulf geosyncline to less than 20 km some 200 km into the continental platform. There are at least five periods of basaltic vulcanism, together with granodiorite laccoliths dated at 1 750 m.y. (Smith, 1976). The geosyncline is now an orogenic belt that has not been created by subduction or plate collision.

Fig. 13.6 Suggested reconstruction of the North Atlantic Basin as a mid-continental (Mediterranean) geosyncline in Late Liassic time (after Van Houten and Brown, 1977).

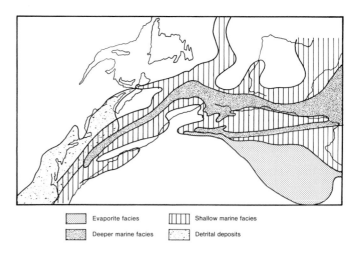

Evaporite facies   Shallow marine facies
Deeper marine facies   Detrital deposits

It should be noted that coastal-plain sedimentation and island-arc sedimentation are both marginal to continents, but many of the classical geosynclines were formed in troughs in the middle of continental masses (Fig. 13.6). Another analogy therefore, is a sea between continents, such as the Mediterranean.

Geosynclines and mountains

Young mountain ranges presumed to be formed from geosynclinal sediments have deep roots, revealed by negative gravity anomalies and by seismic effects which show the base of the crust may be perhaps 60 km down compared to the normal crustal thickness of 30 km. As the mountains are worn down, the chain rises isostatically, lifting mountains further and enabling erosion to continue, leading to further uplift and so on. As time goes by deeper and deeper levels of the geosyncline are exposed, until eventually the deepest metamorphic rocks and granites are exposed. At this stage the crustal thickness has been reduced to the global average and there may be no further isostatic uplift – the remains of the old geosycline are simply part of the bedrock of average continental crust. Such planated ancient geosynclines may consist only of belts of metamorphic and igneous rocks. In Europe the Alps have an isostatic root (not to be confused with the nappe-root zone of metamorphic rock) but the Palaeozoic fold belts do not.

Shields consist of belts of metamorphic and igneous rocks of different ages, and the distinct belts than can be distinguished may be related to different 'orogenies', on the supposition that each belt has gone through something like the geosynclinal history outlined above.

To conclude this chapter it will be very instructive to review the ideas of the originator of the geosyncline concept, James Hall. He studied the Appalachian region for many years before announcing his idea that the Appalachians had been formed through long-continued sedimentation in shallow seas, with sand, limestone and gravel containing fossils and structures such as ripple marks to show its shallow-water origin. The sediments accumulated to thicknesses of 12 000 m showing that subsidence accompanied sedimentation. The rocks could be traced to lesser thicknesses along the margin of the old trough. Hall visualized that folding and faulting accompanied the subsidence of the trough. It seemed to some of his contemporaries that he had a theory of mountain building with the mountains left out. But Hall proposed that mountains as physiographic features were created at a later stage, by erosion after the folded and faulted pile of sediments was uplifted as part of a general continental rise.

Spencer (1966) says 'However, this view has never been widely accepted', but the many examples presented in this book suggest that Hall's pioneer views were correct, or at least nearer to the truth than those of many of his successors.

That geosynclines exist much as Hall described them is not disputed, but the cause of subsidence, the character of the margins of the syncline, the location of modern analogues, the cause and timing of folding, faulting, igneous intrusion and uplift are all hotly debated to the present day.

14 Island arcs, trenches and back-arc basins

General description

The coasts of Asia bounding the Pacific are notable for their arcs of offshore islands that run from the Aleutians in the north to the Kermadec–Tonga trough in the south (Fig. 14.1). Similar island arcs are found in Indonesia (the only arc in the Indian Ocean) and in the Scotia arc and Caribbean (Antillean) arc, the only arcs in the Atlantic. These arcs have an importance in modern tectonic ideas quite out of proportion to the area they cover, and require more detailed consideration of both their topographic features and their tectonic implications. They will bring together many of the aspects of earth structure, gravity, earthquakes and volcanoes that have been summarized in this book so far.

Most 'arcs' are arcuate, forming small parts of small circles on the globe with a radius of about 3 000 km. However, some straight-line features such as the Tonga–Kermadec rise and trench have many of the features associated with arcs and would be regarded by some as 'straight arcs'.

Some arcs are backed by continent (Japan), some by sea floor (Scotia arc), some pass laterally into Cordilleran-type fold belts (Aleutians), and some arcs appear to be in a back-to-back arrangement (Philippines). Most arcs face convex to the east, but a few (Indonesia, New Hebrides) face west. Some trenches are arranged one behind the other but facing the same way (Philippines, Marianas).

Topographically, the front of an island arc is generally rising while the back of the arc is sinking. In New Britain, for instance, the front of the arc is shown to be rising by flights of uplifted coral terraces, while the other side of the island is marked by sinking extinct volcanoes and drowned coasts. Similarly, Papua New Guinea as a whole seems to be rising to the north, where the coastal ranges appear to have been uplifted in the Plio–Pleistocene times, and sinking to the south where the vast deltas of the Fly River and the Papuan Gulf are being built.

Some arcs are single, such as the Kurile arc between Kamchatka and Japan, consisting of a row of active or recently extinct volcanoes behind a trench. Others, such as Sumatra, are double arcs, with an additional line of non-volcanic islands between the trench and the

Fig. 14.1 Trenches and Benioff zones of the western Pacific. Depth to the Benioff zone indicated by 100-km contours (modified from Ringwood, 1974).

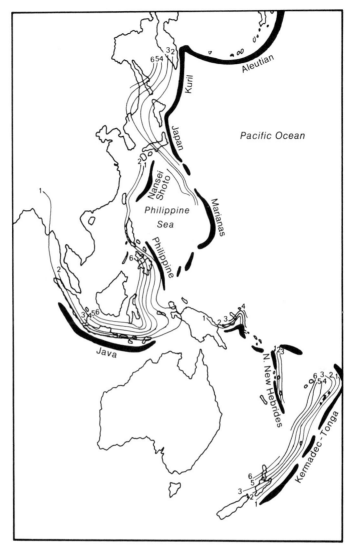

volcanic arc (Fig. 14.2). The line of islands off Sumatra continues as a submarine ridge off Java.

On the ocean side of arcs there is a deep trench in the ocean floor, and indeed the deepest parts of the oceans are always in this situation. The deepest known part of the ocean is the Marianas Deep at 11 035 m, deeper than the height of Mt Everest, and depths of 2–8 km are normal. Most trenches have a steep wall up to 10° on one side and a more gently-curved slope of about 5° on the other. Generally the ocean side is curved, as if bowed down, and the continental side is steep. In the Timor trench, however, the Australian slope is curved down, and the Timor side is steep.

Traverses of many trenches reveal numerous normal faults, and deep sediments that are essentially horizontal (Fig. 14.3), with some mantle bedding where the beds bend slightly against topographic features like a mantle draped over an irregular surface. The Chile trench and the New Britain trench are of this type.

An area of relatively shallow water separates the islands of the arc

Fig. 14.2   The Banda Arc, a double arc with an outer sedimentary island arc and an inner volcanic arc.

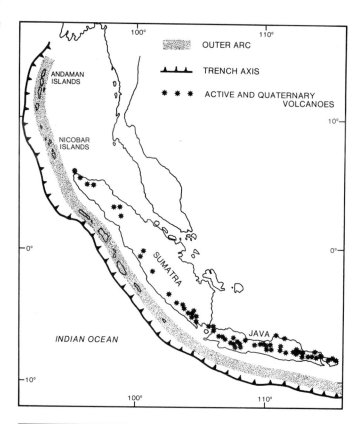

Fig. 14.3   Seismic sparker section across the New Britain Trench, Papua New Guinea (after Finlayson *et al.*, 1976).

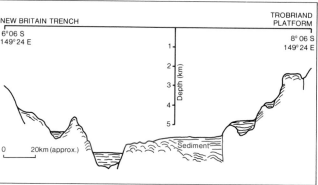

and the trench bottom. This is called the arc-trench gap. It is also known as the fore-arc region, and is becoming increasingly incorporated into plate-tectonic explanations of tectonics, as this is the area where subduction, underplating, or other forms of accretion will most clearly take place.

Those arcs that border continental masses are separated from the continent by shallow seas, known as back-arc basins such as the Sea of Japan. The Caribbean and certainly the Scotia arc have nothing but more ocean behind them, so any general theory of arcs must not require a backing continent. The Banda arc is almost folded back on itself, and never had a continent behind it.

Plate-tectonic explanations

Currently tectonic theories and plate-tectonic theories see island arcs as the clues to mountain-building processes and the sites of future mountain ranges. Many variations on this theme have been proposed, but a fairly general idea prevails, roughly as follows.

To compensate for the formation of new ocean crust by sea-floor spreading, crust must somewhere be destroyed. Crust is subducted at trenches, often at the site of island arcs.

The overriding slab is pushed up to form mountains such as the Andes or the mountain ranges of western North America. In other places the front of an overriding slab emerges as an island arc. In either case volcanoes erupt on the warped upper plate behind the subduction zone, and earthquakes mark the underthrusting of the lower slab.

In ocean–ocean collisions, as in the Scotia arc, one slab of ocean crust is overriding another – the Pacific side is overriding the Atlantic. Here the trench arc and volcanic arc are present, but the dominant sedimentation of continent–ocean arcs are missing.

In continent–ocean arcs sediment from the continent is trapped behind the arc, while the trench is allegedly filled largely by volcanogenic sediments. These parallel belts of sediment may be equivalent to those found in eugeosynclines and miogeosynclines described in Chapter 13.

The downgoing slab is gently arched, and the overriding slab is more rugged and steep. In the trench there may be sedimentation or even volcanic activity, but ultimately sediments are dragged down into the mantle by the subduction. In another hypothesis the sediments are scraped off the downgoing slab to create a crunched-up mass of overthrust sediments. The downgoing slab is thought to be marked by high-pressure/low-temperature metamorphism, with creation of distinctive greenschist and blueschist rocks. This belt is also characterized by ophiolites – an assemblage of distinctive basic igneous rocks including gabbro, pillow lavas and serpentinite. With continued pressure the mass of rocks in association with the arc may be pushed against the continent, forming a new mountain chain, similar perhaps to those of the older mountain chains of the Palaeozoic or Alpine orogenies, adding another strip of land to the growing continents – the process of cratonization.

Further details and problems

The elementary explanation just given has a charming simplicity that does not survive closer examination. Some data fit, some data don't; some data can be explained by elaboration of plate-tectonic mechanisms, some seem to resist any plate-tectonic explanation. The subduction hypothesis has been used as a Procrustean bed to which data have been fitted with varying degrees of violence, and there are literally hundreds of variations on the theme to account for particular sets of data.

In what follows some local details will be interwoven with speculations on plate-tectonic mechanisms, and sometimes with reservations or objections to the ruling theory. At present there is no simple explanation for all the data at our disposal, and it seems important to generate new concepts to rival plate tectonics and allow some measure of the use of multiple-working hypotheses.

Volcanoes

Volcanoes are active on many arcs. These are essentially andesitic and frequently erupt with great violence as at Krakatoa (1882) and Mont Pelée (1902).

Volcanoes make a petrological series of sorts across the island arcs. Whatever the method of producing different rocks, there is a change across an arc region with more alkaline ones nearer the back. There seems to be a good correlation of rock type with depth to the Benioff zone, and the amount of potassium in the lavas has a particularly good correlation (Fig. 8.4, p. 110). Kuno (1966) produced the diagram shown in Fig. 14.4, and other authors have produced many variations on this theme, with different chemical grouping, petrological types and

Fig. 14.4 Some ideas on the relationship between volcanicity and island arcs after the authors cited.

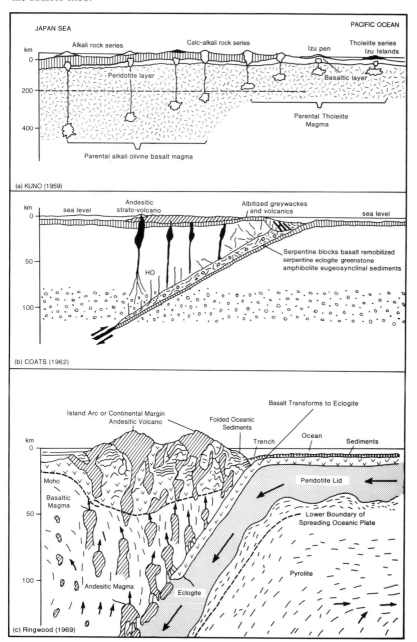

so forth. Several mechanisms have been suggested to derive the varied magmas, with three main themes:

1. Contamination. Mixing of a basic parent magma with siallic crust or sediment.
2. Differentiation. Fractional crystallization of a primary olivine basalt to produce andesites and other rocks. The mechanism requires the addition of water, which may be derived from subducted sediments.
3. Partial melting. Calc-alkaline magmas may be produced by the partial melting of oceanic crust or its high-pressure equivalent carried to great depth (Green and Ringwood, 1968). This method is probably the favourite at present.

It is generally presumed in plate tectonics that partial melting of a downgoing, subducted slab at about 150–200 km depth gives rise to the magmas that rise and are erupted as volcanoes located 150–200 km from the trench axis.

Extension or compression

Simple plate-tectonic theory would suggest that plate boundaries should be dominantly in compression, but there is a great delta of evidence for tension and extension.

Trenches are generally bounded by what appear to be normal faults, and Menard (1964) wrote 'Almost everyone who sees an echogram of the side benches and bottom troughs of trenches believes they are produced by normal faulting ... the topography of the trenches suggests tension'.

The problem was enlarged by Hatherton (1971) who wrote 'the special problem of the trench is the conflict between the theoretically compressional nature of plate boundaries and the apparently tensional nature of the trench itself. Indeed why the collision of two "rigid" plates should produce a trench is so far unexplained.' The arc-trench gap is also rather hard to explain by simple compression.

If we are permitted to think that trenches might really be tensional features, it is interesting to note that older extensional basins are known such as the St George Basin in the Bering Sea behind the Aleutian arc (Marlow, Scholl and Cooper, 1977). The basin is over 300 km long, 30 to 50 km wide, and filled with over 10 km of Upper Mesozoic and Tertiary deposits which have filled it to the brim so that the sea floor is now virtually featureless (Fig. 14.5).

Fig. 14.5   Interpretative drawing of the St George Basin (behind the Aleutian arc) based on seismic reflection profiles (simplified after Marlow, Scholl and Cooper, 1977).

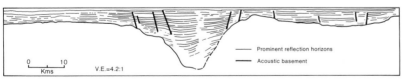

Prominent reflection horizons
Acoustic basement
0   10
Kms   V.E.=4.2:1

Back-arc basins all seem to show that extension is taking place behind the arc, not compression as might be imagined from a simple plate-tectonic model (Karig, 1971). Spreading rates are about 1 cm/y, or 10 000 B. The minor basins appear to spread, heal, and sometimes spread in new direction from new spreading sites. At least in the western Pacific the existence of these spreading sites suggests that cratonization is not taking place at present, and has not taken place for at least tens of millions of years where back-arc basins are spreading. If the island arcs are subduction sites, then subduction is taking place a long way from the continental margin, and the distance is increasing. Some of the back-arc basins are shown in Fig. 14.6.

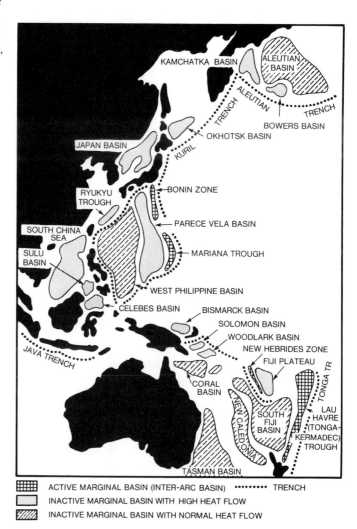

▦	ACTIVE MARGINAL BASIN (INTER-ARC BASIN)	•••••• TRENCH
▭	INACTIVE MARGINAL BASIN WITH HIGH HEAT FLOW	
▨	INACTIVE MARGINAL BASIN WITH NORMAL HEAT FLOW	

In back-arc basins of the western Pacific (Fig. 14.7) Watanabe (1977) found the heat-flow measurements showed the following common features:

1. Less than normal heat flow in the zone from the trench axis to the volcanic zone.
2. A high but variable heat flow over the volcanic zone or island arc.
3. Mean heat flow in the back-arc basin depends on the age of the basin:
    (a) young basins are similar to young ocean ridges, with anomalously high and low values with no clear pattern;
    (b) early to mid-Tertiary basins have flow of 2.0 to 2.2 HFU with low variability;
    (c) Mesozoic basins have mean heat flow close to the world average, with low variability.

To account for back-arc spreading a more complicated subduction model is required, such as that of Töksoz and Bird (1977) who claim 'A necessary consequence of the subduction of oceanic lithosphere is an induced convective circulation in the wedge above the slab. This may play an important role in the formation and evolution of marginal basins' (Fig. 14.8).

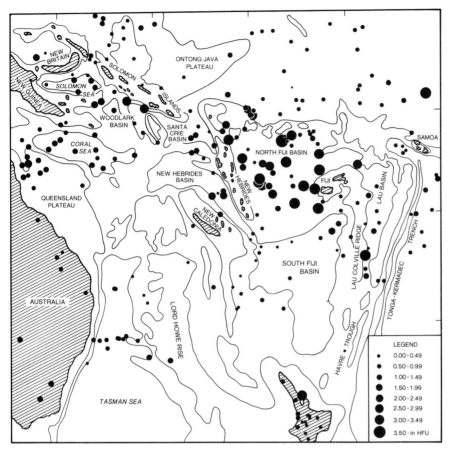

Fig. 14.7 Heat-flow values in the southwestern Pacific (after Watanabe, 1977).

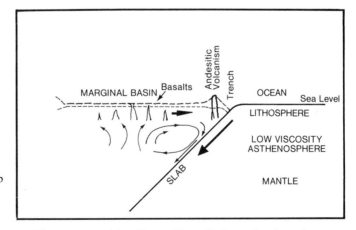

Fig. 14.8 Schematic diagram of convection induced by a downgoing slab and its heating of the overlying lithosphere (after Toksöz and Bird, 1977).

The topographic effects allegedly brought about by compression at island arcs are far from obvious. Where there is ocean–continent collision it is said that the overriding continental slab is pushed up to form mountains. Thus subduction at the Chile trench and overthrusting of South America may push up the Chilean Andes. But Katz (1971) believes the field evidence in Chile shows that the Andes, far from being buckled by subduction and compression, have undergone extension since at least the Miocene, in a belt 300–400 km wide. Similarly behind many island arcs, instead of uplift there is a belt of sea-floor spreading in the back-arc basins.

Kobayashi and Isezaki interpret the northwestern Pacific arc region as shown in Fig. 14.9, with at least three spreading ridges in the basins bounded by opposed subduction zones.

Karig (1971) postulates that all western Pacific island arcs have antecedents that were initially adjacent to continental fragments in the area. These migrated away from the continent, leaving successively younger marginal basins and remnant arcs behind them. Carey (1976) sees the process in reverse, with the continents moving away from the island arcs which are left like a series of moraines as a glacier retreats. Several writers support universal tension. Between the Solomon Islands and Queensland, and the whole area of the Tasman Sea, horst and graben structures recur all the way.

Fig. 14.9 Spreading sites in the northwest Pacific (after Kobayashi and Isezaki, 1976).

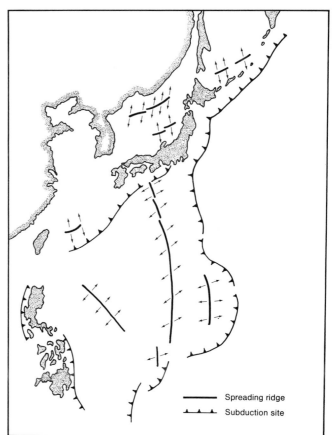

Spreading ridge
Subduction site

Gravity and seismicity

Over, or on the inner side of the trench, is a gravity minimum. On the continental or island side is a positive anomaly.

In the Tonga trench the negative anomaly coincides with the trench, which came to be accepted as 'normal' despite earlier work in Java where the negative anomaly occurs on the concave side (towards the land). In New Zealand the negative anomaly passes through the centre of North Island where there is no trench or tectogene. The negative anomaly is not so much associated with the trench as with the projection of the belt of deep and intermediate earthquakes (Hatherton, 1974). The integrated positive and negative anomaly pair approximate zero, though they are usually considered to be independent: the positive anomaly is ascribed to the higher density of a cool, downthrust (subducted) lithospheric slab, while the negative anomaly is attributed to the accumulation and subsequent underthrusting of sediments in the trench. The distance between the maxima of the positive and negative anomalies is constant at about 115 km (Hatherton, 1974). On the ocean side the negative anomaly recovers to a positive value only half that on the landward side.

Fig. 14.10 Composite cross sections of earthquake centres in various arc-trench locations (simplified after Isacks and Barazangi, 1977).

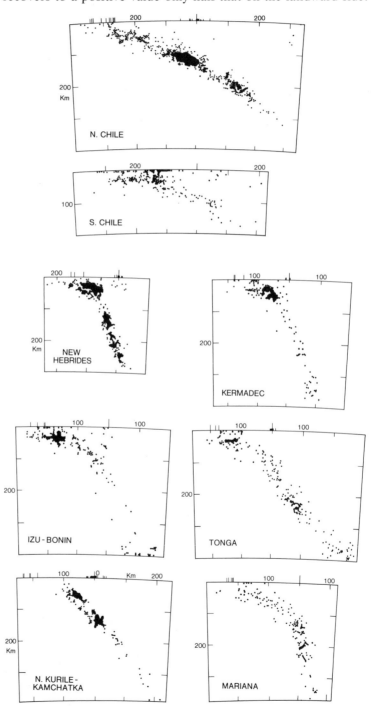

Arcs are characterized by much earthquake activity, and the foci of the earthquakes often appear to fall in a fairly narrow zone called the Benioff zone about 50 km wide and dipping towards the continent at angles between 15° and 90° (Fig. 14.10). Seismicity goes to 700 km, well below the brittle zone where earthquakes are expected, but most is in the top 40 km. The Benioff zone generally appears to be either vertical or to dip under the arc at a steep angle, the deeper quakes showing a steeper Benioff zone than the upper ones in some cases. The Benioff zone itself is many kilometres (often 50 km) thick, and if the earthquakes are related to fracture and fault movement then the faults mark a broad fault zone, not a single major fault.

The relationship between the inclined and shallow seismicity at island arcs is not clear, but varies (Fig. 14.11). In Tonga–Kermadec the Benioff zone intersects the trench and shallow activity is confined to a zone about 200 km wide. In Indonesia and Kamchatka the shallow

Fig. 14.11 Schematic representation of relative positions of trench, negative anomaly, volcanism and seismicity in three active margins (after Hatherton, 1974).

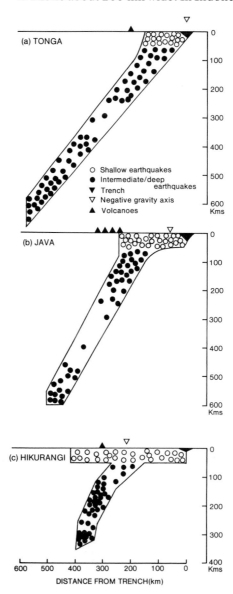

seismicity occupies an area between the trench and the Benioff zone outcrop, which is 150 km from the trench. In the Hikurangi arc the Benioff zone bisects a broad belt of shallow seismicity, and is almost 200 km from the trench. Negative anomalies everywhere coincide with the downturn of the Benioff zone. All the seismicity, both shallow and deep, is on the continental side of the trench. The seismic motions generally indicate tensional motions down to 300 km, and compressional conditions below that. However, Benioff zones seem to become increasingly complex with further work. Figure 14.12 shows a two-layered deep seismic zone under northeast Japan, in which the upper layer events are down-dip compressional, and the lower layer events are down-dip tensional (Uyeda, 1977). There is no satisfactory explanation so far for this relationship.

Fig. 14.12   Two-layered deep seismic plane under the northeast Japan arc (after Uyeda, 1977).

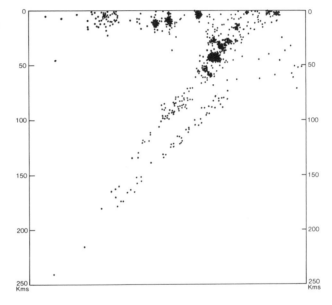

The intersection of a dipping Benioff zone and the curved earth's surface produces an arc. Any oblique plane cutting the curvature of the earth will make a curve, but to provide the curvature of the island arcs a very shallow dip to the plane would be needed, not the steep dip that is actually found on Benioff zones. The Benioff zone must make a conical surface if it dips at a steep angle from the surface trace of the island arc. In Indonesia the Benioff zone has been mapped to show the inclination, and at the end of the Banda arc it does indeed form a conical surface (Fig. 14.2). If the Benioff zone marks a zone of subduction, as envisaged in the plate-tectonic hypothesis, then the material of the subducted plate must be converging at depth towards the base of the cones, which creates various awkward space problems.

Seismic studies of island arcs have placed great stress on the Benioff zone, which stresses in turn the subduction idea. Yet strike-slip faulting is an important feature of many arcs, with lateral displacements amounting to many kilometres along faults roughly parallel to the arc. These faults have direct topographic expression and are a major feature in the geomorphology of arcs such as Sumatra and Java. If the Benioff fault zone is involved in strike-slip movements there would have to be rotation of the half cone it bounds, which is not very

probable. It is more likely that the strike-slip movement is confined t
the upper crust (Hodgson, 1962).

Sediments and rocks

Scholl and Marlow (1974) note that the Chile trench is almost empt
and cannot accept the plate-tectonic suggestion that sediment has bee
lost by subduction. They remain genuinely perplexed as to wh
evidence for subduction or off-scraping of trench deposits is no
glaringly apparent, and in response to the suggestion that the evidenc
has itself vanished down the subduction zone they add 'You can't hav
your trench and eat it too!'

They also report that the typical sediment of trenches include
pelagic sediments, that is sediment of the deep sea as distinct from tha
derived from land, but the coastal mountains fringing the Pacific ar
virtually devoid of oceanic deposits. They consider that the volume o
oceanic debris that would have been scraped off the trenches should b
very noticeable, if it were present at all, but the fold belts typicall
contain greywacke derived from the land, and volcanic rocks. Thes
typical eugeosynclinal sediments are unlike present-day trencl
sediments. Furthermore, the volume of Pleistocene detritus formin,
the bulk of the turbidite sequence in the Peru–Chile trench an
flooding the adjacent sea floor is in good balance with the amount tha
could have been eroded from the nearby continents. Therefore n
significant amount has been removed tectonically. A simila
conclusion is drawn from the Aleutian trench.

The Aleutian trench is tectonically strange. The rate of convergenc
along the eastern and central segments of the trench is severa
centimetres per year. Farther west the angle of convergence betwee
the Pacific and Americas plates becomes progressively smaller unti
west of about 176 °E it no longer exists and the two plates ar
presumably in strike-slip contact (Fig. 14.13). The internal structure o

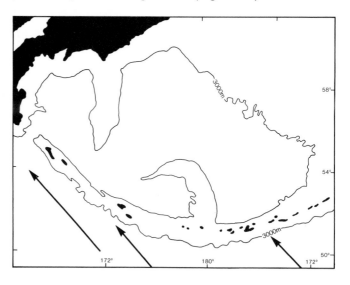

Fig. 14.13 Sketch map of the Aleutian trench and direction of sea floor spreading or subduction.

flat-lying beds and their abrupt contact with the steep inner wall of the
trench are no different here, where little or no underthrusting occurs
than along the central and eastern segments where presumabl
underthrusting is rapid.

The outer Indonesian arc is the top of a wedge of mélange and

Fig. 14.14   Sections through the
subduction system of the Java Trench off
southern Sumatra. (*a*) Major tectonic
components (densities in g/cm³). (*b*)
Mechanism of deformation of the
mélange wedge (after Hamilton, 1977).

imbricated rocks whose steep-to-moderate dips are disharmonic to the gently-dipping, subducted oceanic plate beneath. It seems that this has indeed grown by scraping off of oceanic sediments and basement, with internal imbrication (Fig. 14.14). The seaward side was raised above sea level by mélange stuffed beneath it by Neogene subduction (Hamilton 1977).

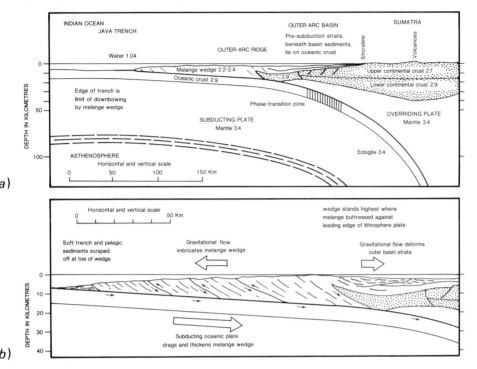

An alternative explanation for the Indonesian situation is given by Worzel (1976). If we consider that the sediment on top of the oceanic basement surface was deposited on that surface, and that isostatic adjustment took place, we can reconstruct the original surface by removing the sediment and returning the basement surface to isostatic equilibrium, as has been done in Fig. 14.15. As can be seen, this returns the basement surface to a nearly horizontal surface, suggesting that no more elaborate mechanism is needed than sedimentation and isostatic compensation.

In the Peru–Chile trench between 36° and 42 °S, the estimated volume of Pleistocene clastic debris derived from the continent requires an average erosion rate over the past 1.2 m.y. of about 500 B. This is the average figure for erosion of highland (p. 249) and implies that a good balance exists between the volume of material eroded on the land and the sediment deposited in the sea. If sediment were lost by subduction over this period the erosion rate on land would have to be significantly increased, giving an erosion rate that is improbably high, though not impossible. Similar calculations have been done for the Aleutian trench and the conclusion seems to be the same (Scholl, Marlow and Cooper, 1977). Perhaps this trench is less active now, but in the past the Aleutian trench is thought to have subducted without trace a large area of sea floor and an actively-spreading ridge system, including a triple junction (Pitman and Hayes, 1968).

Fig. 14.15 (*a*) Line drawing of a multichannel seismic section across the Java Trench. (*b*) The same section with the sediments removed and the oceanic basement restored to isostatic equilibrium (after Worzel, 1976).

(*a*)
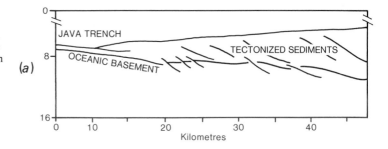

(*b*)
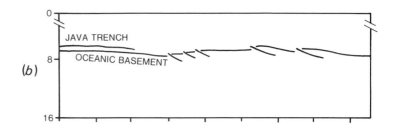

At the present time Eocene ocean floor has reached the Peru–Chile trench. Judging from the amount of pre-Eocene to Cretaceous sea floor preserved on the western side of the East Pacific Rise spreading site, about 7 000 km of sea floor must have been consumed at this trench, yet there is no geological or geophysical indication for any significant accretion of sea floor or its sedimentary cover at the continental margin.

Sediment may either be subducted (consumed) at trenches, or scraped off and accreted. Offscraping is generally thought to create eugeosynclinal deposits around the Pacific. Where subduction occurs the bulk of margins and flanking fold belts will consist of continental sediments deformed in place. Deformed rocks underlie the inner slopes of many trenches, whether or not there is a thick (1–2 km) section of 'scrapable' deposits on the adjacent sea floor.

Most Pacific trenches contain pelagic (oceanic) deposits rather than terrigenous (continental) deposits. Since some 7 000 km of crust are thought to have been underthrust since the Mesozoic there should be large volumes of oceanic beds in coastal mountains, but in fact oceanic deposits are minor.

Such oceanic deposits as *are* found in coastal fold belts are the same age as associated terrigenous beds, but if they were oceanic scrapings some should be as much as 100 m.y. older than the enclosing sediments. Some coastal mountains that have been underthrust during most of Mesozoic and Tertiary time, such as the Andes of Chile and Peru, are underlain by Palaeozoic and older rocks, so all sediment arriving from the ocean side must be subducted, not offscraped. Accretion has not occurred at such margins.

Paired metamorphic belts

Island arcs, especially those around the Pacific, are characterized by paired metamorphic belts of similar age but different type. An inner low-pressure facies with andalusite lies in the volcanic arc, and an outer high-pressure belt with glaucophane (the blueschists) occupies the outer part of the arc-trench gap and near the inner wall of the trench

The pair of metamorphic belts may be used, in a fossil situation, to determine the orientation of any supposed arc, and the presumed direction of subduction. Problems arise with a situation like that in Japan (Fig. 14.16), where Hokkaido has the high-pressure belt on the wrong side, if Honshu is regarded as normal. Okada (1974) has tried to resolve this situation by showing that there are actually four belts in Hokkaido, giving two pairs both sited on westerly-dipping subduction zones.

Fig. 14.16 Three pairs of regional metamorphic belts in Japan (after Miyashiro, 1972).

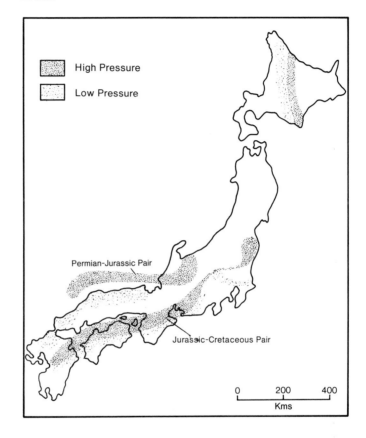

If the theory is correct, the paired belts will not yet be exposed in active arcs, and the best evidence will come from old arcs of Mesozoic–Cenozoic age. Fourteen sets of paired belts have been identified around the Pacific (Fig. 14.17), including New Zealand (Jurassic/Cretaceous), California (Late Jurassic/Late Cretaceous) Washington (Permian/Trias) and Chile (Late Palaeozoic). Clearly the cordilleran situation is being regarded as an arc situation in this compilation.

The high-pressure/low-temperature belts are caused by the tectonic underthrusting in the trench, with a very low geothermal gradient (about 10 °C/km) while the low-pressure/high-temperature belt is associated with the steep geothermal gradient (>25 °C/km) in the vicinity of the volcanic arc associated with the rise of new magmas through the crust (Miyashiro, 1974).

The low P/T type has andalusite and sillimanite, and the high P/T type has glaucophane, jadeite and lawsonite. In the process of underplating successive sheets of mélange are thrust under the

Fig. 14.17   Paired metamorphic belts in the circum-Pacific region. Full lines represent low-pressure metamorphism; broken lines represent high-pressure belts (after Miyashiro, 1973).

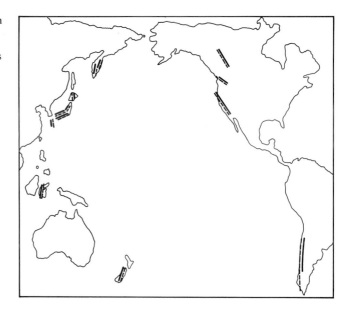

continent. The seaward-side sheets are the youngest, weakly recrystallized ones, while those nearer the continent are older, higher grade and more deeply subducted. The inclination of the plate junction decreases, and this might result eventually in a seaward stepping of the subduction zone (Ernst, 1975).

Alternative hypotheses

Worzel (1976) has challenged the subduction hypothesis, and claims that the geophysical data are consistent with a subsided block at the trench. He has modelled data from many trenches, and Fig. 14.18 shows just one example. The calculated gravity profile on this hypothesis fits very closely to the observed profile, and although other authors are able to produce similar fits from subduction assumptions, Worzel believes that much greater negative anomalies would be produced by an underthrust slab. Worzel also says that his reconstructions are consistent with other evidence such as seismic results, and we have already seen that topographic and sedimentological evidence suggests that many trenches are simple grabens, consistent with the subsided-block hypothesis of Worzel. Other features such as the Benioff zone and the petrological variation of volcanoes are not accounted for in this hypothesis.

Nagumo and Kasahara (1976) show the dynamics of the Mariana arc-trench system as in Fig. 14.19. There is a motion couple with uplift of the ridge due to mantle diapirism, with associated depression of the trench. They suggest that this motion forms a convection cell, and high strain concentrated along the marginal escarpment along the trench generates the high micro-earthquake activity observed there. Vertical motion due to mantle diapirism and its associated local convection seems to be adequate to explain the seismicity of the island arc-trench system.

A totally different interpretation of island arcs and associated features has been put forward by Krebs (1975) in a stimulating paper that relates arc-trench systems to mountain systems and to global *vertical* tectonics instead of the horizontal movements of plate

Fig. 14.18 (*a*) Structure of the Puerto Rico Trench and Arc, depicted by layers of varying velocity and density, showing the trench as a downfaulted block with no subduction. (*b*) Observed gravity anomalies, and the computed gravity anomaly based on the section above (after Worzel, 1976).

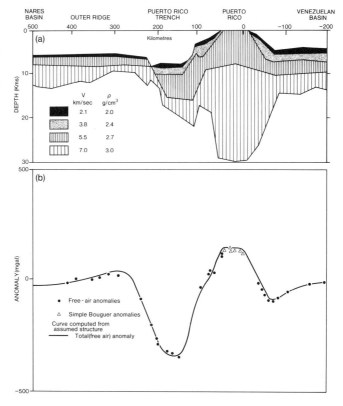

Fig. 14.19 A schematic diagram of the dynamics of the island-arc trench system in the southern Mariana region, with the arc interpreted as resulting from mantle diapirism (after Nagumo and Kasahara, 1976).

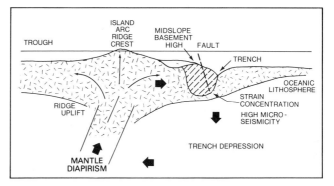

tectonics. According to Krebs if, in the south Pacific, the remnant or inactive trenches and arcs are taken into account as well as the active ones a complex of bilaterally symmetrical structures is revealed. On the axis of the system there is a long inter-arc basin, bordered on both sides by convergent dipping thrust planes or planes of seismic activity. Usually only one side is active at present, and so gives rise to the apparently asymmetrical behaviour of island arcs. The axis is therefore bounded by either active or inactive arcs and trenches. The key areas for understanding the system are the inter-arc basins which represent the top of diapir-like upwelling material from the asthenosphere. These subcrustal asthenoliths are characterized by crustal thinning, extension, high heat flow, deep earthquakes, positive gravity anomalies, extrusion of mantle-derived tholeiite basalts, and intrusion of ultramafic massifs. Krebs certainly gives enough criteria to put his model to the test. Basically vertical forces cause the asthenolith to rise, and then it spreads horizontally under gravity to produce a host of secondary effects.

The interpreted bilateral symmetry of the southwest Pacific is shown in Fig. 14.20, and the more detailed map of New Zealand (Fig. 14.21) shows how bilateral symmetry can be found in New Zealand which is generally regarded as very asymmetrical.

Fig. 14.22 shows Krebs's version of the double arc of the Philippines. A more conventional explanation of the double-arc system is the arc–arc collision already mentioned in Chapter 4. The New Guinea area is complicated (on Krebs's or any other model) and Figs. 14.23 and 14.24 show how it may be interpreted on the bilateral

Fig. 14.20   Tectonics of the southwest Pacific. (*a*) Plate-tectonics version. (*b*) Two-sided symmetrical system (after Krebs, 1975).

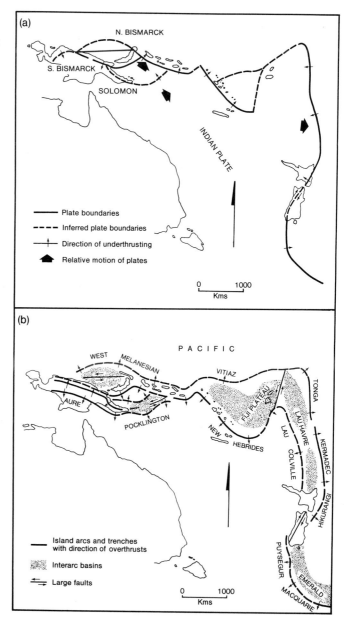

Fig. 14.21   New Zealand interpreted as a symmetrical-arc system (after Krebs, 1975).

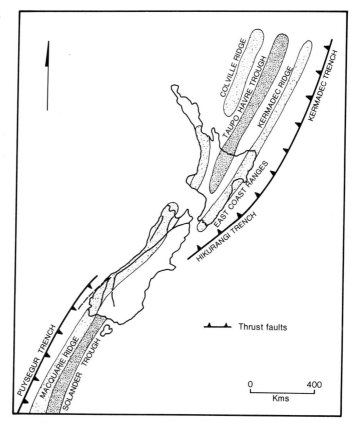

hypothesis. There is convergence on the Solomon Basin, with diverging Benioff zones at the New Britain Trench and the Woodlark Trench. The Woodlark Basin and the Bismarck Basin are regarded as typical inter-arc basins with upwelling asthenoliths. The Bismarck and Woodlark Basins are about 3 km deep with thin sedimentary cover, high heat flow, crustal extension linear magnetic anomalies, spreading centres, tholeiitic basalts, deep seismic shocks and large strike-slip faults; while the Solomon Basin is 5 km deep and has thicker sediments and an upper crust 15–20 km thick.

The Mediterranean may also be of this type, according to Krebs, and other workers have found anomalous structure, with dome-shaped low-velocity channels that might correspond to the asthenolith of Krebs (Fig. 14.25). A spreading origin of the Mediterranean is shown in Fig. 14.26.

In this chapter I have tried to emphasize the gaps in our knowledge of island arcs and associated features, to point out the variations that are concealed in simple generalizations, to expose weaknesses in conventional plate-tectonic explanations, and to present some alternatives that are at least partially successful. Nevertheless it is clear that this is an area of fundamental importance, and future work here is most likely to lead to advances not only in our knowledge of arcs and trenches, but in fundamentals of global tectonics.

Fig. 14.22   The Philippines interpreted
as a symmetrical-arc system (after Krebs,
1975).

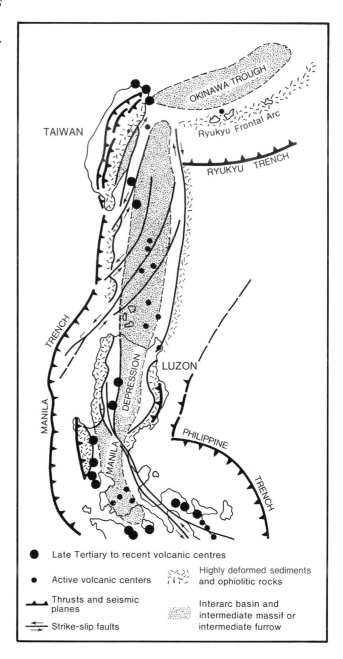

TAIWAN

OKINAWA TROUGH

Ryukyu Frontal Arc

RYUKYU   TRENCH

MANILA
TRENCH

LUZON

DEPRESSION

PHILIPPINE

MANILA

TRENCH

●   Late Tertiary to recent volcanic centres

•   Active volcanic centers

Thrusts and seismic
planes

Strike-slip faults

Highly deformed sediments
and ophiolitic rocks

Interarc basin and
intermediate massif or
intermediate furrow

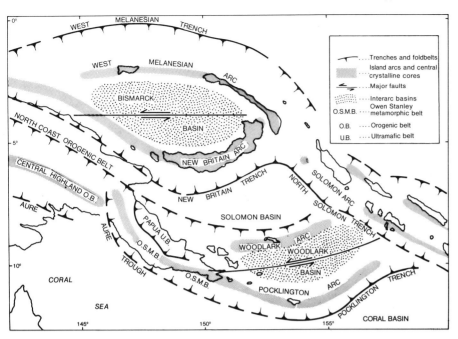

Fig. 14.23 New Guinea interpreted as a
symmetrical-arc system (after Krebs,
1975).

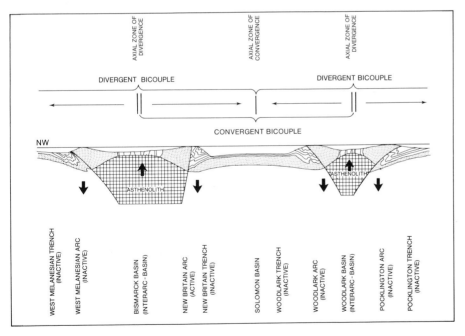

Fig. 14.24 Cross-section of the New
Guinea area to show the tectonic
components of the symmetrical-arc
system (after Krebs, 1975).

Fig. 14.25 Structure of the upper mantle under the western Mediterranean (after Berry and Knopoff, 1967).

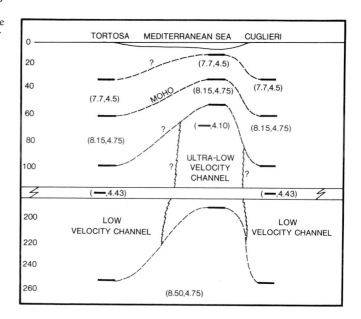

Fig. 14.26 Origin of the western Mediterranean (modified from Glangeaud, 1957). (*a*) Present. (*b*) Miocene. (*c*) Late Jurassic.

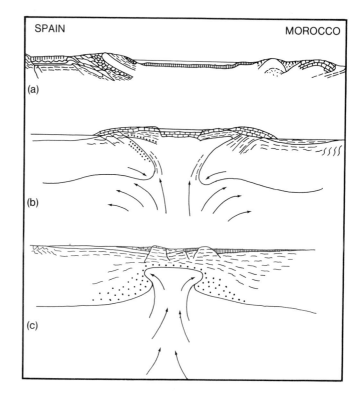

15 Geomorphology and tectonics of the oceans

Topographically the ocean floor can be divided into two basic units, the deep-sea floor and the continental margins. The deep-sea floor consists of the following components:

Ocean rises   The spreading sites, mid-ocean ridges or simply ocean rises are arbitrarily separated at a depth of 4 km from the rest of the deep-sea floor, which gives them an area of about 30 per cent of the ocean.

Slow-spreading rises, like the North Atlantic mid-ocean ridge, have high relief and a distinct graben or rift in the centre. Fast-spreading rises, like the East Pacific Rise, have broadly convex summits without a median rift. Rises are the youngest part of the ocean and have little or no sediment cover. The rises are broken by transform faults, which are marked topographically by scarps or trenches with an amplitude of up to 2 km, with rugged topographic details.

Abyssal plains and hills   Abyssal plains occupy about 42 per cent of the ocean, and occur at depths of 3–6 km. The hills consist of basalt, often in the form of pillow lava, and were formed at spreading sites. As sea floor moves from the ridge it sinks and acquires a cover of sediment which thickens away from the ridge. The sediment is dominantly brown clay, but biogenic silica and carbonate may make a variable contribution depending on latitude and water depth. The plains are simply accumulations of sediment burying hills which occasionally protrude. The deep-sea sediments accumulate very slowly, at a rate of only a few millimetres per thousand years.

Trenches   The deepest parts of the ocean are not in the middle but near the edges, in trenches that follow about half the continental margins. They are twice as deep as average ocean floor. Some follow the continental edge of South America, which is backed by the Andes mountains, but the coastal ranges of North America are not accompanied by a trench. The rest of the trenches are associated with island arcs. Trenches are bounded by abyssal plains on the ocean side, by continental slope or back-arc basins on the other side, but in exceptional places, like the Scotia arc, there is no continental backing.

Seamounts   These are individual volcanoes on the sea floor. Some ar
in chains, marking the trace of ocean plate movement over hot spots i
the asthenosphere, others are (on present knowledge) randoml
distributed.

Guyots   These are flat-topped seamounts, subaerially eroded in th
past since when they have sunk, sometimes by several kilometre
There are over 2 000 known seamounts and guyots.

Aseismic ridges   These are volcanic lineaments such as the Walv
Ridge, the Rio Grande Ridge and the Ninety East Ridge, whic
possibly mark the trace of hot spots beneath the crust.

The continental margins consist of the following units:

Continental shelf   This is the shallow platform seaward from th
continental coast, terminated on the ocean side by a change of gradie
called the shelf break, which is at an average depth of only 130 m b
can reach about 200 m. The average shelf width is 75 km and th
average gradient is 1:500. Local relief may reach 20 m in parts.
  Bloom (1978) considers that the depth of the shelf, very close to th
estimate of full glacial eustatic lowering of sea level, shows that th
shelf break is almost certainly genetically related to Quaternar
sea-level fluctuations, and the variation in depth must have bee
caused by isostatic or other tectonic uplift or depression, glaci
erosion, and post-glacial deposition. Some shelves are rock cut, bt
most have hundreds of metres of sediment. The surface sediment
fluvial, aeolion or estuarine, and much of it is of subaerial origi
modified to some extent by the rapid post-glacial rise of sea level, wit
little evidence of sedimentary gradation across the shelf. Quaternar
shelves are not typical of past conditions, and their form an
composition are of limited use in interpreting ancient marin
conditions.
  On the landward side many shelves are continuous with adjacer
plains on land. Few valleys cross the shelf because of the cover c
modern sediment, but the outer edge is frequently notched by the sta
of canyons.
  Tectonic activity may be active even when sea floors remai
relatively smooth. Some modern continental shelves have smoot
upper surfaces, but fault blocks are present in the basement beneatl
Sellwood and Jenkyns (1975) have described an ancient example fror
Britain, for part of the Jurassic. Jurassic basins and swells wer
characterized by great and negligible subsidence respectively, bt
sedimentation was always rapid enough to maintain a roughly level se
floor. This Jurassic example is not strictly comparable with a moder
shelf, but represents an epeiric sea – a sea that has spread over a wid
area of continent – and is related to the opening of the Atlantic.

Marginal plateaus   These are like shelves in some respects but deepe
They may be downfaulted shelves, but some are probably separate
small areas of continental-type crust that fail to emerge as land area
They are sometimes referred to as 'microcontinents'.

Deltas   The major rivers of the world have built vast deltas man
kilometres thick (the Amazon cone is 11 km thick) and hundreds c
kilometres across, distinctly related to their source area. The Amazor

Niger and Mississippi are well-known examples, the last being a rather special case as it enters the Gulf of Mexico rather than the open ocean.

Back-arc basins   Island arcs are subaerial features that are found mainly in the west Pacific. They are separated from the backing continent by shallow seas, known as back-arc basins, many of which are spreading sites.

Arc-trench gap   This is the unfortunate name given not to a gap but to the stretch of shallow sea between the steep continent-side of a deep-sea trench and its backing arc of islands.

Epicontinental seas   In situations without island arcs, other barriers may isolate small seas bordering the continent. These are the epicontinental seas, mainly found in the Atlantic.

Continental slope   This plunges from the shelf break, rarely deeper than 200 m, to depths of usually 2–3 km at the top of the continental rise. The average slope is only about 4°, but steeper ones may reach 27° as in the Florida peninsula and the slope off southern Australia. In plan the slope is straight or somewhat sinuous, and it varies in width from 15 to 30 km.

Continental slopes are of varied structure and origin. Some are fault scarps – the edges of continents that have drifted apart. Others are debris slopes, and yet others appear to be various kinds of 'dam' (fault blocks, coral reefs, volcanic piles) behind which sediments have been trapped.

Submarine canyons   These are huge canyons, sometimes with tributaries like continental valleys, cut deep into the continental slope and extending headwards onto the shelf. They are V-shaped in section, up to 1 000 m deep, and extend to depths of over 5 000 m. They are found on all coasts, and while some line up with major rivers, others originate on the continental slopes. Many are cut into solid rock, and they are clearly erosional features.

It is known that the canyons act as chutes for turbidity currents, and thought that these might well erode some of the canyons. The canyons of the Mediterranean were eroded by surface rivers when the Mediterranean was a dry tectonic basin in Miocene times (Hsu, 1972).

Continental rise   Continental rises are submarine equivalents of piedmont alluvial plains, found at the base of the continental slope at depths of several kilometres. They have slopes of about 1° and they occupy about 5 per cent of the ocean floor. Seismic studies show they are wedges of sediments, up to 6 km thick and extending as far as 600 km from the base of the continental slope.

Some have apices at the mouths of submarine canyons, others do not, but all are clearly derived from continental sedimentation. Meandering levee-bounded valleys meander across the fans, formed from turbidity currents. A valley on the Monterey fan, for instance, is 36 km long, with levees up to 370 m high.

Table 15.1 gives some further descriptive details of submarine topography. Table 15.2 gives the dominant geological and geophysical character of major submarine physiographic provinces.

Table 15.1 Typical dimensions of major features of submarine topography (after Heezen and Wilson, 1968)

Feature	Typical dimensions	Comments	Example
Continental shelf	Width: few to > 300 km Relief: < 20 m Depth: < 200 m Gradient: < 1:1000	Seaward boundary, or shelf break occurs where gradient changes to > 1:40	Northeastern United States
Marginal plateau	Similar to shelf except depth: between 200–1 200 m	No seaward barrier or sill	Blake Plateau
Epicontinental sea	Similar to shelf except depth: between 100–1 500 m Relief: often > 40 m	Seaward barrier or sill	Gulf of Maine
Continental slope	Width: few to 150 km Relief: locally > 2 000 m Depth: drops from 100+ to 2 000+ m Gradient: > 1:40 (3–6°)	Upper boundary is shelf break; high relief is associated with canyons	Northeastern United States
Marginal escarpment	Width: few km Height: 2 000–4 000 m Depth: begin at 1 000–3 000 m Gradient: > 1:10	Precipitous slope	Blake Escarpment
Landward slope of trench	Height: several thousand metres Depth: drop from 500–5 000 m Gradient: > 1:40	Always associated with trench	North of Puerto Rico
Continental rise	Width: few to > 300 km Relief: < 40 m Depth: 1 500–5 000 m Gradient: 1:1000 to 1:700	Seaward limit is change to < 1:1000, although 1:2 500 (and 1:50) can occur	Northeastern United States
Outer ridge	Width: > 150 km Height: 200–2 000 m	Lies seaward of a basin or trench	North of Puerto Rico
Marginal basin	Variable dimensions Depth: < 5 000 m Gradient: often < 1:1000	Defined by position *re* outer ridge, continental slope	Blake Plateau area
Marginal trench	Width: 30–100 km Length: 300–5 000 km Depth: 3 000–10 000 m Relief: > 2 000 m	Narrow, steep sided	Puerto Rico Trench
Continental borderland	Similar to shelf except relief: 100–1 000 m Depth: up to 2 000 m Gradient: 1:1000 1:40	Rises and depressions are common	Off Southern California
Submarine canyon	Width: 1–15 km Relief: 20–2 000 m Depth: 20–2 000 m Gradient: < 1:40	Most commonly cut continental rise; length to 500 km	Hudson Canyon, off New York
Abyssal hills	Width: 100 m–100 km Height: few to 1 500 m	Found seaward of abyssal plains, usually in belts	Western Atlantic Abyssal Hills Province
Abyssal plains	Width: few to 1 000 km Depth: usually > 3 000 km Gradient: < 1:1000 (to 1:10,000) Relief: Nil	Small plains occur in trench bottoms; plains are very flat, gentle	Nares and Sohm Plains, North Atlantic
Oceanic rises	Width: 300–500 km Height: up to 5 000 m	Aseismic features with highly variable relief (smooth slopes - rough scarps)	Bermuda Rise
Aseismic ridges	Width: to 150 km Length: to 4 000 km Height: up to 4 000 m	Strongly asymmetrical	Walvis Ridge

Table 15.1 (*continued*)

Feature	Typical dimensions	Comments	Example
Seamounts	Width: 2–100 km Height: > 1 000 m Depth: 0–2 000 m (top surface)	Submerged volcanoes; often wave truncated and sub-merged (guyots); may have atolls	East Pacific
Oceanic plateaus	Dimensions similar to oceanic rises Gradient: < 1:1 000 on top; > 1:40 on sides	Also termed 'Microcontinents'	Mascarene Plateau Indian Ocean
Mid-Oceanic Ridge	Width: 2 000–4 000 km Length: 35 000–40 000 miles Height: 1–3 km Depth: 0–5 000 m	Includes flank and crest regions. World's greatest mountain range	Mid-Atlantic Ridge
Ridge flank	Width: 500–1 500 km Relief: up to 1 000 m Depth: > 3 000 m	Includes fractured plateau, and steps on flank	Mid-Atlantic Ridge
Ridge crest	Width: 500–1 000 km Relief: 2 000 m Depth: 2 000–4 000 m	Includes rift zone, which may be 20–50 km wide	Mid-Atlantic Ridge

Table 15.2 Dominant geologic and geophysical characteristics of some major submarine physiographic provinces (after Heezen and Wilson, 1968)

Province	Surface geology	Geophysical characteristics (N.B. all areas are isostatically compensated, except trenches)	Probable genesis
Continental shelf	Unconsolidated terrigenous sediments of all sizes	Aseismic, with magnetic anomalies over buried volcanoes, structures	Represents filled-in ridges and troughs
Continental slope	Unconsolidated terrigenous silts and clays; in places lithified and truncated	Aseismic, with major magnetic anomaly associated with abrupt edge of continent	Represents structural edge of continent
Continental rise	Unconsolidated silts and clays often deposited by turbidity currents and reworked by contour currents	Aseismic, with smooth magnetics (modern trench areas are seismic and isostatically uncompensated)	Sedimentary prism, perhaps filling in old trenches
Abyssal plains	Unconsolidated silts and sands of turbidity current and pelagic origin	Aseismic with magnetic anomalies revealing linear patterns unrelated to obvious surficial structures	Sedimentary plain overlying abyssal hill terrain
Abyssal hills	Alkali basalt and gabbro	Similar to abyssal plains	Older Mid-Oceanic Ridge crust
Mid-Oceanic Ridge flanks	Similar to abyssal hills	Aseismic; minor linear magnetic anomalies paralleling boundaries	Older Mid-Oceanic Ridge crust, brought to surface in crest area and moved from crest with time
Mid-Oceanic Ridge crest	Alkali basalt and gabbro (serpentine)	Highly seismic with large magnetic anomaly over rift High heat flow	Zone of introduction of mantle material which becomes oceanic crust and moves away from ridge crest

Coral reefs

Coral reefs are limestone accumulations consisting of the skeletons of corals and other creatures, especially algal and worm remains that cement the coral together. Coral reefs bring together problems of sedimentation and tectonics, and provide data that are very helpful solving such problems.

There are several factors that limit coral growth. They need light, usually only grow down to 50 m, though in exceptional conditions the may reach 100 m. Their preferred temperature is around 25 °C b they can tolerate temperatures down to about 18 °C. Normal marin salinity is preferred, and fresh water or excessive salinity in closed basin is inhibiting. They require a solid foundation to build on, a supply food brought in by waves, and little siltation so they are absent off riv mouths.

Bores have shown that some coral islands consist of great thickne of limestone – 1 400 m at Eniwetok in the Marshall Island and perha 1 500 m on Bikini atoll. The coral limestone could not start growing such great depths, and the only explanation is that the base of the re has continued to sink, from an original position within the depth coral growth, and that this growth has kept pace with the subsidenc This would take a long time and these limestones go back to t Tertiary.

In contrast there are some coral reef areas like Bermuda where t whole archipelago stands on a platform at −75 m. The Great Barri Reef of Australia is on a platform about 140 m deep.

There are some interesting geographic differences between reefs the present day. Reefs in the Indian and Pacific Oceans have a pi algal ridge which forms a breaker zone and offers considerab protection to the island behind. The Atlantic reefs have no pink alg ridge, so no breaker zone, and the coral islands tend to be small a easily devastated. This fact is of environmental interest today, but al suggests that reefs in the past may have varied in some properties fro those of today, so we cannot push uniformitarianism too far.

Coral reefs are classified into various groups depending on the relation to backing land masses. Fringing reefs actually grow on t basement island and form the shoreline. There is no lagoon betwe the fringing reef and the backing island.

Barrier reefs are separated from the mainland by a lagoon. Atol are rings of coral reef with a lagoon in the middle, but no backir island.

The growth of coral reefs, and the creation of the different types, closely related to sea-level changes. With a falling sea level a fringir reef is the kind most likely to grow. It can grow seawards for only short distance before the sea falls.

With a rising sea level the reef grows upwards and outwards ar forms a barrier reef with an inner lagoon. As Darwin realized in 183 barrier reefs can only form with a rising sea level.

Atolls represent a special case of growth with a rising sea leve when sea level rises so high that it covers the original island that ma have been in the centre of the ring-shaped barrier reef.

If sea level rises too fast it is possible that coral growth cannot ke pace, and it may be killed off if it reaches too great a depth. Ma guyots are capped by old coral reefs killed off in this way. Darw suggested general subsidence for the formation of the range observed reefs, and his theory is still intact in general principle.

Other workers noting that many lagoons had a rather uniform dep at about 100 m suggested that Pleistocene sea level changes we responsible for the range of coral types. In a time of low sea level, ree would emerge, the corals die, and the reefs be eroded. When sea lev rose again the corals would grow upwards to the present level, and form a young barrier reef. Objections to this outline as a general theory can be made on various grounds – the lagoons are not so uniform in depth, the islands behind the reefs are not eroded, and so on – but there are no doubt genuine effects of sea-level change.

In fact sea level can be studied very well in coral reef areas, especially those where tectonic uplift takes place. The finest example is the Huon Peninsula in New Guinea (Fig. 15.1), studied by Chappell (1974) and others, where coral terraces can be traced to 2 500 m above sea level. This is also an area of good exposures, so details of the coral structure can be observed. Rising sea levels would be marked by upward and outward growth of coral, a stillstand would be marked by outward but not upward growth,with some burial of lower reefs, and a fall in sea level is marked by the start of a new reef growing on the old forset beds of the previous reef. By making observations along these lines a table of rises and falls of sea level can be drawn up. By dating the limestone by uranium series dating the sea level variations can be placed on a time scale.

Now it is well known that in many parts of the world the sea level has risen to its present position since the last glaciation, and has almost reached the level of the last interglacial sea level of 120 000 years ago. There is a variation of perhaps plus or minus 4 m, but this level is the best touchstone in the past. In New Guinea the 120 000 year terrace is at an elevation of about 240 m, and if uniform uplift is assumed, the rate of uplift over this period is 2 mm per year (2 000 B). By plotting the sea level variation data against this assumed constant uplift, it is possible to get a sea level curve, as shown in Fig. 15.2 for the past 120 000 years.

Raised coral reefs are found in many parts of the world, and the method can be extended to derive relative tectonic movements in different places. Some groups of islands show differential uplift, like the Trobriands with one uplift in the western island of Kaileuna, two uplifts in the central Kiriwina, and about five uplifts in eastern Kitava. Yet other islands are capped by coral terraces sequences that have since been faulted. Coral reefs thus offer a tool for tectonic analysis with a usefulness quite out of proportion to the small area they cover.

Tectonics and coastal classification

Many classifications of coasts are possible and may be based on energy expenditure, the nature of coastal materials, or other features. We are interested here in classifications that relate to structure and tectonics.

One of the oldest classifications is into Atlantic-type coasts, with rock structure oblique to the coast, and Pacific-type coasts in which rock structures parallel the coast.

Other classifications depend on emergence or submergence of coasts, with associated regression (retreat) of the coastline, or progradation of sediment. Johnson (1919) produced a classification with four main groups: emerged, submerged, neutral and compound. Unfortunately nearly all coasts turn out to be compound, so the classification is of little use.

Shepard (1973), recognizing that present coasts are mainly submerged by eustatic rise of sea level since the last glaciation, divided coasts into two kinds: *primary*, those drowned but otherwise unchanged, and *secondary*, those modified by erosion and deposition.

Valentin (1970) devised a system (Fig. 15.3) that uses submergence

Fig. 15.1   Raised coral terraces on the
Huon Peninsula, Papua New Guinea.

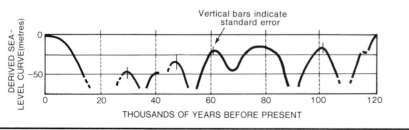

Fig. 15.2   Sea level curve derived from
study of dated raised coral terraces of the
Huon Peninsula, Papua New Guinea.
Turning points are heights above sea level
of very shallow water facies or tidal
platforms (after Chappell, 1974).

and emergence together with shoreline erosion (retrogradation) and progradation as factors which determine whether a coast has advanced seawards or retreated landward.

Fig. 15.3   Genetic types of coastal configuration (after Valentin, 1970).

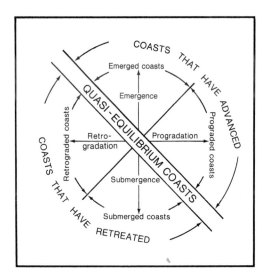

Inman and Nordstrom (1971) have presented what is probably the most tectonically-based classification of coasts. In the first place they categorize coastal features into three orders:

First-order features are those associated with tectonic plates, with linear dimensions along the coastline of about 1 000 km, dimensions across the coastline including continental shelves and coastal plains of about 100 km, and vertical dimensions from ocean floor to mountain summits of about 10 km.

Second-order features are associated with processes of erosion and deposition, and have length, width and height dimensions of about 100, 10 and 1 km.

Third-order features are those dependent on wave action.

A coastal zone is defined in terms of the first two orders, and a shore zone is associated with third-order features.

In tectonic terms the coastal zones can be classified as follows:

1. *Collision coasts*
   (a) Continental collision coasts, where a thick continental plate collides with a thin plate, as on the west coasts of the Americas.
   (b) Island-arc collision coasts, where thin plates collide, as in Indonesia, the Philippines and Aleutian island arcs.
2. *Trailing-edge coasts*
   (a) Neo-trailing-edge coasts formed at recently-formed separations such as the Red Sea and the Gulf of California.
   (b) Afro-trailing-edge coasts, such as the Atlantic and Indian Ocean coasts, of Africa, where opposite coasts are both trailing.
   (c) Amero-trailing-edge coasts, which are trailing-edge coasts on continents with a collision coast on the other side, such as the east coasts of the Americas.
3. Marginal sea coasts, which are coasts protected from the open ocean by island arcs, such as Vietnam and southern China.

The world distribution of tectonic coasts is shown in Fig. 15.4 and

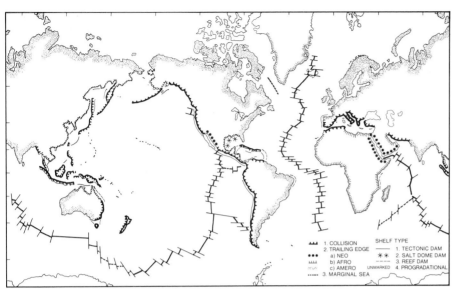

Fig. 15.4 World-wide distribution of tectonic coastal and shelf types (after Inman and Nordstrom, 1971).

Table 15.3 Statistical distribution of first-order features of the world coastal zones (see Fig. 15.4) (after Inman and Nordstrom, 1971)

Continent	Total coast length (10^3 km)	1. Collision coast		2. Trailing-edge coast						3. Marginal sea coast	
		10^3 km	%	a. Neo-		b. Afro-		c. Amero-		10^3 km	%
				10^3 km	%	10^3 km	%	10^3 km	%		
Europe-Asia	75.2	9.7	12.9	7.8	10.4	1.7	2.3	31.4	41.7	24.6	32.7
Africa	24.7	1.6	6.5			23.1	93.5				
North America	43.4	11.4	26.3	2.4	5.5			24.2	55.8	5.4	12.4
South America	27.3	9.0	32.9	2.4	8.8			12.5	45.8	3.4	12.5
Antarctica	24.5										
Australia	14.9	2.5	16.8	1.3	8.8			5.7	38.2	5.4	36.2
Large islands (> 2 500 km²)	136.1	82.3	60.5			14.5	10.6	39.3	28.9		
Small islands (< 2 500 km²)	93.0	60.3	65.0					32.7	35.0		
World excluding Antarctica	439.7	171.1	89.1	18.9	4.3	29.8	6.8	155.4	85.4	38.8	8.8

the statistical distribution of tectonic coasts shown in Table 15.3. This classification has its problems. For instance, India would have had an Afro-trailing-edge coast while it was drifting north, but when it collided with Asia and became located against the Himalayas it became an Amero-trailing-edge coast, although most of the coastal features were unchanged. In Fig. 15.4 the European coast of the Mediterranean is shown as a collision coast, though a case could be made as to regard it as a trailing-edge coast. For these and other reasons Inman and Nordstrom propose another classification based on less speculative characteristics, the morphology (Fig. 15.5).

1. Mountainous coast. Shelf width is less than 50 km, coastal mountains 300 m or higher. Shore is usually rocky and cliffed.
2a. Narrow shelf, hilly coast. Shelf width less than 50 km, coastal hills of 300 m or less. Shore of headlands and bays.
2b. Narrow shelf, plains coast. Shelf width less than 50 km. Plains may be somewhat elevated.
3a. Wide shelf, plains coast. Shelf width less than 50 km. Low-lying coastal plains bordered by a wide shore zone.
3b. Wide shelf, hilly coast. Shelf width less than 50 km. Grades into plains coast.

4.   Deltaic coast. Usually refers to a bulge extending 50 km or more along the coast resulting from sediments deposited where a river enters the sea.
5.   Reef coast.
6.   Glaciated coast. Dominated by the erosional effects of glaciers.

The world distribution of these morphological coastal types is shown in Fig. 15.5, and the relationship between this classification and the earlier tectonic classifcation is shown in Table 15.4.

Fig. 15.5   World-wide distribution of morphological coastal types (after Inman and Nordstrom, 1971).

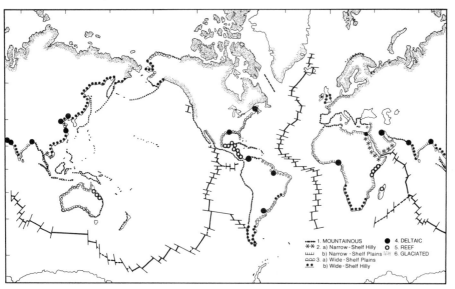

In the same way that Potter's classification of rivers (p. 178) and Audley-Charles's classification of deltas put emphasis on tectonic location, the Inman and Nordstrom classification enables coasts to be classified on a tectonic basis.

Table 15.4 Comparison (in %) of morphological classification of coasts with the first-order tectonic classification (after Inman and Nordstrom, 1971)

Morphologic class	*First-order tectonic class*				
	1. Collision coast	*2. Trailing-edge coast*			*3. Marginal sea coast*
		a. Neo-	*b. Afro-*	*c. Amero-*	
Mountainous coast (1)	97.2	8.0			2.5
Narrow shelf, hilly coast (2a)		75.1	14.1		5.6
Narrow shelf, plains coast (2b)		15.9	46.2	1.5	
Wide shelf, plains coast (3a)			4.0	89.3	3.1
Wide shelf, hilly coast (3b)				2.2	77.4
Deltaic coast (4)		1.0	3.4	1.3	5.8
Reef coast (5)			3.0	1.9	5.6
Glaciated coast (6)	2.8		29.3	3.8	
Total	100.0	100.0	100.0	100.0	100.0
Percentage of world coastline excluding Antarctica	39.0	4.6	7.5	35.2	8.1

Tectonic aspects of submarine geomorphology

There is general agreement on the origin of deep-sea topography. The middle of the ocean rise is a spreading site, which may be marked by a rift, and from this the sea floor drifts away and sinks (possibly because of cooling) to form the abyssal plains.The sea-floor topography is varied by volcanic eruptions and tectonic scarps along transform and other faults.

In plate-tectonic explanations there is a distinction between active and passive coasts. The Atlantic coasts are passive, and as the ocean spreads the continents move apart. Accumulation of sediment derived from the bordering continents builds a typical shelf, slope and continental rise betwen the deep sea and the continent. Large rivers such as the Amazon and Niger build large deltas, which may cause isostatic compensation but are not affected by subduction or other horizontal movements, for the sea floor on which they are deposited is moving at the same rate as the continents from which they are derived. In general the slopes on passive coasts will be gentle, but not always. The South Australian continental slope, for example, is very steep although passive. The slope is regarded as a little-modified scarp slope formed when Australia split from Antarctica, and the slope has experienced little sedimentary build-up because few large rivers drain to it, much of the hinterland is arid, and the continent is very flat with limited erosion.

Along passive coasts there is a build-up of sediment in marginal basins that dates back to the break-up of the original supercontinent. The thick wedge of sediment off North America, the many basins of deposition around Africa, and the basins of the North Sea and Irish Sea off Europe are typical, and often date back to the Cretaceous. A large basin off Western Australia has accumulated sediment almost continuously since the Ordovician (Fairbridge and Finkl, 1978).

Tectonically-active coasts have assemblages of island arcs, trenches, back-arc basins, and related features which are much more controversial. The southwest Pacific is especially complicated with its wide complex of seas, trenches and ridges in the triangle between New Guinea, New Zealand and Tonga. Subduction is a favoured explanation, but as explained elsewhere (p. 197) it is not entirely acceptable. Some trenches are empty, some have horizontal sedimentary layers, and some have highly-disturbed sedimentary deposits, so some elaboration of subduction is required to account for the observed range. Back-arc basins also vary in age, spreading history, heat flow, sediment thickness and other properties, so again an over-simple explanation is inadequate. It seems reasonable to suppose that the controversy over much of these matters will be resolved in the next decade or two when accumulation of further data constrains the present rather free speculation.

The origin of ocean water and its implications

No matter whether the earth originated by cooling of a hot body or the accretion of cold planetesimals, the atmosphere and hydrosphere are derived from volcanic emanations. A hot earth would lose any early atmosphere to space, and would only be able to retain water when it had cooled to 400 °C, and the amount gained by condensation would be small. Further water would then be added from the lithosphere in the form of volcanic exhalations. One might expect volcanic activity to be more frequent, and therefore exhalation greater, in the early part of the earth's history because more radioactive elements with short half-life would be present. A cold-accretion earth would, of course, have to derive all its water from emanations from the lithosphere.

Volcanic activity and the production of water for the accumulating hydrosphere are therefore clearly related. Periods of increased water accumulation should correspond to periods of increased volcanism.A study of the origin of the hydrosphere should therefore throw light on the history of vulcanism.

When did the earth's water accumulate, and what is the balance of gains and losses through geological time? Numerous possible hypotheses have been proposed at different times, including the following:

1. A large original ocean was formed by condensation when the earth cooled, and has been slowly shrinking by losses to space. This now seems so unlikely that further discussion is unnecessary.
2. The ocean accumulated early in the earth's history and has since remained static. However, we know that volcanoes make some additions of juvenile water, so we have the next models.
3. Largely static ocean, with small increments from volcanoes.
4. Continuous growth of oceans through geological time, which may be divided into several variants.
   (a) Uniform growth of oceans through time.
   (b) More rapid growth in early stages because of radioactivity of short half-life isotopes producing more volcanic emanations.
   (c) On geological evidence several authors have suggested a rapid increase in water production in the Jurassic.

Some of these hypotheses are shown on Fig. 15.6. A fuller account of the origin of ocean water is in Rubey (1951) and Mason (1966).

Fig. 15.6   Variation in the amount of ocean water during geological time (after Mason, 1966).

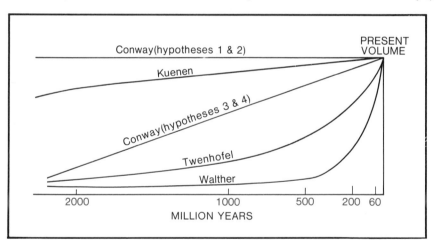

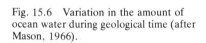

In Rubey's classical paper on the origin of water (1951) he concluded that the volume of water must have increased through geological time. In an era before continental drift was accepted, Rubey thought that the extra water might be contained by increasing the depth of the ocean basins. Carey thinks the increased capacity needed can be attained by increasing the area of the oceans, by an expanding earth. Using a plate-tectonic model Deffeyes (1970), on the basis that the creation of new sea floor is 2.5 ± 0.2 km²/y, calculates that the rate of water released (if juvenile water causes serpentinization) is 2.9 km³/y. This is seven times Hess's estimate (1962, p. 613) of 0.4 km³/y, the revision depending entirely on different spreading rates. The new rate would produce all the present volume of seawater in post-Cambrian time.

The origin of the earth's water is only part of the story; we must also consider the origin of ocean basins, where most water is held. First we will review the main properties of ocean basins that must be accounted for.

1. Ocean basins are fundamentally different from continents, being floored by sima and bounded by the edges of siallic continents. Only shallow seas overlap onto the continental shelves.

2. To express the matter simply, the oceans appear to be full to the brim. Fairly small increases in amount of water would flood large areas of land, and fairly small water losses would soon reveal continental slopes.

3. Although the borders of seas have repeatedly changed through geological time, there is no interchange of continents and deep oceans. Former deep oceanic sediments are not found on the continents.

4. No sediments older than Jurassic have been found in ocean basins, and no dated ocean floor is older than Jurassic. The last item is perhaps the most puzzling of all, suggesting either that all older ocean basins have been lost or that deep ocean basins did not exist before the Jurassic.

The generally preferred hypothesis at present is that older oceans have been lost by subduction. A system of plates that move together, collide, then break along a new suture and drift apart with creation of new ocean and subduction of old ocean is envisaged. But it is not evident why this mechanism would consume *all* the older sea floor and it therefore fails to account satisfactorily for the total lack of ancient sea floor, even though it accounts best for present knowledge of the migration of continental plates.

A second and less orthodox explanation is that the earth is expanding by creation of new sea floor, the continents remaining more or less constant in area. This explanation is propounded most fully by Carey (1976). It involves the creation of all modern ocean basins, except for an area called the Eopacific, since Mesozoic times. A case for more limited, but still significant expansion has been made by Steiner (1977). There is little argument that the new sea floor area has been created since the Mesozoic, but most geologists prefer to think of subduction of an equal amount of sea floor, on an earth of constant radius. The expanding earth hypothesis is considered further in Chapter 18.

If the earth is indeed expanding, the origin of water needs to be reconsidered, especially if expansion only started in the Mesozoic.

The first problem is where was the water in pre-Mesozoic times. If the ocean water had already evolved into something like its present volume, then the same amount of water on a smaller earth would cause the average depth of water to be deeper than at present. This does not seem to be borne out by geological evidence, for there were certainly many land areas in pre-Cretaceous times, and many seas were shallow.

If the ocean basins suddenly appeared in Cretaceous time there should be a fall in sea level, as the same amount of water was spread onto a larger earth, and in particular the water would fall into the newly-made ocean basins. In general, however, the Cretaceous was a time of flooding, so if the ocean basins did suddenly appear, water would have to be produced at a rate sufficient to maintain or even raise sea levels. If, as we believe, water can only be provided by volcanic emanations, then a period of increased water production should correspond to a period of increased vulcanicity. The Cretaceous Parana plateau basalts of Brazil cover an area of over 750 000 km² and the mainly Jurassic Karroo basalts of South Africa cover over 140 000 km² and were once much more extensive. But over the world as a whole it is not evident that volcanic activity was vastly more extensive on land, and we are reduced to producing water from the submarine volcanoes in the newly-formed rift oceans.

If a Mesozoic start to continental drift and earth expansion is accepted, then a new model for water production must be postulated. In the early history of the earth, water was produced fairly rapidly, but later the quantity probably increased only slowly. At this stage water was held in shallow seas and occasional deeper troughs on the earth's siallic crust, as the ocean basins had not yet appeared. In Jurassic and later times the crust fragmented, ocean basins appeared, but an increase in water production has kept the ocean basins full up to the present time. Presumably volcanic activity and emanation increased at the same time.

The key point in assessing the possibilities of the various hypotheses so far considered is therefore this: the amount of drift, water production and vulcanicity are interrelated, and allow at least qualitative checks to be made on the mutual relationships contained in any hypothesis.

Before examining the question further, we might consider what kind of vulcanicity would produce the water. In all probability when an ocean is expanding it is the production of new basalt along the rift that brings most of the water; in other words the contribution of the familiar terrestrial volcanoes is probably rather small, and submarine eruption is most important. It is well known that below a certain depth explosive eruption is impossible but quiet eruption of basalt could possibly add water to the ocean. This would be extremely hard to detect or estimate, but if it happened at all it would profoundly affect attempts at quantifying water production.

Some quantitative estimates of water production are very clearly presented by Holmes (1965, p. 1026). The average annual output of water from volcanoes is estimated to be 1.4×10^{16}g. The total volume of oceans and seas is 1.4×10^{24}g. It therefore appears that volcanic emanation at the present rate could fill the ocean in 100 m.y. This length of time appears to be too short, and some water is evidently being recycled. To perform the calculation accurately it is necessary to know the amount of *juvenile* water produced each year, and in particular what amount of water is added by the volcanoes of sub-oceanic ridges where contamination or recycling is improbable.

Since the average depth of oceans is 3.8 km, every new square kilometre of ocean floor produced by an expanding earth requires 3.8 km³ of water to cover it if the present average sea level is to be maintained. At first sight it seems unlikely that water can be produced at this rate.

However, in the expanding earth hypothesis the problem of the growth of water volume cannot be separated from the growth of mantle volume. An average square centimetre of the earth has to account for 2×10^5 g of water and 10^9 g of mantle, which implies at least 0.2 per cent of water in the pre-differentiation mantle. Even if this quantity had to be increased by a factor of ten it would still not look unreasonable.

If water production fails to keep up with expansion, in the long-run there should be falling sea levels to accompany expansion of the ocean basins. The history of rise and fall of sea level may perhaps throw light on this problem.

16 Sea level changes

There is abundant evidence of relative changes in the levels of land and sea. Many of the rocks on the continents, including those in mountain regions, were laid down in the sea and subsequently uplifted. There are also places where rocks or sediments deposited on land, such as sand dunes, are now inundated by the sea.

Many coastal features such as beaches and sea cliffs are now found at considerable heights above sea level, and in tropical areas there are many coral terraces, now hundreds of metres above sea level, that grew as reefs below the waves. Rivers can only cut down to sea level, but there are many drowned valleys where the old river courses are hundreds of metres under the sea.

The evidence for changing sea levels has now been studied for a long time in many parts of the world, and the story of sea level changes has been partially deciphered. Some changes are of limited local extent, and presumably reflect local tectonic movement, resulting in local uplift or subsidence. Other changes act over large areas, possibly the whole world. World-wide simultaneous change in sea level is known as eustatic change.

There seems little doubt that the eustatic changes of the past half-million years have been dominated by the growth and decline of ice caps: these are glacio-eustatic sea level changes, and they may go back well into the Cenozoic.

Late Cenozoic glaciation

Glaciers existed in Antarctica at least as far back as 27 m.y., as indicated by sub-glacial volcanics of that date (Le Masurier, 1972). By 25 to 30 m.y. ago Australia had crossed latitude 50°S as it moved northwards, the circum-Antarctica Current had developed and the large East Antarctic plateau was probably ice covered (Kennet et al., 1974).

The modern global atmospheric and oceanic circulation patterns date from that time. By 9 or 10 m.y. ago valley glaciers appeared in the mountains of North America (Denton and Armstrong, 1969), and the Greenland ice sheet may have formed 3 m.y. ago. It is clear that substantial areas were extensively glaciated long before the traditional 'ice age' which started 1.8 m.y. ago with the beginning of the Pleistocene.

The Riss, Saale or Illinoian glaciation began about 0.5 m.y. ago, and was followed by the last interglacial (the Riss–Würm, Eem or Sangamon). Numerous botanical, pedological and geological lines of evidence suggest that the last interglacial was warmer than the present, which is consistent with greater melting of ice and slightly higher sea levels than those of today.

After the interglacial, temperatures fell and the interval from about 115 000 to 10 000 years ago is the last glaciation (Würm, Weichsel or Wisconsin). It may be sub-divided into smaller periods of glacial advance (stadials) and periods of retreat (interstadials). The maximum advance of ice seems to have been about 15 000–18 000 years ago.

Even during the glacial maximum the cooling was not extreme, and the average change in summer sea-surface temperature was only $-2.3\,°C$ (CLIMAP 1976), though some areas had more severe falls like the North Atlantic ($-18\,°C$).

The subsequent retreat of the ice was very rapid, and by 6 500 years ago sea level was within a few metres of its present level and the climate was approximately the same as that of today. Between 6 000 and 3 000 years ago there was the so-called Hypsithermal Interval when many areas were warmer and drier than today.

The volume of water abstracted by glaciers has a eustatic effect, and estimates of lowering during the last glaciation range from 80 to 140 m, with present 'best estimate' about -130 m. Lowering exposed about half the continental shelf, so rivers extended seawards and cut down. The rising sea level of the last 15 000 years has drowned all coasts except those with rapid glacial-isostatic or other tectonic uplift. During the most rapid phase, between 10 000 and 7 000 years ago, sea level may have risen at a rate of 10 000 B. At present nearly all rivers enter the sea either in estuaries or across deltaic fills, both features of the glacial eustatic history, and today's valley mouths cannot be considered 'normal'.

The sea is now only a few metres below its level of the last interglacial, about 120 000–140 000 years ago, and many coastal features that seem at first sight to be in equilibrium with present conditions are relics from the last interglacial and only recently reoccupied by the sea. The sea has been at its present high elevation for only about 10 per cent of the past 2 m.y.

There is complication brought about by tectonic feedback resulting from the weight of the water. When a large area of land is inundated by sea, the added weight of the water may cause it to sink slightly, and such sinking may be compensated to some extent by a broad uplift some distance inland. If a depth of sea water is removed from a land area, the loss of load may lead to isostatic uplift of the emergent land. the edges of continents are therefore not passive markers on which changes of sea level are recorded, but flip up and down in response to changes in water load.

Apart from isostatic feedback, a stable area might be expected to give a good record of eustatic changes, but the derivation of the record has been somewhat elusive. Many arguments have arisen over what areas are really stable, and different allegedly stable areas have given different sea level histories.

Winkler and Howard (1977) studied the shorelines of the southern Atlantic coastal plain without the assumption that the area was stable.

Topographic shoreline features were mapped and differentiated into age groups on the basis of progradational discontinuities, contrasts in the state of preservation, and changes in coastal morphology. They discovered three well-preserved shoreline sequences, which permitted palaeogeographic reconstruction, and the results suggest that all three shoreline sequences have been deformed (Fig. 16.1). Warping in the Pleistocene followed persistent Cenozoic structural trends.

Fig. 16.1   Tentative curves of shoreline warping of the Atlantic Coastal Plain of the United States. Each curve represents maximum height of transgression inferred for each shoreline sequence. Dashed lines indicate uncertainty about sea levels (after Winkler and Howard, 1977).

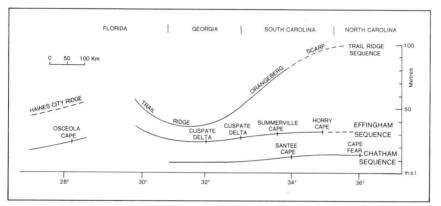

Early eustatic changes

Further back in time changes in sea level are recorded in the sedimentary record. In Cretaceous times, for instance, many areas of the world were inundated by the 'Cretaceous flood', a rise in eustatic sea level that left shallow-water Cretaceous sediments over large areas.

Some sea level changes have regional extent, like the Tertiary marine deposits of various sedimentary basins around Australia, where the sequence reveals the following general pattern:

1. At the base are littoral sandstones and shales of Palaeocene to Eocene age.
2. Marine conditions extended into the Upper Eocene.
3. The sea withdrew in the Oligocene.
4. The sea extended even further in Late Oligocene and Early Miocene times.
5. The sea withdrew in the Late Miocene.
6. After a Pliocene transgression of limited extent the sea again withdrew.

Growth and shrinkage of ice caps have caused the sea level changes in the Upper Tertiary, but it seems impossible for this cause to have been responsible for the Lower Tertiary and Mesozoic changes in sea level. Some tectonic cause seems to be indicated.

Table 16.1

Transgression	Date (m.y.)	Withdrawal
Pliocene	0–5	
		Late Miocene withdrawal
U. and M. Miocene	12–15	
M. Miocene	18–23	
		Oligocene withdrawal
U. Eocene	40–42	
M. Eocene	45–48	
Paleocene	51–58	
Uppermost Cretaceous	68–70	
mid-Upper Cretaceous	81–86	

Offshore exploration of thick sedimentary sequences by very high energy seismic methods reveals bodies of sediment, dated by fossils, that indicate a series of pre-Quaternary marine transgressions. Transgressions occurred at the dates shown in Table 16.1.

Vail and Todd (1977), who describe the technique of seismic stratigraphy, also report the very significant fact that single eustatic cycles are asymmetrical, with a gradual rise to stillstand and an abrupt fall.

This technique cannot be carried beyond the Mesozoic as there are no older deposits in the ocean basins, but other stratigraphic clues have been utilized to postulate eustatic sea level changes right back to the Cambrian and even earlier. Some changes may be glacio-eustatic, associated with ice ages in the Eocambrian and Permo–Carboniferous, but others require a tectonic cause.

McKerrow (1979) has suggested that depth-related benthic communities, if accurately described and correlated, can give good indications of relative changes in sea level, and provides examples from the Ordovician and Silurian. After a transgression in the Caradoc, there is a very widespread Ashgillian regression in many parts of the world, which might be linked with the growth of an ice sheet in North Africa at that time. In Europe and America a slow transgression in Llandovery time is widespread. In the Welsh Borderland a progressive shallowing through the Wenlock is followed by a sudden deepening in the early Ludlow and a regression in the late Ludlow, but synchronous Upper Silurian changes are not obvious elsewhere in Europe or North America.

The Carboniferous in northwest Europe shows more than 25 large transgressions and intervening regressions up to the middle part of the Westphalian, and these may be of general occurrence. Each large transgression consisted of pulses, each pulse reaching further than its predecessor (Ramsbottom, 1979).

Donovan *et al*. (1979) described seven transgressions in the Lower Lias on the northern flank of the London Platform in the southern midlands of England, each followed by a period of stability. The transgressions involved an average sea level rise of 25 m and they occurred at intervals of about 1 m.y. If we assume the transgression took half the interval and the stable period half, the rate of transgressive rise is 25 m in 500 000 years or 500 B.

Tectonic causes of transgression and regression

Tectonic causes of sea level changes have often been postulated, and for sea level changes of the Mesozoic, when there could be no appeal to glacio-eustatic changes, it seems essential. But what are the tectonic processes involved, and how do they affect sea level? Before considering geological aspects of the problem it is necessary to sort out some simpler geometrical concepts.

Fig. 16.2 depicts a simple sea level system, such as a fish tank with clay on the bottom. If the clay is moulded into a bump and depression there is no total change in volume, and the level of the water remains the same.

We cannot make a hardware model on a sphere, but can conceive of a sphere of solid material surrounded by a concentric layer of water. If we imagine a simplified earth with no continents, and the ocean waters lying on the top of the sima – what we might call a simasphere – we can do analogous deformations. As in the fish tank, deformation of the

Fig. 16.2 (*a*) A fish tank floored with clay. (*b*) The same tank, with the clay moulded into hills and depressions. Tectonic warping of the floor does not cause changes in water level unless the hills project above water level. (*c*) An ocean on a spherical simasphere. (*d*) The same globe with tectonic deformation of the sea floor. Unless the sima changes volume or projects above sea level there will be no effect on sea level.

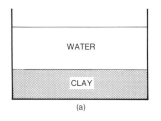

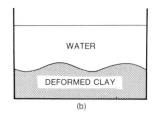

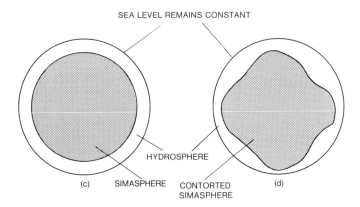

sima into bumps and depressions will not alter the sea level if the deformation is entirely under water. There are two ways in which deformation *can* affect sea level:

1. If the system is not isovolumetric some bumps may arise by increase in volume (perhaps by heating). If, for example, the mid-ocean ridges were created by a simple volume expansion or heating, there would be a genuine rise in sea level.

   This is postulated by Dietz (1961) who suggested that during a period of slow spreading the basaltic sea floor would cool in place, become less buoyant, and sink. This increases the volume of the ocean basins, and causes a regression as oceans drain of the continents.

2. If the deformed simasphere breaks the surface of the water there will be a general change in sea level. This is of little significance on the hypothetical simasphere, but on a real world with emergent continents, the relationship of the emerged and submerged parts of the system is most important.

In considering the effects of density changes of sima on sea levels, it is important to take into account the sphericity of the earth. Returning to a rectilinear model (Fig. 16.2), if the sima increases in density its surface level falls, and the surface of the overlying water falls with it. But on a spherical earth, increase in density of the simasphere causes it to contract into a smaller sphere, with a smaller surface area than the original. With the same volume of water on a smaller area, the depth of the water must increase.

A simple explanation of transgression can be supplied by the elementary geomorphic process of erosion and deposition, though it may be quantitatively inadequate, but for regression some sort of tectonic process is necessary.

On an earth of constant volume, with oceans of constant volume, if the mountains are worn down and the eroded material is deposited in the sea, sea level should rise by displacement, and since the mountains have been worn down conditions for transgression are favoured.

Conversely, times of mountain building would correspond with times of sea level fall and marine regression. Sediments are removed from the ocean and welded onto the craton (perhaps by subduction as in the Andes, perhaps by continental collision as in the Himalayas), resulting in a thickening of the continental crust and an increase in the volume of the ocean basins.

In brief, periods of continental erosion correspond with transgression, periods of mountain building correspond with regression.

Bearing in mind the asymmetry of eustatic cycles described earlier, with a gradual rise of sea level to a stillstand level followed by an abrupt fall, this explanation leads to very interesting conclusions. A long-continued period of erosion would correspond with deposition in ocean basins and slow rise of sea level, and would result in partial planation of the continent. At this stage erosion would be very much reduced and a stillstand might be recorded in the marine record. The rapid fall of sea level suggests that the next phase of mountain building (whether by subduction, plastic flow under the continents, or any other mechanism) is abrupt. From the tectonic point of view this is important evidence in favour of synchroneity of orogeny around the world. From the geomorphic viewpoint there is a tectonic basis for the cyclical ideas of such writers as Davis and King, with major periods dominated by erosion separated by distinct, short periods of uplift.

Another way in which tectonics may affect the world's marine transgressions and regressions is postulated by Valentine and Moores (1970) and Hays and Pitman (1973). During periods of sea-floor spreading the emergent active ridges have a considerable volume (2.5×10^8 km^3 at present) and these topographic highs displace ocean water, causing transgression. If spreading slows down or stops the ridges subside, causing an increase in the volume of the ocean basins and therefore a marine regression.

As an illustration, the mid-Cretaceous was a period of particularly rapid sea-floor spreading in both the Atlantic and Pacific Oceans, and is matched by a corresponding marine transgression that almost doubled the area of continental shelves. In the Lower Palaeocene rates were like those of the present and there was a corresponding regression.

If the ridges *could* simply subside to the level of deeper sea-floor, sea level would fall by about 0.5 km. However, this argument seems to neglect the sphericity of the earth, and will only work if the volume of the ridge is brought about by local heating-induced volume change, not by circulation within the earth with subduction balancing uprising at the ridge. This model also appears to be too coarse to account for the many transgressions and regressions since the Cretaceous.

Valentine and Moores further suggest that continental assembly or collision favours regression, and that fragmentation favours transgression.

On an expanding earth, with generation of new water, a quite different model is required. If water production exceeds the growth of sea-floor area then sea level will rise. If the growth of sea-floor area is ahead of water production there will be a fall in sea level.

A long-standing controversy in tectonics is whether mountain building is episodic or continuous. Some authorities such as Rutten (1969) think that the periodicity of mountain-building periods –

Caledonian, Hercynian, Alpine and so forth – has been much overstressed, and that mountain building takes place virtually all the time though in different places. Others distinguish distinct orogenic periods, with quiet periods between, and may even find a cyclic pattern in orogenic periods. The answer to this problem may help to resolve some arguments about marine regressions.

To restate an earlier point, periods of erosion and deposition should produce partially-planated continents and marine transgressions, periods of mountain building should cause uplift to get ahead of erosion, and so marine regression. Now if mountain building is going on all the time somewhere, it should have very little effect on world-wide sea level changes. But if there really are distinct periods of mountain building they should coincide with periods of marine regression. The Oligocene withdrawal and Late Miocene withdrawal may well correspond to active mountain-building periods, and even the current glacio-eustatic changes may be superimposed on a general withdrawal associated with Pleistocene mountain building. The asymmetry of eustatic cycles, with slow transgressions and rapid regressions, also offer some support for the notion that mountain building occurs in distinct and short bursts.

Transgressive deposits, eustasy and tectonics

The cause of marine transgressions seems at first sight to be elusive, because they may be caused by rise of sea level, subsidence of continents, or a combination of both and the geological evidence – transgressive deposits – is the same in each case. This apparently insoluble problem has been tackled by Bond (1976) by comparing actual sediment thicknesses with those calculated as possible from a maximum sea level rise, making full allowance for isostatic compensation of the sedimentary load.

The procedure is in three parts:

1. Assume the transgression was caused entirely by a rise of sea level.
2. Estimate the maximum sea level rise that is required.
3. Calculate the maximum thickness of transgressive strata that can accumulate in a water depth equal to the sea level rise.
4. Compare the calculated thickness with observed thickness. If the actual thickness exceeds the calculated thickness then subsidence of the crust is required to provide space for the excess thickness.

If a transgression is caused by sea level rise over vertically stationary continents then the area covered by transgressive deposits depends only on the highest elevation to which the sea rises, and the slope (hypsometry) of the submerged continents.

For instance, to find the sea level rise in the Late Cretaceous transgression, it is necessary to plot the area of the transgression on an appropriate hypsometric curve. Problems lie in estimating the area of the transgression correctly, and choosing the right hypsometric curve – rather basic problems, but nevertheless problems that have been tackled by Forney (1975), and Bond (1976) who describe some of the details of the calculation.

The only hypsometric curve available is the modern one (Kossinna, 1933, p. 882). An earlier one, say the Late Cretaceous one, would have significant differences because of post-Cretaceous mountain building and epeirogenic uplift. The modern curve should be steeper than earlier ones, so a calculation based on areas flooded on the modern hypsometric curve will overestimate the sea level rise.

As described on p. 31 (isostasy), as sediment accumulates on a continental margin, the weight of the sediment causes the depositional surface to subside isostatically, and the thickness that can accumulate as the result of a rise in sea level is greater than the initial depth of water produced. A simplified formula for the thickness of accumulated sediment is

$$t = 2.4\,h$$

where h is the rise of sea level, t is the thickness of sediments. For example, if the Late Cretaceous sea level rise was 310 m, then the sediment accumulation would be about 740 m; if sea level rise was 200 m, the sediment could accumulate to a thickness of 480 m.

The Late Cretaceous transgression in North America provides a good example. A sea level rise of about 310 m is required. The sea level rise for the Late Cretaceous transgression turns out to be 390 m, but when the transgression began about 10 per cent of the continents was already submerged, corresponding to a sea level elevation of 80 m. Therefore a better (though still overestimated) Late Cretaceous sea level rise is 310 m. The maximum thickness of a trangressive deposit produced by a 310 m rise of sea level, with allowance for isostatic compensation, is only about 700 m.

Bond (1976) calculated the thickness data from North American strata of Albian to Palaeocene age (the total time span of the Late

Fig. 16.3 Isopach map of Albian to Palaeocene strata in North America. Numbers are thicknesses in kilometres; dots in areas where thickness exceeds 700 m; shading in areas where thicknesses are less than 700 m; lines mark Late Cretaceous land areas. Data for westernmost North America are not shown (after Bond, 1976).

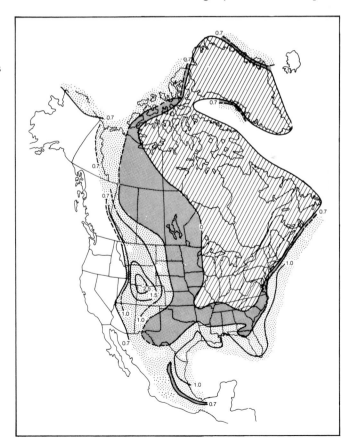

Cretaceous transgression and regression according to Hallam, 1963) and produced a map of isopachs (Fig. 16.3). If the transgression was caused entirely by a sea level rise over continents that subsided only in response to sediment loading, the overestimated sea level rise of 310 m would produce a deposit with a thickness not exceeding 700 m. But as the isopach map shows nearly half of the Upper Cretaceous transgressive deposits are thicker than 700 m, so extra tectonic subsidence is required. This approach gives the minimum area of tectonic subsidence and it is possible that some of the area with less than 700 m of sediments could also have subsided to some extent.

Using his sediment-loading technique, Bond (1978) showed that a combination of eustatic rise of sea level and tectonic subsidence had been active in North America through the Cretaceous. Thicknesses of greater than 480 m are taken to indicate tectonic subsidence. Areas underlain by thicknesses greater than 480 m in the Western Interior Belt expand progressively eastward, suggesting progressive bending of the crust. In the Gulf and Atlantic coasts the area exceeding 480 m shifts rapidly inland in the Early Cretaceous to a position that then remains more or less fixed for the rest of the Cretaceous. This suggests Airy-type loading of a basin bounded by faults, possibly formed by rifting during the opening of the Atlantic and Gulf ocean basins.

Bond (1978) extended the method to other continents and other times. By preparing maps of the shorelines of different periods for different continents (Fig. 16.4), it is possible to calculate the

Fig. 16.4 Palaeogeographic maps. Dots = Albian shoreline; dot-dash = Maestrichtian shoreline; dashes = Eocene shoreline; solid line = Miocene shoreline (after Bond, 1978).

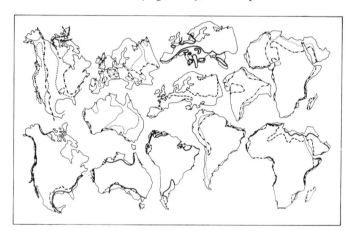

percentages of flooding on different continental hypsometric curves (Fig. 16.5). In this form the data are difficult to assess, but may be re-plotted as in Fig. 16.6 where the elevations of sea level for each continent are plotted individually against the four time-intervals involved.

The points on this figure are scattered, and the African points especially are significantly displaced from the rest. The large displacements suggest that continental hypsometries have changed over large areas, but it is not immediately clear when the changes occurred, or whether sea level might also have changed.

However, it is noticeable that some groups are clustered, such as the points for all but Africa in the Miocene. If the Miocene cluster indicates the approximate rise in sea level then Africa has been substantially uplifted since Miocene time. If so, the amount of Africa's post-Miocene uplift must be removed from its hypsometry before

Fig. 16.5 Percentage of continental area flooded at various times in the geological past plotted on corresponding continental hypsometric curves. Open squares = Albian; solid squares = Late Campanian to Early Maestrichtian; triangles = Eocene; open circles = Miocene; Af, Africa; Au, Australia; Eu, Europe; In, India; NA, North America; SA, South America; Eu adjusted is the curve for Europe with the area south of the Alpine zone excluded (after Bond, 1978).

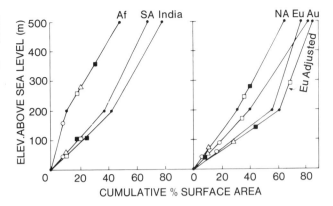

Fig. 16.6 Sea level elevations corresponding to the percentages of flooding that are plotted on the hypsometric curves of Fig. 16.5. (*a*) Distribution of uncorrected points. (*b*) Distribution of points after corrections for uplift or subsidence. Points bounded by dashed lines are those that form clusters from which the approximate sea level for each time interval can be inferred. Continental abbreviations as in Fig. 16.5 (after Bond, 1978).

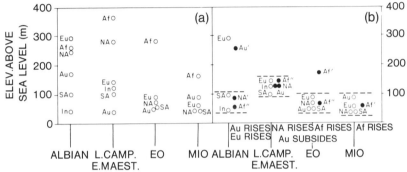

considering the Eocene, and earlier time-intervals. To bring Africa back to the middle of the Miocene cluster a correction of 90 m is required. This correction must next be applied to Africa in Eocene times, but again Africa remains out of the cluster, and a further correction of 120 m is required to get Africa into the middle of the Eocene cluster. Now the sum of the Miocene and Eocene corrections (210 m) must be applied to the African point in the late Campanian to early Maestrichtian time-interval, which brings Africa into the cluster. However, the points for North America and Australia lie outside the cluster for this interval and must be corrected. The downward movement required to correct the North American point implies that the North American continent rose between latest Cretaceous and Eocene times, whereas the upward movement of the Australian point implies that Australia subsided in the same time-interval. Finally, the uncorrected data for the Albian show no clusters, but when the North American and African corrections are applied a new cluster of four continents appears. However, the Australian correction moves Australia well above the cluster, indicating post-Albian to pre-latest Cretaceous uplift of Australia. For the first time Europe lies outside the cluster, and uplift of Europe in the same time-interval is indicated.

Much more testing of data and the procedure are required, but the technique certainly provides interesting and thought-provoking results, and perhaps some direct conclusions. The elevations of the clusters of points (Fig. 16.6) give a rough estimate of the real rise of sea level above the present level during the four time-intervals. The Late Cretaceous transgressions probably were the maximum inundations of the entire Mesozoic era, so the range of 150 to 200 m is a rough estimate of the maximum real sea level rise over the continents during Mesozoic times with a 'best estimate' of 170 m. This value was obtained by plotting the total percentage of flooding on all of the continents on a single composite hypsometric curve. Sea levels obtained by this method are too high because the composite curve is uncorrected for the post-Cretaceous continental uplifts.

Once a sea level curve has been determined, the effects of sea level change can be subtracted from the elevation of continents at specific times to get the actual direction of net movement relative to sea level.

Applying this method, Bond (1978) found that substantial net uplifts relative to other continents are indicated for Europe and Australia between the Albian and Turonian; for North America in the Cenozoic; and for Africa between Eocene and the present. It is suggested that the net uplift in Australia between the Albian and Turonian was largely an isostatic recovery following an episode of tectonic subsidence in the Albian.

Simple allowance for isostatic loading of ice and water and sediment is apparently not good enough, and Farrell and Clark (1976) have presented a more complex model. They discuss an 'exact model' for calculating the changes in sea level that occur when ice and water masses are rearranged on the surface of a non-rotating earth which exhibits instant elastic and delayed viscoelastic response. They find that if a quantity of ice equivalent to a uniform 100-m rise in sea level melts from the Laurentide and Fennoscandian ice sheets, then in the South Pacific the instantaneous rise can be as large as 120 m, while in the North Atlantic the instantaneous rise is always less than 100 m. There is a zone in the North Atlantic with almost no sea level change, and near Greenland and Norway the sea level falls rather than rises, by over 100 m. A thousand years after the melting a bulge migrating towards the ice loads causes water to flow from the South Pacific into the North Pacific, so raised beaches should be formed in the South Pacific. The gravitational attraction of an ice mass upon a nearby ocean tends to hold sea level high in the vicinity of the ice, and this extra load near the ice may have a further effect on post-glacial isostatic adjustments.

Fig. 16.7   A possible constant freeboard global tectonic model illustrating feedback mechanisms that maintain the volume of ocean basins equal to the volume of ocean water (after Wise, 1974).

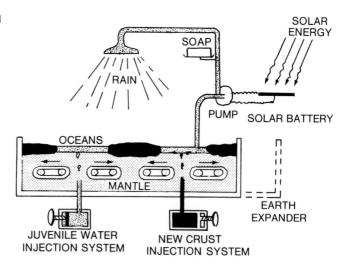

The relative elevation of continents with respect to sea level may be termed the continental freeboard. Wise (1974) proposes a constant freeboard model, and claims that for over 80 per cent of Phanerozoic time freeboard has remained within 60 m of a normal value that is 20 m above present sea level. He thinks Egyed's interpretation (1956) of steady continental emergence is invalid, and his ideas conflict with those of Bond that have just been described.

Over the past 2 500 m.y., he argues, there was neither net accretion nor retreat of the continents. In this case the margins should be considered as the focus of operation of a constant volume system, the 'conservation of continents' system mentioned in Chapter 1. The margins are the place where escaping sediments are welded back onto continental rafts of constant area and constant thickness.

As Wise points out, a constant freeboard with time seems anomalous on a globe that might be evolving both surface water and crust with time, eroding continents at variable rates, and dumping variable amounts of sediments in the oceans to displace seawater. It is obvious that continents are not getting submerged with time, so a mechanism is necessary to push back eroded material, regulating the volume of ocean basins to coincide with the volume of ocean waters, an adjustment that Wise suggests has been maintained to within 2 per cent of volume over most Phanerozoic time.

He provides a model of the Rube Goldberg/Heath Robinson variety, which can be used as a basis for tinkering with various changes to see if feedback brings the system back to normal (Fig. 16.7).

For example, to take what Wise thinks is the most extreme change, we can try global expansion by moving the tank wall. This increases the area and hence the volume of ocean basins, leaving the continents standing higher. Increased continental elevation increases exposed area and stream gradients. This increases regional rates of erosion, and sedimentation extends the continents laterally at sea level, displacing ocean waters and raising sea level. Rising sea levels reduce rates and areas of erosion. The end result is a return to equilibrium at constant freeboard, with thinner, broader continents and shallower, wider oceans having the same ratio of continental depth to oceanic depth as before (in this overly-simplified isostatic system), so that the volume of ocean basins again equals the volume of ocean water. Essentially this system is one of dynamic equilibrium, and is independent of shape or number of continents or ocean basins. Its operation is controlled by tectonic processes of the mantle which drive the deeper half of the equilibrium towards wide ocean basins and thicker, smaller continents, and surficial processes which push the system towards thin and widespread continents separated by shallow ocean basins.

17 Rates of landscape erosion and crustal movement

We have seen that landscapes result from the interplay of erosion and uplift: if erosion dominates, mountains are worn down and eventually a plain may be formed; if uplift dominates then mountains or plateaus are formed; if there is a balance between uplift and erosion some sort of steady-state landscape may be attained, though such a balance is improbable. To understand what might happen to landscapes in theory, and to know what has happened in reality, it is very useful to have some idea of the relative rates of erosion and tectonic movement.

Process rates are very variable, and so are the units they have been measured in. In the case of erosion some workers have used the volume of rock removed, others the weight of rock removed, others the amount of surface lowering; some use metric units and some do not; and rates may be expressed per year, per hundred years, per thousand years, or for other time spans. It is extremely helpful to have all rates in the same units, and in this book I shall express rates in *millimetres per thousand years*. This unit has been called the Bubnoff unit, symbolized B, and it can be applied equally well to erosion, uplift or horizontal movement rates (Fischer, 1969).

Rates of erosion

Erosion rates can be measured directly. A series of stakes can be embedded on a valley side, buried sufficiently deeply to prevent their movement. Marks can be made on the stakes to record the position of the ground surface. If the ground surface is lowered over the years then the succession of marks made, perhaps annually, can be used to see the rate of surface lowering. This simple technique has many flaws and more elaborate methods are used in reality, but the principle is the same.

Perhaps the surface is not lowered by a simple planing-off. It may be that all the upper metre or so of soil is moving downslope. A series of pins placed in a vertical sequence in the wall of a pit dug in the soil will record such motion and if the pit is re-excavated after a suitable period of time the pins will have moved differentially downslope (Fig. 17.1). This measuring pit is called a Young pit, after its inventor. Table 17.1 shows some results obtained by Williams (1973), examining granite

Fig. 17.1 Method of determining downslope volumetric creep from movement of rods in a Young pit (after Williams, 1973).

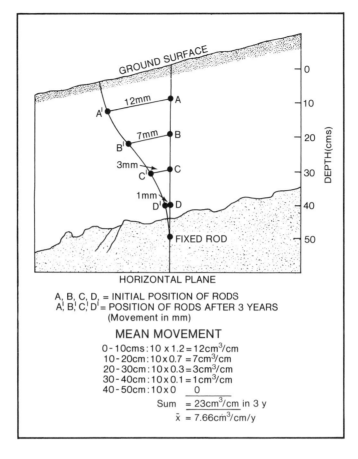

A, B, C, D, = INITIAL POSITION OF RODS
A¦ B¦ C¦ D¦ = POSITION OF RODS AFTER 3 YEARS
(Movement in mm)

MEAN MOVEMENT

0 - 10cms : 10 x 1.2 = 12cm³/cm
10 - 20cm : 10 x 0.7 = 7cm³/cm
20 - 30cm : 10 x 0.3 = 3cm³/cm
30 - 40cm : 10 x 0.1 = 1cm³/cm
40 - 50cm : 10 x 0   0

Sum   = 23cm³/cm in 3 y
x̄ = 7.66cm³/cm/y

Table 17.1 Mean surface lowering by slopewash and soil creep (after Williams, 1973)

Northern granite		Northern sandstone		Southern granite		Southern sandstone	
Slopewash	*Creep*	*Slopewash*	*Creep*	*Slopewash*	*Creep*	*Slopewash*	*Creep*
54	18	56	11	54	7	103	15

and sandstone regions in both northern (tropical) and southern (temperate) Australia. His results show that slope wash is about five times more important than creep in these areas, but other workers have found the reverse situation, such as Young (1960) in the Pennines and Kirkby (1967) in Scotland.

Repeated surveys of a hillslope may show the rate of incision of gullies, from which erosion rates can be computed. If gullies have formed on an initially smooth slope in known time the same sort of calculation can be performed. The volcano, Vulcan, in New Guinea had smooth slopes when first formed in 1937. When resurveyed in 1967 the gulleys showed an erosion rate (averaged over the 30 years of the volcano's existence) 18 000 B (Ollier and Brown, 1971). Using a similar method but a longer time-scale (650 000 y) Ruxton and McDougall (1967) showed that the rate of erosion of the Hydrographer volcano of Papua New Guinea varied from 80 to 750 B.

A quite different technique measures the amount of material eroded or deposited in a known time. Suppose a reservoir is built in 1950. By 1980 it will have collected a certain volume of sediment which can be determined. The area of catchment from which this was derived is known, the time span is known, changes in density can be estimated, and so the rate of erosion (surface lowering) can be calculated. Rates of river erosion can also be calculated from measurements of river flow together with measurements of bedload (which is difficult), suspended load and dissolved load.

Some figures for large rivers of the USA are as follows:

Mississippi	52 B
Missouri	39 B
Colorado	58 B
Ohio	52 B

These figures are similar to those of other large rivers such as the following:

Amazon	50 B
Congo	15 B

The Amazon gets the bulk of its load from the Andean catchments in its headwaters. The figures for American rivers are currently higher than what might be called the geological norm because of erosion resulting from farming and other human activity.

It is said that the average rate for the USA as a whole is only 34 B, so some smaller rivers must have significantly lower erosion rates than those quoted above. However, some small basins have been found to have very high rates, including the following:

small basin, Nebraska	6 700 B
small basin in loess, Iowa	12 800 B

Garrels and Mackenzie (1971) have provided figures for all continents, simplified in Table 17.2.

Table 17.2 Denudation of the continents (after Garrels and Mackenzie, 1971)

Continent	Chemical denudation (B)	Mechanical denudation (B)	Total denudation (B)
North America	13	34	47
South America	11	22	33
Asia	13	12	25
Africa	9	7	16
Europe	17	11	28
Australia	0.8	11	11.8

The all Australia total denudation is the lowest of all continents, and the Western Australian craton has perhaps the lowest denudation rate yet recorded, 0.2 B.

In limestone areas where solution is dominant, the rate of erosion may be calculated from the amount of carbonate carried away in rivers. The Caucasian karst area has an erosion rate of 75–145 B. In South Wales the rate on limestone is 40 B. In Northern England a surface lowering of 40 B was determined from carbonate measurements and checked by direct methods, but there was also the equivalent of another 43 B by sub-surface solution, so the total landscape lowering rate is 83 B (Sweeting, 1966).

The greatest factor affecting erosion rates is generally thought to be relief. Mountainous areas generally have higher erosion rates, including:

European Alps	400–1000 B
Himalaya (Kosi River)	975 B
Himalaya (Bengal fan catchment)	720 B

The mountainous tributaries of the Amazon have relatively high rates (Ucayali, 184 B; Maranon, 138 B) compared with the lowland tributaries (Javari, 32 B; Jutai, 18 B; Jurua, 33 B; Tefe, 6 B; Xingu, 2 B).

Some estimates of average erosion rates of the mountainous regions of the world are:

92–800	(Corbel, 1959)
400–750	(Fournier, 1960)
92–750	(Young, 1969)
300–430	(Ahnert, 1970)

These may be contrasted with estimates of denudation of the world's plains:

12–58	(Corbel)
20–81	(Young)
16–195	(Ahnert)

However, Slaymaker and McPherson (1977) have suggested that estimates from mountainous regions have been exaggerated, and get the lower figures from western Canada as shown in Fig. 17.2.

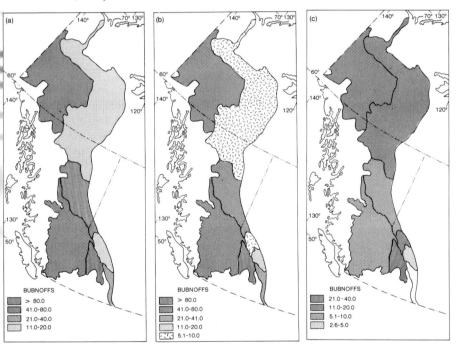

Fig. 17.2 Mean annual denudation in Canadian Cordillera. (*a*) Suspended plus dissolved solids. (*b*) Suspended sediment. (*c*) Total dissolved solids (after Slaymaker and McPherson, 1977).

Climatic factors also play a part. Langbein and Schumm (1958) found that in the USA the maximum erosion was found in semi-arid areas – drier areas had insufficient run-off to cause enough erosion, and wetter areas had more vegetation which has a protecting effect.

Glacial erosion is not remarkably different in rate from fluvial erosion. An average rate for Antarctica is 50 B. Baffin Island varies from 50 to 200 B. An average for glacial and periglacial erosion calculated by Corbel is about 600 B. The average rate of surface lowering by rock glaciers in Alaska during the Holocene is about 900 B.

Detailed work by Wilson showed that neither the Langbein–Schumm relationship, nor any other yet published, is valid on a world-wide basis. He found that land use, not mean annual precipitation, was the dominant factor in determining sediment yield. Another important conclusion he reached was that factors that control sediment supply are more important than those that control transport.

Marine cliff retreat is very limited in area of activity, but is relevant to some geological discussion, especially the formation of plains. Some rates of marine cliff retreat in temperate marine areas are shown below. These are all on soft rock, and eastern England is also subsiding, so these rates are very high. A few thousand B is probably a reasonable estimate on hard-rock coasts. On the limestone coasts of Norfolk Island and Western Australia the rate was 1 000 B.

Limestone, marl	4 000–6 000	Gotland, Sweden
Chalk	67 000	Sussex, England
Till	100 000–200 000	Lake Vattern, Sweden
Till	1 000 000	Eastern England
Till	300 000	Eastern England

If theories of pedimentation are correct the retreat of major escarpments becomes significant for models of world landscape evolution. King (1953) estimates scarp retreat in Mozambique, which might be typical of African scarp retreat, as one foot in 150–300 years or about 5 000 B.

On geological time-scales estimates of erosion may be made from the volume of sediment in a sedimentary basin, the length of time involved, and estimates of the area of catchment for the basin – the last factor is the most difficult and often makes this sort of calculation impossible. Using such techniques Gilluly *et al.* (1970) found that the erosion rate in the Rocky Mountain area in the Lower Cretaceous was 183 B. In contrast the active Ventura anticline area, also in the Lower Cretaceous, had an erosion rate of 1 524 B. The eastern USA in Triassic times was low and arid, and had a surface lowering rate of 25 B.

Various estimates of the world average rate of erosion have been made including:

- 24 Judson
- 30 Corbel
- 84 Ritter
- 69 Holeman
- 35 Stoddart

Even the lowest rate would be sufficient to reduce all the world to sea level in 25 m.y. if other processes did not intervene. Another approach is to estimate the total amount of sediment added to the world's oceans from the continents, which is said to be 9.6×10^9 tons per year, of which 9.3×10^9 is brought by rivers. This gives a world average erosion rate of 24 B.

Many of the estimates in the literature are undoubtedly very inaccurate, and some are only partial. Investigators may neglect the loss in solution when working out reservoir data, they may neglect wind erosion when using river data, they may neglect mechanical losses when working out the lowering of limestone landscapes from water analysis, and so on. Young (1974) stresses that if the maximum contribution to knowledge is to be derived from denudation-rate studies, improvements in experimental design are needed, especially the use of recording sites that are based on a system of stratified area sampling. This ideal has not been achieved anywhere, so far as I know, so we must make do with the scrappy information that has been accumulated on a less systematic basis. Nevertheless enough rates have been estimated for some generalizations to be made.

Schumm (1977) suggests the following averages:

Plains	72 B
Mountains	915 B

Young (1969) suggests the following:

Normal relief	46 B
Steep relief	500 B

If the lower estimates of the average world erosion rates given above are correct, these regional figures must still be too high. However, they will serve as rough guides, and for future discussion in this book the following rough estimates of erosion rates will be used:

Lowlands	50 B
Highlands	500 B

Rates of uplift and subsidence

Rates of uplift or subsidence can be determined by repeated accurate surveying. This has been carried out in many parts of the world, and increasing accuracy of methods in recent years will make this a valuable and routine method in the future. Some examples will give an idea of the results obtained. The results will again be expressed in Bubnoff units.

In Costa Rica recent levelling shows uplift of 2 500 B. In the same area geological studies have shown that since the Upper Miocene the average rate of uplift in the Central mountains was only 100–200 B.

The Carpathian Mountains show varying rates of uplift, with 13 000 B in the west, 11 000 B in the east, and 4 000 B in the south.

In the Caspian region rates change rapidly, increasing from west to east. In the Crimean Mountains the rate is 2 000–4 000 B, in the western Caucasus it is 6 000–8 000 B, and in the axial zone of the central and eastern Caucasus it is over 12 000 B.

In the Swiss Alps several methods have been used to determine uplift rates. Repeated surveying shows a present uplift of 1 mm/y or 1 000 B. By a fairly elaborate method using fission-track dating of apatites, the deformation of isotherms shows an uplift of 0.3–0.6 mm/y (3 000–6 000 B) from 6 to 10 m.y. ago (Schaer, Reimer and Wagner, 1975). The uplift of the gipfelflur (p. 156) took place since Pliocene times, which gives an uplift of about 5 000 m in 5 m.y., or 1 000 B.

In the British Isles estimates of rate of uplift are based on rather flimsy evidence, but after assembling the evidence Walsh *et al.* (1972) produced a graph (Fig. 17.3) with a gradient of about 1 m in 15 250 years, or 65 B, since mid-Tertiary times. In Southern Italy the rate over the past 170 000 years has been 100 to 200 B, and uplift in Poland has been estimated at rates between 500 and 1 000 B.

The Adirondack Dome has an anomalously high rate of uplift (compared with its surrounding) and levelling over 18 years showed an uplift rate of 40 mm in the centre, and a subsidence of 50 mm at the northern end (Isachsen, 1975).

Re-levelling techniques show that Nova Scotia is subsiding at a rate of 5 000 B. The Black Sea is sinking irregularly with an average rate of

Fig. 17.3 Graph to show the possible rate of uplift of the upland areas of the British Isles during the Neogene.

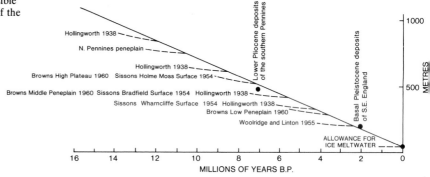

about 1 000 B and locally reaching 6 500 B. Depressions in the western Carpathians reach a maximum subsidence rate of 5 300 B.

The greatest rate of movement established by re-surveying techniques is probably an uplift in Japan of 76 000 B.

Stratigraphic data can be used to get uplift rates. Thus in the European Alps Upper Cretaceous marine rocks are found at 2 500 m, giving an uplift rate of about 30 B. Similarly Upper Cretaceous rocks at 8 400 m on Mount Kanchenjunga give an uplift rate of 90 B. Clearly these are minimal rates, very much less than most uplift rates measured over shorter periods.

The uplift of the Ethiopian plateau (where Jurassic marine limestones were progressively raised to an altitude of 2 500 m, in the Cenozoic) occurred at a mean rate of about 100 B. This was a broad movement acting on areas of hundreds of thousands of square kilometres. The collapse of the central rift during the same period indicates negative movement at a rate of 200 B.

Subsidence of the Caroline Islands region was at a rate of 4 000 B since 4 100 B.P. and at a rate of 1 900 B between 4 100 and 6 500 B.P. This affected a region of about 2 500 000 km², so is thought to be a measure of eustatic sea level change rather than local tectonics.

Chappell (1974) has measured rates of uplift revealed by uplifted coral terraces in New Guinea (Fig. 15.1) which were dated by uranium-series dating. The terraces are warped, indicating different rates of uplift along the coast, but rates average 2 000 B with a maximum of 4 000 B. The same technique has been used in Timor. Three sites on the North Coast gave an uplift rate of 500 B, similar to the rate of 470 B found on the nearby island of Atauro, but a fourth site on the north coast near Dili gave an uplift rate of only 30 B. In the New Hebrides a rate of 670 B was found from coral uplift on the island of Malo, and on Efate rates ranged from 550 to 770 B.

Coral reef growth can usually, but not always, keep pace with subsidence. The Alacran Reef in Mexico has the thickest Holocene section, 33.5 m, known from either Atlantic or Indo–Pacific reefs. The rate of accumulation is a maximum of 12 m/1 000 y (12 000 B) and is attributed to accumulation of the open framework constructed by the rapidly growing *Acropora cervicornis* (Macintyre, Burke and Stuckenrath, 1977).

Many areas of glacio-eustatic movement have series of old shorelines enabling accurate dating and measurement of uplift rates, both overall and for specific periods. In Fennoscandia the maximum rate over the past 7 000 years was 16 000 B; present rates are shown in Fig. 17.4. In the Lake Superior region rates of 5 000 B have been determined. In the Baltic Shield a maximum of 6 000 B was found. In Scotland rates vary from 9 000 to 300 B. Britain appears to be tilting, rising in the west and sinking in the east. A subsidence rate measured in East Anglia was 1 000 B.

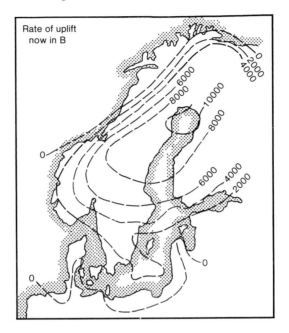

Fig. 17.4 Present rate of uplift in Scandinavia in B (after Spencer, 1965).

Archaeological methods may occasionally be used to determine rates of earth movement. In Iran an ancient canal of known age, presumably nearly level when functional, was bent by an anticline that grew across the canal with an uplift rate of 3 000 B.

Lees and Falcon (1952) found a maximum rate of uplift of an anticline in Iran at 12 000 B, while also in Iran Vita-Finzi (1979) found a maximum uplift rate of 7 400 B. From his uplift rates he also attempted to calculate rates of crustal shortening, though evidence for shortening as opposed to purely vertical uplift is not presented.

Some areas exhibit slow and sustained region movement, but in others there are short sharp movements associated with faults and earthquakes. These give a direct measure of the amount of local movement, but being virtually instantaneous they cannot be generalized into a meaningful long-term average rate. During the 1964 earthquake of Anchorage, Alaska, coast levels rose over 2 m and in other places subsided almost 2 m. The uplift and subsidence affected an area of 90 000 km².

In New Zealand the greatest rate of uplift, near the Alpine Fault, is 10 000 B, contrasting with the rate for the east coast of South Island which is only 300 B.

The Black Forest and the Vosges which bound the Rhine graben have been uplifted to a maximum of 2 500 m in post-Middle Eocene times, giving a rate of about 55 B.

In several areas maps of uplift and subsidence have been made. The form of the maps depends on the resolution of the survey data, and perhaps on the ideas of the investigator. Some areas seem to have simple domes and basins of rather rounded form (Fig. 17.5). Others appear to fall into several compartments, as if several fault-bounded

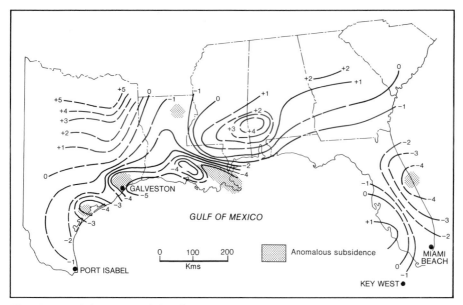

Fig. 17.5   Preliminary rates of elevation
change in south eastern US (after Meade,
1975). Isolines are rate of vertical
movement in mm/y.

blocks are jostling together with greatest movement across the fault
boundaries (Fig. 17.6).

In the eastern United States modern movements appear to be
related to earlier Phanerozoic trends, but the rates of modern
movements are much larger than average rates over the past 130 m.y.
(Brown and Oliver, 1976). Movements are thus either episodic or
oscillatory about a long-term trend.

Fig. 17.6   Map of recent crustal
movements in the Soviet part of the
eastern Carpathians (after Somov and
Kuznetsova, 1975). Isolines are rate of
vertical movement in mm/y.

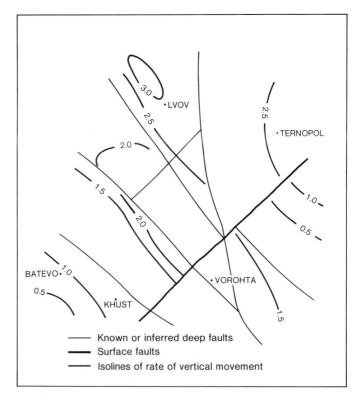

The Atlantic Gulf coastal plains are tilting down away from the continental interior, as might be expected with loading of the shelf and erosion of the continent. The Appalachian Highlands are rising relative to the coast at up to 6 000 B. Maxima correlate strongly with topographic highs, and there is a suggestion of correlation of modern seismicity with extremes of relative velocities in the Appalachian Highlands province. The Interior Plains are tilting downwards. The wavelength between successive zones of relative uplift in the Appalachian Highlands is about 300 km, suggesting tht the entire thickness of the lithosphere and probably the asthenosphere are involved.

Granite intrusion and gneiss domes

Pitcher (1975) indicates that times of uprise and freezing of individual pulses in the intrusion of batholiths are of the order of a million years. The construction of a composite batholith takes some 50 to 70 m.y. and although distinct episodes may be recognized, overall it is likely to be a continuous process. As Gilluly has pointed out (1973) if drift is continuous then so is magmatic intrusion, so rates of emplacement should be several centimetres per year, say 50 000 B. Igneous intrusion leads to uplift, and an estimate of the rate of uplift of a granite batholith by Fyfe (1970) is 2 cm/y or 20 000 B. Stephansson (quoted by Osmaston, 1977) has calculated the rate of intrusion of granite diapirs as likely to be about 1 km/m.y. (1 000 B). The Adirondack Mountain massif is a dissected elongate dome with a long axis of 190 km and a width of 140 km, where Palaeozoic cover rocks have been eroded from a metamorphic core. Repeated levelling has shown an uplift of 40 mm in the centre and a subsidence of 50 mm to the north between 1955 and 1973. This differential uplift, particularly with subsidence at the northern end, argues for a tectonic rather than a glacio-isostatic mechanism. Kahle and Pawlowicz (1977) found that the Adirondack Dome rate of uplift during the 18 years of observation amounted to an average of 3.2 mm/y for 18 years, or 3 200 B. Fletcher (1972) has suggested the growth rate of mantled gneiss domes is up to 5 km in 30 m.y. which works out at 1 mm in 6 years or 166 B. Ollier and Pain (1980) studied gneiss domes associated with Plio–Pleistocene granites and found an uplift rate of about 1.5 mm/y or 1 500 B, about ten times faster than Fletcher's estimate.

From all these measurements, and many others, it seems that tectonic uplift probably averages 1 000 B. Areas of subsidence are much smaller than uplifted areas, and usually closely bordered by larger areas of uplift that more than compensate for the subsidence.

Submarine uplift and subsidence

Submarine topography is subject to rises and falls just like continental topography. Some movement may be systematic, some may be merely local. It has been suggested that the sea floor is pushed up at mid-ocean rises, and sinks as it moves away from the spreading centre. This accounts for the major topography of the oceans, and implies a sinking rate of about 15 B.

Individual rates come from subsided seamounts and guyots. Rockall Island, a fragment of siallic rocks in the North Atlantic, sank after the break-up of the northern land mass when the Atlantic opened. It sank 1 400 m in 55 m.y., at an average rate of 25 B. Coral reefs on a submerged ridge off Brazil are now at a depth of 3.5 km and

are about 30 m.y. old, so sank at an average rate of subsidence o
116 B. It seems that uplift and subsidence in the oceans is at a rat
about an order lower than on continents.

Horizontal movements

Horizontal relative movement of portions of the earth can sometime
be directly measured after an earthquake by displacement of rivers
roads, fences, or other features that enable easy measurement of th
offset. Displacement in metres are not uncommon, and reached 21 n
on the San Andreas fault during the earthquake of 1906. On a longe
time-scale, right lateral motion of the San Andreas fault is moving th
Mendocino triple junction northward relative to North America a
about 60 000 B (Atwater, 1970).

Repeated fault movement may be recorded in river terraces a
where the Wairau Fault crosses the Branch River in New Zealand
where periods of river terrace formation alternate with faul
movements and show the sequence of Fig. 5.12. The total offset i
20 000 years was 65 m giving an average rate of 3 450 B.

The horizontal displacement of geological features on opposit
sides of faults enables the rate of movement over much longe
time-spans to be deduced. For example the Bocono Fault is a majo
fault in northwestern South America, with right-lateral strike-sli
displacement. Movement has been estimated at between 15 anc
100 km since the Mesozoic and between 69 and 100 m in th
Holocene (Schubert and Henneberg, 1975). These estimates giv
average rates of 1 000 B in the long term and 10 000 B in th
Holocene. Other rates determined in this manner include th
following:

Java	5 000 B
Sumatra	12 000 B
Israel (Dead Sea)	7 500 B
Arabia	8 000 B
Ethiopian rift	12 000 B
New Zealand	14 000–18 000 B
Carpathians	23 000 B

Sea-floor spreading rates are another kind of horizontal movemen
of the earth. Plates move from the ridge at about 3 cm/y or 30 000 B
This is one of the fastest of geomorphic rates and must surely have
many significant geomorphic effects. An inferred rate of convergenc
at a subduction site west of Peru was calculated as about 100 km/m.y
or 100 000 B (Noble and McKee, 1977).

Other tectonic rates are consistent with the sea-floor spreading rate
The rate of movement of the Hawaiian chain hot spot was at ar
average rate of 110 000 B, and a similar rate has been found in the
Marqueas, Society and Austral Island chains (Duncan and McDougall
1976).

Movement of the Australian continent over hot spots that createc
Cenozoic volcanoes was at a rate of 66 000 B, which is reasonabl
close to the rate of 56 000 B for the northward drift of Australia a
calculated from sea-floor spreading data from the Southern Ocean.

Crustal extension was measured in the Rio Grande rift nea
Alberquerque by interpretation of fault-dip decrease. It appeared to
be extended 8 km in 26 m.y., giving 0.3 mm/y or 300 B (Woodward
1977). This compares with rates of 100 B for the Rhine graben (Illies

1972) and rates from 400 to 1 000 B for East African rift valleys (Baker, Mohr and Williams, 1972). In Iceland, where extension is taking place by dyke intrusion, the rate is about 5 000 B, comparable with sea-floor spreading sites and not with continental extension. Quite a different type of horizontal movement is that associated with the movement of nappes (Ch. 9). Elter and Trevisan (1973) showed that the average velocity of a 50 km wide nappe was about 10 000 B. The rate of movement of some nappes in the Pyrenees had a velocity of sliding of about 5 000 B according to Choukroune and Seguret (1973). Gravity sliding in the Taconic zone of North America has been estimated at 1 000 to 10 000 B (Voight and Cady, 1978).

In future the development of advanced surveying techniques such as lunar ranging (Bender and Silverberg, 1975) and very long baseline interferometry (VLBI) are expected to measure intercontinental distances to an accuracy of one centimetre and so give direct measurements of horizontal displacements of continents and deformation within continents.

Rates of landscape formation

To summarize, the average rates affecting landscape formation approximate to the following values:

Erosion of lowlands	50
Erosion of highlands	500
Uplift rates	1 000
Sideways movement	10 000

The discrepancy between these figures clearly needs an explanation. Any world model should explain the significance of the actual figures, and the relationship between them.

Erosion rates are fairly straightforward, and the difference between highlands and lowlands is to be expected. The absolute rate is of interest, for at this rate it is possible to erode all continents down to sea level in about 25 m.y. if uplift does not intervene. Clearly the continents have not been removed, even though geological time available would enable them to be lowered many times over. There has to be some law for the conservation of continents. Isostatic uplift provides part of the answer, but is not sufficient by itself. If the continent simply kept rising to keep pace with erosion then the total thickness of siallic crust would be consumed before very long and the underlying sima exposed. This does not happen. Direct observation shows that continental debris is deposited around the edge of the continents, and if these deposits are somehow consolidated they might become part of the continent. But if continents grew in this way they would get ever thinner to account for the increased area of continental material. Furthermore, the very good fit of continents suggests little accretion since the Jurassic at least. Some kind of sub-continental flow or subduction seems to be necessary if continents are to maintain their thickness despite the relatively rapid rates of erosion.

The rate of uplift is in fact about double that of erosion, and on face value suggests that mountains are growing. We may be living in a period when mountain uplift is in reality outpacing erosion. It is also possible that the rates are biased because erosion measurements relate to whole regions, whereas uplift rates relate to high peaks, to individual ranges, or to more restricted areas.

The vast majority of horizontal movements in the earth are related to sea-floor spreading, and the areas of high movement are almost entirely related to actions at plate boundaries. Since sea-floor spreading takes place at a high rate, it is inevitable that plate interactions will often be at a high rate too.

The high rate of sea-floor spreading means the creation of a lot of new sea floor, which must either be destroyed at subduction zones (at an even greater rate as subduction zones are very much shorter than spreading sites) or the earth is expanding. Other movements result from adjustments within plate boundaries, and also to the creep of nappes under gravity tectonics.

The high rate of horizontal movement on the earth is very relevant to problems of continental drift, reconstruction of past geography, location of 'lost' land masses and similar problems. It is important to realize that the mobility of the earth is much greater in the horizontal plane than in a vertical direction, and relative displacement of fragments of crust in plan can be much faster than mountain building

18   The expanding earth

There are four main possibilities in global tectonics.

1. A contracting earth. This was propounded by Elie de Beaumont in 1829 and had a long vogue, but it seems to have no supporters at present.
2. A steady-state earth with fixed continents and oceans. A minority of present day geologists maintain this view, including Meyerhoff and Meyerhoff (1972).
3. A steady-state earth with plate motion. This is the overwhelming favourite at present, and its supporters are too numerous to list.
4. The expanding earth. A view supported by a minority of geologists, including Carey, Steiner, Owens, Shields and others. This group present various lines of evidence that are awkard for plate tectonics and have not been convincingly refuted to the satisfaction of the expanders. It may be that the expanding earth today is in the position that continental drift was in 20 years ago – generally disbelieved despite the evidence – so it is worth reviewing here. It also provides a valuable antithesis to the uncritical acceptance of all the plausible but undemonstrated aspects of plate tectonics.

Evidence for an expanding earth

1. Gaping gores, or the petal effect   If a balloon is covered with paper and then inflated, the paper will split and triangular gapes appear between the triangular paper fragments. A number of triangular 'petals' appear to be separated by intervening, paper-free balloon surface. The analogy is the tapering southern continents, at a simple level. The southern continents resemble the petals of a flower on a swelling bud. But the gape argument goes further. If the continents are fitted together, as in the Bullard fit, there are small failures in the closure between Africa and South America (two petals) because they are being fitted (allegedly) on a globe that is slightly too large. But the gape between eastern Europe and Africa is excessively gaping (if these two were ever in contact) and further east the gape increases to produce the very large Tethys Sea, which is, according to some palaeogeographers, much wider than the real Tethys.

2. The fit   Just as the fit of opposing continents provides convincing evidence for the opening of the Atlantic, so the fit of other continents seems to be too good to be a coincidence. But they only fit together with the maximum matching if they are assembled on a smaller globe than the present one. The global fit involved putting Australia against South America, virtually closing the Pacific, so Pangaea is a continent that covers the entire globe (Barnett, 1962).

Barnett (1962) cut rubber templates from a 4½-in globe and assembled them on a 3-in globe. Atlantic fits were conventional and the Pacific was closed by bringing West Antarctica against the southern Andes, eastern Australia against Central America, and the northern margin of Australia against North America. Barnett wrote 'it is difficult to believe that chance alone can explain this fitting together of the continental margins'.

Creer (1965) prepared plastic shells of the continents on a 50-cm globe and remoulded them to 37-cm and later 27-cm globes, and like Barnett found the fit too good to be due to coincidence.

But the assemblage used by Creer would find little support from most geologists, and it might be argued that since different fits are apparently equally good at filling all the available spaces on a smaller globe, the method by no means proves that the fit is more than coincidence. It might be merely a measure of the patience and ingenuity of the fitter.

Brosske (reported in Jordan, 1971) has presented another whole-earth compaction of continents (Fig. 18.1).

Fig. 18.1 Reconstruction of the earth with all the continents reassembled (after Brosske, 1962).

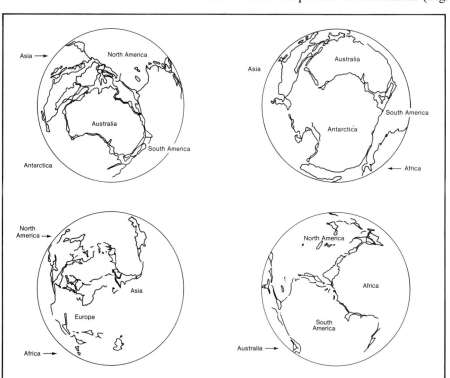

A more detailed attempt has been presented by Shields (1979), who assembled much biogeographical information to show that the Pacific was formerly closed. His assemblage is shown in Fig. 18.2 which shows

Fig. 18.2 Morphological fit with Pacific closure proposed for Early Jurassic times (after Shields, 1979).

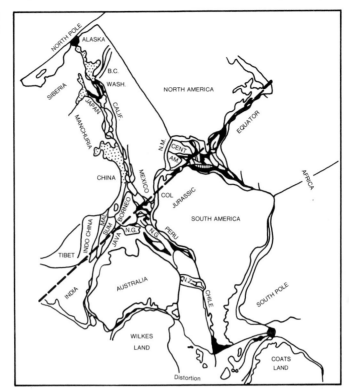

Australia in about the same position that Barnett and Brosske envisaged, but with a large land mass between which became the various ridges of the southwest Pacific – New Caledonia, New Hebrides, Lord Howe Rise, New Zealand and others. There are some geological and geomorphic advantages of such a fit (the western source of South American sediments (Katz, 1971, p. 186), and the eastern catchments for the beheaded Australian rivers (p. 174). There is other biogeographical support for a smaller Pacific (Axelrod, 1976), but very little effort has been expended, so far, on contemplating, let alone testing and evaluating trans-Pacific relationships. Nearly all the effort in demonstrating continental drift relates to trans-Atlantic and trans-Indian Ocean fits.

Shields presents a series of maps and a fairly-detailed scenario for continental displacement, including the surprising detail that the Pacific started to open later than the Atlantic. The next few years will put these ideas to the test by assembling details, at least to the level of the trans-Atlantic data.

3. Absence of ancient sea floor   On the plate-tectonic model, new sea floor is created at spreading sites and old crust is destroyed at subduction sites. The spreading sites are actually rather complex, with changes in the position of spreading sites, cessation of spreading at some sites and fresh starts elsewhere, and so on. On this model it seems that straightforward conveyor-belt motion from spreading sites does not take place, and it is therefore not likely that *all* old sea floor should be subducted. If all the ocean floors were created since the Mesozoic crustal expansion, there would of course, be no older sea floor.

4. Palaeogeography of seas   If the oceans were created since the beginning of the Phanerozoic, and the water on the earth were constant, the same amount of water should be covering a larger area and there should be a general regression of the oceans from the land. This might be measured by finding how much of the continents was covered by seas at various periods of earth history. An expanding earth should show a general regression, a shrinking earth should show an overall transgression, and a steady-state earth should show no systematic change.

Egyed (1956) has carried out this experiment. He measured the areas covered by seas at different geological periods, using two separate palaeogeographic atlases, prepared by Termier and Termier (1952) and Strakhov (1948).

These atlases were not prepared with this experiment in mind, but they both showed the same general trend – a consistent regression, consistent with an expanding earth. This seems to Egyed to be the correct conclusion, even allowing for inaccuracies in the raw data. However, Wise (1974) has reconsidered the evidence and believes the errors are too great, the emergent trend is an artefact, and he finds that a steady-state earth provides the best model. Carey seems to attach little value to this evidence because it is much more probable that new ocean water would be produced during expansion, and the assumptions of the method are not valid.

5. Pacific convergence   The earth can be divided into two unequal parts – the Pacific, bounded by its many alleged subduction sites, and the expanding hemisphere which has many spreading sites but few if any subduction sites.

All the continents around the Pacific are moving towards the Pacific, and quite irrespective of any subduction, these continents are moving towards each other. The Pacific rim is less than a great circle, so the continents around have moved from the other hemisphere, over the great circle concentric with the Pacific rim, and are now converging. They should therefore be compressing each other laterally, or at least getting nearer to each other. But the reverse is found.

Meservey (1969) has tackled this problem with topological thoroughness, and his argument is as follows:

Figure 18.3(b) is an equidistant projection of the earth's surface with the origin of the projection in the central Pacific on the equator at 165 °W. This projection preserves undistorted the radial direction and distance along the great circles passing through the origin. The perimeter represents the point in Africa antipodal to the origin. The circle formed by the meridians 105 °E and 75 °W forms the circumference of the earth half-way between the origin and its opposite point.

On this projection the present position of the continents are shown. If the continents drifted to their present position from an arrangement roughly like that in Fig. 18.3(a), as is generally accepted, then they can be conceptually drifted back along the arrows as shown in Fig. 18.3(b). Since the paths are generally radial they are not significantly distorted.

Points A to J mark the perimeter of the Pacific. Since the Pacific covers only 0.35 of the earth's surface, the perimeter must be expanded as the continents are returned to their original position, so that they can fit over the circumference of the earth fomed by the 105 °E and 75 °W meridians.

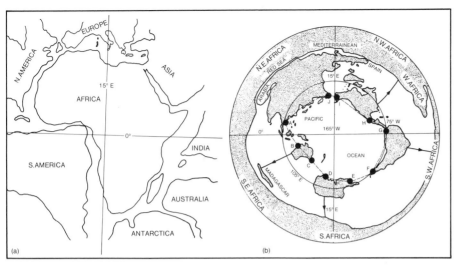

Fig. 18.3 (a) Original relative positions of the continents with respect to Africa. (b) Equidistant equatorial projection of the present earth showing the perimeter of the Pacific and the motions of the various continents to achieve the reconstruction shown in (a) (after Meservey, 1969).

To change the length of the Pacific perimeter, assuming the plates are rigid, any change must take place in the links, AB, CD, EF, GH and IJ. To get the continents back over the circumference of the earth these links must be stretched. In other words, as we go back in time the links must be longer than they are now. But the evidence of sea-floor spreading shows, without any ambiguity, that they were shorter in the past and are at their longest now.

To achieve an arrangement like Fig. 18.3(b) the perimeter of the Pacific must have at some time enclosed at least half of the earth's surface in order to pass over the earth's circumference (as we go back in time) and be assembled on the opposite side of the earth.

There is no topologically possible transformation of the continents on an earth of the present size that can take the continents back to the position of Fig. 18.3(a) if the constraints on the perimeter are accepted.

The perimeter argument seems inescapable. All the data show that Australia drifted away from Antarctica (CD) and was in contact with it 50 m.y. ago. South America drifted away from Antarctica and the Scotia arc was formed between; South America drifted from North America, with the formation of the Caribbean arc between. IJ is the site of the Alaskan orocline with perhaps little change in length, and link AB contains many areas of back-arc spreading and has almost certainly extended.

The topological argument is independent of any possible subduction.

6. Arctic convergence Palaeomagnetic data show the Permian equator crossed North America through Texas, but is now on the Amazon, so North America has moved north about 40° since the Permian. In Europe the Permian equator was about the south of France, now about latitude 40 °N, so Europe has moved north. Greenland is about 50° nearer the pole than it was in the Permian, Siberia 17° and Pacific Asia about 25°. Even allowing for considerable error in measurement or estimate, it seems beyond argument that all the continents have moved north by a considerable amount, since the Permian. Yet the Arctic is a spreading site, where nobody has ever suggested subduction. Carey (1976, p. 199) suggests that the only possible solution to this paradox is global expansion, which allows the parallels of latitude to sweep over the continental blocks (Fig. 18.4).

Fig. 18.4 Migration of parallels of latitude across continents during asymmetrical expansion (after Carey, 1976).

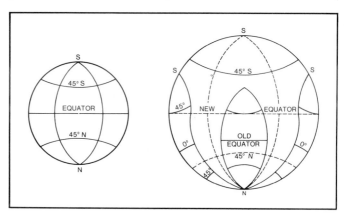

7. *The double-equator paradox* The Triassic equator, derived from palaeomagnetic measurements of European rocks passes through the Mediterranean. African rocks put the Late Triassic to Early Jurassic equator through the Cameroons and Aden (Irving, 1964, Figs. 9.8 and 9.47), about 30° to the south.

Plate-tectonics explanations regard the Mediterranean zone as a collision zone between Africa and Europe, creating the Alpine mountain chains, and presumably losing large amounts of crust by subduction. On an expanding earth Africa retreats from Europe, with repetition of latitudinal zones.

The same paradox is found in the Americas, where the Triassic palaeomagnetic equator as determined from North American rocks runs from Baja California to Florida, but the equator determined from South American rocks runs from northern Peru to the mouth of the Amazon (Irving, 1964, Figs. 9.27 and 9.72), though here convergence is not suggested on plate-tectonic theory.

Although the Arctic paradox and the double-equator paradox depend, at least partly, on palaeomagnetic evidence, many experts consider that palaeomagnetic results as a whole present a cogent argument against the expanding earth, and positively demonstrate that the earth has maintained constant size.

McElhinny and Brock (1975) have demonstrated that palaeomagnetic measurements provide a geometrical picture of the world which is inconsistent with expansion, at least by more than a very small amount which might be within the limits of palaeomagnetic accuracy. Their palaeomagnetic radius determinations based on African data suggest a Triassic palaeoradius of 1.08± 0.015, a Cretaceous palaeoradius of 1.03 ± 0.19 and an overall Mesozoic palaeoradius of 1.08 ± 0.13, results which they rightly point out do not support any expansion. Carey (1976) remains unconvinced, and believes the method of measuring magnetic dips does not give true palaeolatitudes on an expanding earth.

8. *Absence of subduction sites* If it could be demonstrated that there are no subduction sites, the plate-tectonic model would lose half of its mechanism and the remainder – spreading sites alone – would mean an expanding earth. It is not surprising therefore that Carey (1976) has

marshalled together all the evidence that might be used to explode the 'subduction myth'.

Trenches have the appearance of tensional graben, being bounded by normal faults.

Many of them have horizontal sediment, with no sign of compression or subduction.

Some are almost empty, as described by Scholl and Marlow (1974) who also point out that the nature of the sediments that are found in trenches are not the same as those on the neighbouring continents.

Carey points out that detected motions on Benioff zone, parallel to the zone, could be interpreted as shear movements related to a strain elipsoid with the major stress vertical.

Not all advocates of the expanding earth are against subduction, and some believe in limited subduction. They would base their conclusions partly on the next argument.

9. Mismatch of spreading and subduction sites   Spreading sites are about twice as long as subduction sites, so it seems intuitively that there is a problem with a steady-size earth. The areas produced at spreading sites must be subducted at a greater speed to consume the same area, and with a continuity of flow there should be some parts of the ocean floor that are gaining speed as they approach the subduction sites.

On a quantitative basis Steiner (1977) has measured the amount of crust generated at spreading sites at different times, using atlases of sea-floor spreading prepared without this use in mind. He has also estimated the amount of subduction.

He found that during the past 165 m.y. sea-floor generation by spreading has been estimated at 431×10^6 km^2 and global subduction as 325×10^6 km^2 – sea-floor spreading exceeds subduction by 33 per cent. With an expanding-earth model this works out to a palaeoradius of the earth of 6 668 km $\pm$ 13 per cent at Jurassic time, or 0.89 of the present radius.

By working out the palaeoradii and the rate of change of palaeoradius for various Periods and Epochs Steiner showed that spreading and subduction rates have increased with the passage of time.

10. Estimates of palaeoradius   If the earth were to expand there would be an increase in the length of a day, and a decrease in the number of days in a year.

The observed increase in the length of a day is about 2 ms/100 y. Spencer Jones determined that an increase in the earth's radius of 6 inches would cause the day to lengthen by 5 milliseconds. The lengthening of the day by 2 ms/100 y corresponds to an expansion of two-fifths of 6 inches, about 60 mm/100 y or 600 B.

It must be noted that the increase in the length of a day may also be caused to some extent by tidal friction.

The number of days in a year can be estimated from growth lines in corals. Living corals secrete carbonate in such a way that they have growth lines rather like tree rings, each indicating a day. There is a further zoning related to seasonal variations so it is possible to count the growth lines and see how many days there are in a year. Present-day corals give an average of about 360 days, less than the real figure presumably because on a few cloudy days they fail to secrete enough carbonate.

Wells (1963) has found that some Upper Carboniferous corals indicate about 390 days in a year, and Middle Devonian ones about 400. This is consistent with a shorter day and a smaller earth in the past.

With the Middle Devonian average of 400 days in a year the average rate of lengthening of the day works out at 2.2 s/100 000 y, corresponding to an average increase in the earth's radius of 0.66 mm/y, or 660 B.

The Upper Carboniferous 390 days gives a radius increase of 0.6 mm/y, or 600 B.

It seems from these figures that the earth was expanding in the Palaeozoic at about the same rate that it is expanding now.

An extreme approach to palaeoradius determination can be made if it is assumed that the sialic parts of the present crust originally covered the entire sphere (with the same thickness and area), and that all the ocean basins have appeared as a result of expansion. If the area of sialic crust is 177×10^6 km^2 and the area of the whole globe is 510×10^6 km^2, then the increase in radius works out at 2 620 km.

This figure is the maximum amount of expansion. The rate of expansion depends on how much time the expansion has taken. If expansion has been taking place since Early Precambrian times, over say 4 500 m.y., the rate is 0.58 mm/y, or 580 B.

If expansion has been taking place since the start of the Palaeozoic, over say 600 m.y. the rate is 4.35 mm/y, or 4 350 B.

If expansion has occurred only since the start of the Mesozoic, say 200 m.y. ago, then the rate is 13.1 mm/y, or 13 100 B.

An annual increase in radius of 1.3 cm would lead to an increase in circumference of 8 cm, which on most great circles would be less than the amount of sea-floor spreading, but is of the right order. It is of course about double the rate determined from day length and days-per-year calculations.

11. Condensed palaeoclimates   Spjeldnaes (1973) pointed out:

Fig. 18.5   Trans-Atlantic Ordovician latitudes and palaeogeography (after Carey, 1976).

'fossils and sediments ... put certain constraints on the amount of movement you can construct. If you get coral reefs at the poles or widespread glaciations at the equator, this is a sign that there is something seriously wrong with the reconstruction. To put the record straight – the Bullard fit for North America and Africa is absolutely incompatible with the biological and sedimentological evidence from the Lower Palaeozoic. The problem he refers to is the closeness of strongly-contrasted Ordovician faunas across the Caledonian–Appalachian axis, and the Ordovician tillites in North Africa close to the tropical faunas of Ordovician limestones in the United States.

Carey points out that on a smaller globe (Fig. 18.5) the climatic belts conform to the fossil and glacial data, and also to the position of the equator as determined palaeomagnetically.

12. Hot-spot separation   According to Stewart (1976) a comparison of distances between pairs of hot spots and their traces for different ages generally indicates less separation further back in time. The apparent changes in great-circle distances between hot spots may be consistent with constant separation, with an increase in the earth's radius of up to 12 per cent over the past 120 m.y.

13. Palaeontological continuity   As an alternative to the collision hypothesis in which India travels a great distance from an original position in the south to bump into a convenient embayment in Asia, it is possible that Peninsular India has always been near to those parts of Asia to which it is now attached. This is compatible with the opening of the Indian Ocean by sea-floor spreading, but incompatible with an earth of constant dimensions: it implies expansion. Such a hypothesis also takes account of palaeontological relationships that cannot be reconciled with the postulated original wide separation of India and the presently neighbouring countries. The vast oceanic Tethys is probably an artefact, and it was probably a much smaller epicontinental sea (Crawford, 1979).

14. Expanding universe and expanding earth   Astronomical results clearly indicate an expanding universe, a vastly bigger undertaking than merely an expanding earth. If the universe is expanding, for whatever reason, it might be a reasonable supposition that everything in it is also expanding.

This does not appear to be true. There is no evidence that the moon is expanding, nor any of the solar planets except the possibility of slight expansion on Mars. The volcanically active moon of Venus, Io, *may* be expanding, but we have no indication of it. The earth seems to be the odd one out, if it is expanding, but may nevertheless be related in some way to the expanding universe.

The cause of expansion

Almost all attempts to refute the expanding earth start with an attack on the lack of any mechanism to bring it about. This is to some extent irrelevant, and the first thing to do is to find out if the earth is expanding as an empirical fact. If it is, then some speculation on the cause might be in order, but it is not necessary to have a causal explanation to determine empirical facts. We don't really know the cause of mountain building, but we know empirically that mountains are there, and some mechanism or mechanisms must be responsible. We have no reasonable explanation for the reversals in the earth's magnetism, but they are empirical facts.

Another example is provided by Carey. At the end of the nineteenth century Kelvin terrorized geologists into accepting a time scale far shorter than their evidence suggested, because he had the immutable laws of physics on his side. Unfortunately for him he proved wrong, and the discovery of radioactivity allowed the geologists to reinstate their empirically-derived time scales. A nineteenth-century geologist, armed with all the physics of his day, would not have been able to provide a physical explanation for his problems, but he would have been wise to stick to his geological facts. At a later time continental drift was rejected because there was 'no mechanism' known to physics, but when the physicists finally proved continental drift by palaeomagnetism the 'mechanism' objection disappeared. In fact various mechanisms are put forward such as the convection hypothesis and the pull of a sinking slab, but these all have their problems (see p. 52). The failures of Kelvin and Jeffreys have given geologists new confidence in their own results, and they no longer feel it necessary to get the physicists' permission to propose geological theories.

But if a mechanism *must* be provided, there are three main contenders:

1. Phase change at constant mass   Numerous writers have suggested phase changes, starting with initial high specific gravity (many thousands as in white dwarfs, or 17 suggested by Egyed). The core of the earth may be a pressure paramorph of ferro-magnesian silicates that can turn into other phases, including typical mantle material and basalt.

2. Secular decrease in G   The gravitational constant, G, might not be so constant. Cosmologists have proposed secular decrease in G on cosmological grounds quite independently of geological evidence. It is even possible to calculate the rate of earth expansion from any given rate of decrease of G. The estimated average rate of Hoyle and Narlikar (1971) corresponds to a circumference increase of about 2 mm/y (2 000 B) or an increase in radius of 320 B. This is about an order too small, but not so far off as to suggest the concept be discarded.

3. Secular increase in mass   One of the first exponents of an expanding earth, Hilgenberg (1933), and more recently Neyman and Kirillov (quoted in Carey, 1976, p. 454) have proposed that the change in volume was due primarily to a growth in mass, derived from the conversion of energy.

Tests

At least two lines of evidence from those listed in this chapter have the standing of crucial tests. The convergence of continents on an expanding Arctic, and the convergence of continents around the Pacific while demonstrably separating, are only possible on an expanding earth. But two new tests have been devised that have the advantage of predicting rather than merely explaining observations. 1. Two corner-cube reflectors have been placed on the moon.

Observatories in Canberra, Honolulu and Tokyo will measure the distance from their telescope to the corner cube. Repetition will enable the observatories to get their relative position to within a few centimetres.

These three stations are approaching each other at several cm/y on the plate-tectonic hypothesis, but are separating at a few cm/y on the expanding-earth hypothesis. Re-measurement after a few years should establish the truth.

2. Very long baseline interferometry (VLBI) is rapidly reaching the stage where it will be possible to measure intercontinental movements. Once the technique is established it should only take a few years of observations to discriminate between plate movements and earth expansion.

3. A postulated though seemingly impossible test is to send a pulsed beam of electrons through the earth from Chicago to Cocos Island (Indian Ocean) and measure the length of the chord to an accuracy of a few centimetres. If this could be done, and repeated in a few years, it should indicate whether the earth is expanding or not.

19 Theories of mountain and plateau formation

Orogeny is a word meaning the genesis of mountains, and when proposed it meant just that. Unfortunately in later years the idea of folding and mountain building being the same thing became entrenched, and the term orogeny came to mean the folding of rocks. Nowadays we may think that folding and mountain building are two different things, but orogeny is now used to refer to the folding of rocks in fold belts. It does *not* mean mountain building, despite its etymology. We shall have to use the longer term 'mountain building' to be clear. If an authority is needed for this practice, King (1969) wrote in his influential paper 'In this account, and on the legend of the "Tectonic map of North America", "orogeny" is therefore used for the processes by which the rock structures within the mountain chains or foldbelts are created.'

In contrast with orogeny. early geologists used *epeirogeny* to mean the uplift of broad areas, as opposed to the narrow fold belts of mountain chains. Gilbert (1890, p. 340) coined the phrase epeirogeny and was one of the first to use orogeny, so it is useful to get his views: 'The process of mountain formation is orogeny, the process of continent formation is epeirogeny, and the two collectively are diastrophism'. Epeirogeny is still a valid term. As we shall see many mountains result from the erosion of areas that have been uplifted epeirogenically.

The paradox was noted long ago by Stille (1936) who expressed it thus:

As a matter of fact, orogeny in the tectonic sense generally fails as an explanation for the existence of the topographically great mountains of the earth, such as the Alps of Europe or the Cordilleras of North America. These mountains exist – or still exist – as a result of post-orogenic *en bloc* movements, for the most part still going on, and belonging to the category of epeirogenic processes. Thus arises the terminologic contradiction, that the mountains as we see them today owe their origin not to what is called orogeny, but to an entirely different type of movement that is to be strongly contrasted with the orogenic process.

To make matters worse, many of the theories about the origin of mountains are based on dubious assumptions, and many hypotheses purporting to be about mountain building are in fact concerned with geosynclines, plate tectonics, or the origin of mobile belts – belts of deformed rocks which may or may not be coincident with mountain belts or former mountain belts.

Theories come in all varieties of sophistication, but many fail to distinguish clearly between the folding of the rocks and the formation of mountains, and we must include here various ideas that are really concerned with fold belts but because they are called theories of orogenesis they imply a relationship with mountain chains.

Two main concepts have been used to explain the creation of mobile belts:

(a) The lateral (tangential) compression hypothesis, or vice concept, and

(b) The vertical (radial) tectonic concept, with vertical uplift of broad areas.

There are then two related assumptions:

(a) Assumption that the strata are shortened and that the two opposite sides of a sedimentary packet move towards each other during folding.

(b) Assumption that there is little shortening and there may even be extension (dilation).

Finally, we may mention assumptions about earth volume, which include the following:

1. Hypothesis of a contracting earth. This was an old favourite, last presented seriously by Lees (1952), who wrote: 'Mountain building is the consequence of contraction of the interior of the earth and crustal compression from this cause has been dominant throughout revealed geological time.'

2. Hypothesis of an earth of constant volume. This is by far the commonest assumption.

3. Hypothesis of an expanding earth, most seriously proposed by Carey (1976).

4. Hypothesis of oscillating earth volume. So far as I know nobody has seriously proposed this, and it is included here merely to complete the list of possibilities.

These concepts can be pieced together in various ways. Badgley (1965) for instance favours the vertical tectonic concept for the medial portions of mobile belts and believes that both shortening and stretching occur but at different tectonic positions in the mobile belt. Van Bemmelen uses *bi-causality* as a principle, distinguishing primary vertical movement of the crust which create potential or topographic relief energy, and secondary tectogenesis which includes all the gravitational reactions to the primary uplift.

Lateral (tangential) compression theories

This is virtually the simple theory of compression as envisaged on a shrinking earth – the shrivelled apple theory. The rocks in a geosyncline are seen to be folded, so what is easier than to think they are folded by squashing, as if in some great vice. The blocks on the sides of the geosyncline are thought to be rather rigid, and they move together like the jaws of a vice, and it is the mobile contents of the geosyncline that become folded and uplifted into mountains.

Associated with the idea of lateral compression is the idea of crustal shortening. In theory it should be possible to unravel the folds of a fold belt or orogen and derive the length of the sedimentary basin before folding and the amount of crustal shortening. Cox and Cox (1974) for instance state quite plainly: 'The crustal shortening that accompanies the formation of fold mountain may be on the order of tens of miles, as in the Appalachians, or hundreds of miles, as in the Alps.' The unfolding process may lead to difficulties, as Brock (1972) realized when he wrote: 'The Tectonic map of Africa, which labels every exposed bit of shield granite as an orogeny, tacitly demands starting with an Africa of twice its present linear dimensions to accommodate all the crustal shortening demanded by the 'roots of old mountains'.

There are some problems with lateral compression when modern geosyncline equivalents are considered for they do not seem to be located between two rigid blocks. The prism of sediment off the east coast of North America lies between continent and ocean, and so do the island-arc analogues. This vice idea had been more or less superseded before it was realized that the Palaeozoic geosynclines of Europe and North America were indeed in just this position of a trough between two continental masses in the pre-drift assemblage.

But the main objection to the theory, stated simply, is from the problems of rock mechanics, discussed in Chapter 9. Nevertheless some of the modern theories of plate tectonics, despite the sophistication of argument about plate movement, resort to crude compression to form fold belts and mountain chains. As a modern example, Thomas (1977) in discussing the Appalachian–Ouachita folding suggests that both the distribution of sediments and the outlines of later compressional folding reflect an originally zig-zag continental margin. He suggests a compression, not parallel to the movement during rifting, to create folding and uplift, and within re-entrants he believes that compressive stress has been transmitted through the width of thick *incompetent* (my italics) clastic sediments. On the promontories, in contrast, he thinks compression has been transmitted directly through continental crust, because here the pre-Appalachian basement is involved in the folding. Another example is provided by Cogne (1979) who wrote that the collision of northward-moving Gondwanan continental masses, against the then sub-equatorial Europe, occurred at the end of the Devonian, resulting in the development of the Hercynian chain characterized during the Carboniferous by very major crustal thrusting and shearing and the production of new granites by anatexis – remelting of the basement.

Kulm and Fowler (1974) claim that the compressional thrust model offers the best explanation for the Late Cenozoic evolution of the Oregon margin. Pleistocene abyssal plain and fan deposits are being thrust beneath the earlier Cenozoic rocks that underlie the continental shelf. These abyssal deposits have been uplifted more than 1 kilometre and incorporated into the lower and middle continental slope. The stratigraphic position of abyssal deposits on the continental slope, and their age relationships strongly suggest imbricate thrusting of thick slices of sand turbidites typical of submarine fans, which alternate with silt turbidites characteristic of abyssal plains. Underthrusting is thought to have produced uplift of the lower continental slope at an average rate of 1 000 B. Compaction reduces the rate of uplift of older deposits to only 100 B.

Even though the ideas of vice-like compression may be abandoned, compression is evident locally, even in gravity slides. It is also possible that in a compressive situation, as in the Himalayas where India collides with Asia, or in the Pyrenees (Fig. 6.13, p. 88), compression may be accommodated by rock flow in the plastic zone leading to a thickening of the crust, with gravity slides in the surface zone. Thus lateral compression does not form folds directly, but it may cause the crustal thickening that leads to both gravity sliding and mountain formation.

The tectogene

The tectogene idea was very similar to that described above, but was a little more specific. It was based to some extent on geophysics, and was proposed by Vening Meinesz to account for the intense negative gravity anomalies that he found along the deep-sea trenches of Indonesia, as well as to explain mountains. Griggs (1939) extended the idea to a general theory of mountain building (Fig. 19.1).

Fig. 19.1 The tectogene theory of mountain formation. (*a*) Plastic crust overlying a fluid mantle. In modern terms this would be lithosphere over asthenosphere. (*b*) Slow convection leads to the formation of a geosyncline. (*c*) Faster convection causes a major downwarp, folding the geosynclinal sediments and development of a mountain 'root'. (*d*) Melting of the lower part of the tectogene forms granites. Convection ends and a buoyant rise of the mass of folded, thickened and intruded crust (after Garrel and Mackenzie, 1971).

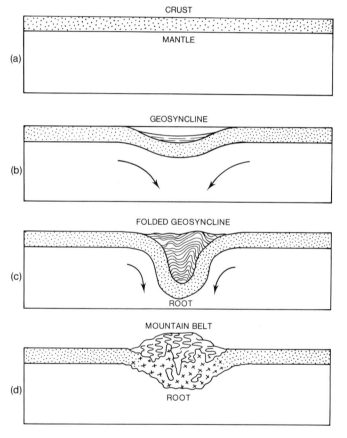

Briefly, it is proposed that a trench develops in the earth's crust, and the down-buckled area becomes filled with sediments, and sinks further, intensifying the geosyncline. The force to form the down-buckle is provided by convection currents. Eventually the deeper parts become sufficiently hot to melt and form igneous intrusions, and the geosyncline is everted, accompanied by the squeezing out of contorted rocks, nappes and the formation of mountains.

In the middle would be a root zone of highly-altered, squashed and heated rocks, while the less-altered rocks would be found in the furthest-travelled rocks. Few real world situations seem to match this model, though the geology of Sulawesi, with divergent fold belts on either side of a central 'root' appears to fit the bill fairly well. With a very asymmetrical development of the tectogene it was also thought that the model could account for the European Alps, using an older model of the Alps which related the nappes to the core zone.

Today the gravity anomalies are accounted for in different ways, and geological structures are generally thought to be inconsistent with the tectogene model.

The Chevallier and Cailleux hypothesis

Chevallier and Cailleux (1959) produced a unique variant on the fold-mountain hypothesis, in which they envisage shrinking continents but not a shrinking earth.

They unfolded all the tectonic folds, and found that the enlarged continents just covered the entire earth surface. They suggest that a primitive sial layer covered the earth, and was then fractured, the cracks corresponding to mid-ocean ridges and earthquake epicentrebelts – a sort of precursor to plate tectonics. The continents then continued to shrink and rotate, but the actual mechanism of mountain formation is left vague.

Foreland folding

Mountain ranges are usually highly asymmetric. Folded rocks are consistently overturned towards one side of the range, and most thrusts travel in the same direction. The side toward which the thrusts have moved and the folds have been overturned is called the 'foreland'. The side from which the surface rocks have moved is the 'hinterland'.

The foreland-folding model takes into account the asymmetry of most geosynclinal belts. The system consists of a land mass, such as a craton or shield, bounded by a trough in which geosynclinal sediments are laid down. At a late stage the contents of the geosyncline are somehow pushed towards and upon the foreland along great thrusts. A modern review is provided by Dott (1978).

The Alpine–Himalayan system, for example, can be thought of as resulting from the deformation of sediments deposited in a composite set of geosynclines developed between the northern foreland of Europe and Asia, and the southern foreland of Africa, Arabia and India. To the north foreland folding produced the Alps, Carpathians, Caucasus and Hindu Kush; to the south the Atlas, Apennines, Dinaric Alps and the Zagros Mountains. Sometimes there are two bordering mountain belts with a median mass in between. The Zagros and Elburz Mountains for example border and thrust away from the Iranian plateau. This demonstrates rather well that foreland folding does *not* imply thrusting brought about by movement of the block behind the fold belt, for the Iranian plateau block could not be thrusting simultaneously in opposite directions. Either the forelands must approach the median area, or the median area has been uplifted and the outward thrusting results from gravity sliding.

The theory of foreland folding has the advantage of incorporating the observed asymmetry of many mountain regions and fold belts. It has the disadvantage of requiring pushing of rock masses in a way that is mechanically difficult, and it does not say what is on the other side of the geosyncline, or what causes the pushing.

Pinch folding

Jordan (1971) relates mountain building to an expanding earth. Expansion necessarily decreases the curvature of the continental mass, and in doing so mountains may be created (Fig. 19.2). Jordan believes this mechanism is the main cause of mountain building, and quotes several supporters including Matschinki and Haber, who introduced the term 'pinch folding' (Quentschfalten).

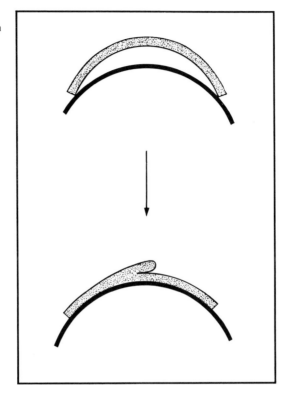

Fig. 19.2 The formation of mountains by pinch folding. On an expanding earth the continent comes to have too small a curvature, and as it collapses onto the broader curve of the substratum it 'pinches' to create mountains (after Jordan, 1971).

Rickard (1969) argued that on an expanding earth, compression would occur along the margin of the continental craton, where a geosyncline and orogen could develop, complete with volcanic belt and Benioff zone, but without consumption of crust by underthrusting.

These theories probably fail because they start with an initial elevation of the centre which cannot be attained, because adjustments in the asthenosphere are too fast. These models are also based on the idea that mountain building is a compressional phenomenon.

Plate tectonics

Plate tectonics were introduced in Chapter 4. Geosynclines and mountain building are explained as possible features of convergent plates. Some possible mechanisms are shown in Fig. 19.3.

A simple story would apply where oceanic crust is being subducted beneath a continental plate. This boundary may, for a while, be marked by a trench in which (geosynclinal) sediments accumulate. Convergent movement between the plates might squash the sediment into folds. Alternatively the sediments might be thrust upon the continental mass, as in foreland folding, or as fault wedges. In the Coast Ranges of California the Franciscan complex is the one most clearly related to presumed subduction. It occurs in a strip about 100 km wide and consists of a series of slices of east-dipping metasediments and ophiolitic mélanges that are older and most metamorphosed to the east. The slices are thought to be accreting

Fig. 19.3 Some possible mechanisms for creation of mountains by plate tectonics. (*a*) Continent–continent collision (Himalayan type). (*b*) Continent–ocean collision with buckling up of the continent and subduction of the ocean floor (Andes type). (*c*) Obduction of the sea floor over the continent, with later isostatic rise to form mountains (Cyprus type).(*d*) Thrusting of marginal sediments on to the continental plate with foreland folding at the front and thrust faulting at the rear (Appalachian type). (*f*) Crustal thickening resulting from plate collision, possibly accompanied by gravity sliding of rocks near the surface.

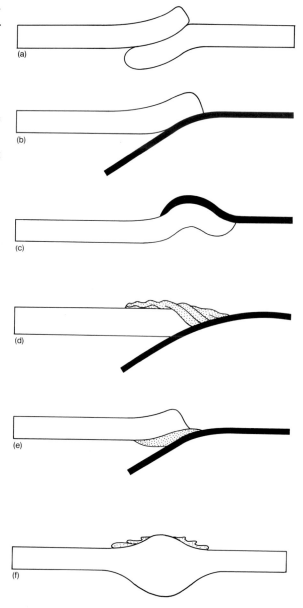

wedges of trench sediments of Jurassic age. A third possibility is that the sediments might be thrust under the continental plate (subducted) and add to the continental thickness by 'underplating'. The increase in crustal thickness of light material could lead to isostatic rise, so the continental edge might rise to form mountains though without direct compression of the rocks of the upper crust. Subducted geosynclinal sediments might also be melted to form andesitic magmas, and the andesitic volcanoes and large intrusive complexes of some mountainous areas might also be accounted for.

Another possible situation is that which purports to explain the origin of the Himalayan mountains. Here geosynclinal sediments are laid down between two continental masses, Asia and India. The northward movement of India compresses the sediments and eventually leads to thrusting and finally to uplift of the mountain belt.

Yet another situation is where sea floor has been thrust over a continental mass. The overridden continental rocks, being lighter, then rose isostically in a large dome, creating the mountains of Cyprus. The old sea-floor rocks are the ophiolites. Many mélange deposits are also associated with this situation. Another place where an overthrust sea floor is found is in eastern Papua.

In many island-arc situations there is a close juxtaposition of mountain ranges and deep sea trenches, which can be accounted for in plate-tectonic terms. But major compressional mountain systems should form only at or very close to converging plate boundaries. If the plates descend at an angle of about 45° it is unlikely that the upper plate will be deformed by compression beyond 1 or 2 km from the toe of the thrust, yet many mountain ranges are hundreds of kilometres from the toe, and some major mountain systems such as the Rocky Mountains, Pyrenees, Brooks Range, Tien Shan, Kun Lun Shan and Nan Shan have formed within continents, remote from any obvious plate boundaries.

There are many more complicated situations, and one of the features of plate tectonics which may be regarded as either a strength or a weakness, is the readiness with which it can be modified to explain, or rather rationalize, a variety of different situations. Thus downgoing slabs can be steep or at moderate angle, going to various depths, at various rates, and in some versions the direction of underthrusting can reverse completely ('flip'), and it is even possible to have convergent Benioff zones (Fig. 4.9, p. 46). Certainly, many of the 'explanations' of geological phenomena seem very facile, and because of the great variety of possible interactions postulated at plate edges the theory would seem to have limited predictive possibilities. To give a specific example, plate-tectonic models of the Alps differ to the extent of having opposite direction of subduction. Dewey and Bird (1970) believe the subduction zone dipped south; Oxburgh (1972) has it dipping north. Estimates of crustal shortening range from nil to 1 000 km in the Alps, but even this is many times less than the shortening implied by plate tectonics. It is more important to note that subduction does not provide a general theory for all mountains. There are many mountains that are remote from any subduction zones at all, and many where the formation of mountains (not geosynclines) seems well separated in time from any activity at plate edges. This is not to say that plate tectonics may not be a valuable concept in some mountain areas, but it cannot be the only explanation.

In the words of Smith (1976):

to assert that a particular orogenic belt is of plate-tectonic origin is in most cases a working hypothesis that may or may not be true. Even if the working hypothesis is accepted, it is still possible to interpret the evolution of the best-known mountain chains in the world by self-consistent but diammetrically opposed plate-tectonic models. This fact, more than any other, illustrates the present understanding of the relationship between plate tectonics and orogeny. Although it has completely reoriented tectonic thinking, it has not solved the problem of how orogenic belts form.

Exactly the same might be said about plate tectonics and mountain building.

Vertical tectonics and gravity tectonics

A group of tectonic explanations depends on vertical uplift, with associated subsidence. Vertical tectonics is now a common term for this idea, and is satisfactory except that no convenient adjective can be derived from it. Cymatogeny has only been used by one author, probably because he invoked geomorphic associations that other tectonicians were uncomfortable with. Vertical uplift may itself create plateaus, domes or mountains, but it is commonly a precursor to lateral movement of material that slides off the uplifted area under the force of gravity. Gravity tectonics utilize the body force of gravity to account for tectonic features at many different levels, from small landslides to major features of the earth.

Landslides, nappes, and diapirs have already been described in Chapters 10 and 11 and it was shown that gravity is the only force available to provide the forces required. In some way, it is postulated, large 'tumours' rise in various parts of the earth. Any surficial sediments on these tumours or domes may slide off, or the dome itself may tend to spread, giving rise to tectonic features such as folds and nappes.

With domal uplift in uniform material one might expect the structures caused by sliding or spreading to be radially distributed. On an uplifted ridge slides on opposite sides might be found. In some instances this appears to be true, but more often there is marked asymmetry, which might be accounted for by the rise of a dome to one side of a depositional basin.

The problem of scale is important in gravity tectonics, and the simple example of a salt diapir provides a useful illustration. If a diapir rises to a height less than about 1 km below ground level the salt in the top part will be heavier than the surrounding rock, but the diapir may continue to rise. This is because the top part is now merely being pushed up by the buoyant driving force of the larger mass of salt below the 1-km level. The system as a whole is losing energy, but the top of the diapir is actually gaining energy. The same reasoning may be applied to diapiric granite intrusions.

At a large scale gravity tectonics will extend below the level of the threshold between brittle rocks and rock flow (described on p. 142) and provide the force, and determine the direction of force in metamorphic, flowing rocks. Where the crust is thickening and rising, metamorphic flow is vertically upwards and here the vertical gneisses will be found.

Several hypotheses of mountain formation are variations on the theme of gravity tectonics.

Cymatogeny

Lester King (1962) coined the term 'cymatogeny' for a kind of broad arching of the earth's surface. He claimed that radial, or vertical, displacement appears to be a common and fundamental activity of the planetary crust, revealed at its simplest in the crustal arches, called *cymatogens* which affect both shields and fold belts. The formation of these arches is called cymatogeny, which King defines as a mode of major vertical (radial) deformation of the earth's crust wherein the fundamental structure induced at intermediate depths is a steeply-inclined tectonic igneous activity. At the earth's surface it is expressed by arching sometimes thousands of metres in height and hundreds of kilometres wide. Rift valleys due to tension and wedge uplifts due to unbalanced forces are frequent minor attributes. The uplifted zones are followed by zones of negative gravity anomalies, and

King maintains that the two sets of data fit so closely as to leave no doubt that the two phenomena are interrelated; he think the gravity anomalies are linked not with the distribution of rock masses but with the topographic deformation. It is, of course, possible that all three features – topography, rock mass and gravity anomaly – are interrelated.

King believes that mountain formation is fundamentally due to vertical movement, entirely independent of lateral (tangential) movement. He points out that although the South American continent drifted west through Cretaceous time, the Andes were not uplifted until later.

Lateral movements of earth materials within a few kilometres of the surface are due primarily to gravitational stresses self-induced in elevated crustal masses. Continental drift, sea-floor spreading, and similar major movements are the only later movements that involve the whole crust and reflect differential 'streaming' of the sub-crust, and these do not make mountains according to King.

It is interesting to note that King worked mainly in South Africa, where there are no Alpine fold belt mountains, and gained supporting evidence from Australia, Brazil and similar places where the deformation of the shield created the cymatogens which, when eroded, became mountain ranges. His emphasis is naturally on uplift of the shield.

The undation hypothesis

The undation hypothesis was proposed by Van Bemmelen (1954) and concerns wave-like motion of double bands of subsidence and uplift. The subsided belt collects sediment, like a geosyncline, and the uplifted band is an arc or tumour of possibly mountainous proportions which is eroded down, adding debris to the trough. Furthermore, very large masses may slide from the tumour as nappes, filling the trough with a series of folded rocks. The distinctive feature of this hypothesis is that the belts migrate. The uplifted arc will be eroded to lowland, and eventually the geosynclinal depression, being a large body of light sediment, will rise to form a new arc, with a corresponding new depression on the sea floor in front. With this mechanism one should find a succession of fold belts getting consistently younger towards the present trough. This is the situation that is found in Malaysia–Indonesia, and the European Alps can also be explained by this mechanism. More recently Van Bemmelen (1976) has reviewed the undation hypothesis in relation to plate tectonics.

What is the driving force for gravity tectonics? According to some workers, such as Press (1973), the earth's crust is heavier than the underlying asthenosphere, a difference that creates widespread instability and the swells and sinks are part of a sequence of changes acting to restore a more stable situation. Van Bemmelen's distinction between the primary vertical uplift and the secondary movements of materials sliding from the uplifted areas is worth emphasis here.

In the theory as summarized so far the uplifted area is a very broad tumour. Folding can also result from the vertical uplift of an individual block, such as a fault block, from an igneous intrusion, or the intrusion of a salt dome. Of particular interest are the folds associated with large fault blocks, which demonstrate that the block spreads out as it rises. This again emphasizes the bicausality of mountain formation – there is something to cause uplift, and the uplifted material then spreads under the influence of gravity.

Resurgent tectonics

Butzer (1976) distinguished between young fold mountains, in which deformation and volcanism continue today, and old fold mountains which have been eroded since the first phase of mountain building and then uplifted broadly. This is probably not a helpful distinction because it depends on observation of present processes, such as geodetic survey, to reveal continuing movement and prove a mountain or mountain range is young. Furthermore some of the mountains normally regarded as young, such as the Alps, have indeed been worn down and undergone broad uplift since the Pliocene, so would count as old mountains.

Yet another complication arises from the fact that renewed movement sometimes takes place in the same areas, sometimes along the same faults, over a vast length of time. Renewed uplift along ancient lines is called resurgent tectonics.

In northern England the post-Triassic resurgent tectonics reactivated faults leading to uplift and tilting of the Askrigg and Alston Blocks, gentle doming of the Lake District and warping along the Pennine axis (Moseley, 1972). Movement along the Craven fault resulted in uplift of the Askrigg Block, but must be visualized in the broader context of the downwarp of the Irish Sea (Fig. 19.4).

Fig. 19.4 Section from the Irish Sea to the Askrigg Block to show Tertiary uplift by step faulting (after Moseley, 1972).

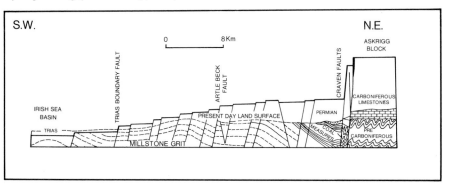

The Neotectonic Map of the USSR was based on theoretical assumptions expressed by Nikolaev and Schultz (1959). The Russians use the term 'neotectonics' to refer to the tectonics of the Neogene, that is for the time since the mid-Tertiary. They believe that strong tangential movements created the Alpine chains in earlier times, that the mid-Tertiary was a period of maximum stability and planation, and a neotectonic phase followed and lasted to the present.

Nikolaev and Schultz believe that the neotectonic phase has the following characteristics:

1. Movements were omnipresent.
2. They differed in intensity from place to place.
3. Movements were predominantly vertical (either up or down) rather than tangential.
4. Upward and downward movements alternate from place to place.
5. Faulting occurred along both new and old lines.

They further claim that neotectonic movements created the present configuration of the land surface and the ocean bottoms, which seems to overstate the case and ignore such processes as erosion, and sea-floor spreading – though of course the latter was not known in 1960.

The most significant feature is the emphasis on vertical tectonics, but this is perhaps not surprising in continental Russia. Some more recent writers use 'neotectonics' to refer to any recent or present-day tectonic activity, without implying support or dissent from the earlier Russian ideas.

Precambrian tectonics

The Precambrian orogenies of Africa are mapped from the directions associated with gneiss foliation and other lineaments, which define distinct belts with distinct ages. Although the metamorphic rocks may be similar in many ways to the so-called root zones of younger mountain chains such as the Alps, they may have quite a different origin and be unrelated to mountain building. They may be metamorphic events only, as described on p. 100).

Brock (1972) considers it unrealistic to have the map of Africa covered by a matting of conflicting orogenies, broad alleged orogenic zones meeting at angles up to 90°. It is tectonically inconceivable that such wide and highly compressed belts could just suddenly end. One explanation is that older orogenies underlie younger ones, but the problem then is how to compress the upper, younger orogenic belt without destroying the lineaments in the underlying material. Another possibility is that the various domains are metamorphic events and represent metamorphic alteration that affected limited areas (see p. 100) converting the lineaments but not representing the roots of old mountains at all.

Some people regard the Pan African 'orogeny' as a metamorphic event, for example. There is undoubtedly a metamorphic event, producing rocks that give radiometric ages of 650 to 500 m.y., but there are no distinct sedimentary basins and no specific 'orogeny'. Others, such as Gass (see p. 000) have tried to explain the Pan African orogeny in plate-tectonic terms. Wynne-Edwards (1977) has suggested that in the Precambrian the rocks were essentially ductile, so the brittle tectonics of the Phanerozoic are absent. This might in part be due to the erosional removal of ancient brittle crusts. England and Richardson (1977) present some evidence to show that this upper crust has indeed been removed. Blueschists develop on the low-temperature end of the metamorphic geotherm and are succeeded in exposure at the surface by greenschist or amphibolite facies rocks. The time scale for this process is consistent with the virtual absence of Precambrian blueschists.

Mountain building in time

There is substantial disagreement amongst geologists about the apparently simple matter of whether orogeny takes place at distinct times or not. Some writers (e.g. Stille, 1936, 1955) think that there are clear episodes of orogeny separated by periods of general quiet; others think that orogeny goes on more or less continuously, and that while one part of the earth is experiencing orogenic change, other parts may be quiet. The question of correlation of orogeny from one place to another is debatable.

In Europe the idea of distinct orogenies was introduced by early writers and is still maintained by many. According to common usage Phanerozoic time in Europe is marked by four orogenic times – the Assyntian (Baikalian), Caledonian, Variscan (Hercynian) and Alpine. More detailed work has distinguished as many as 40 orogenies during this time, mostly thought to be phases of the broader orogenic times. Stille (1936) supposed that each was world wide in its effects. Rutter (1969) does not subscribe to what he calls the 'Stille codex', but it is very hard to dispel the old ideas. It is certainly useful as a shorthand description to be able to refer to 'the Caledonian orogeny' or 'the Alpine orogeny', but such usage should not prevent the realization that orogeny may in fact be less clearly defined in time than these terms might suggest.

In the United States King (1969) doubts the correctness of the proposition that orogeny has been nearly continuous through the life of the fold belts; instead, orogeny that produced actual rock structures (rather than ephemeral mountainous topography) was concentrated in a succession of episodes during each orogenic phase. Individual episodes were not world wide or even extensive. Orogenic times are the natural groupings of these episodes, and to these the traditional names like Taconic, Appalachian, Laramide, etc. are appropriate. These are comparable in scope, but not in age, to the European orogenic times such as Caledonian, Hercynian, Alpine and so forth. A world-wide correlation is therefore not supported by King.

In this discussion I have used the word 'orogeny', but there is an associated idea that actual mountain building might also be periodic. Indeed it might, but as stressed elsewhere it may be completely separate in time from the orogeny associated with the folding of rocks, and many mountains of the present day are very much younger than the last orogeny on the same site.

Plateaus

Plateaus may be divided into two main groups – those associated with mountain chains, and those remote from mountain chains. Examples of the former are the Tibetan plateau north of the Himalayas, the

Fig. 19.5   Possible mechanisms for plateau uplift. (*a*) Accretion (addition) of low-density material to the base of the crust from mantle. (*b*) Isochemical phase change with volume increase in the lower crust or upper mantle. (*c*) Intrusion of low-density plutons in the base of the crust. (*d*) Tectonic thickening of the crust and/or upper mantle. (*e*) Hydration or other metasomatic alteration involving volume increase in the lower crust or upper mantle. (*f*) Heating of crust and upper mantle resulting in volume increase. (*g*) Uplift due to thickening of lithosphere by underthrusting or by float-up of a previously subducted block (after Hobbs, Means and Williams, 1976).

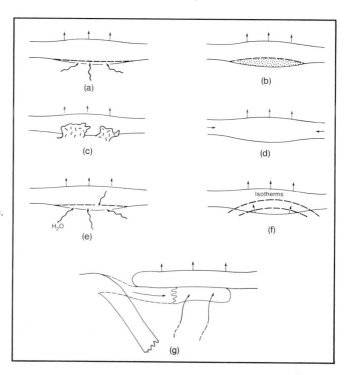

Colorado Plateau west of the Rocky Mountains, and the Altiplano associated with the Andes. Plateaus remote from mountain chains may be further subdivided into those with associated volcanism, such as Tibesti, Jos Plateau, Ahaggar, all in Africa, and those without volcanism such as Serro do Mar, Adirondacks, and Black Hills.

Hobbs, Means and Williams (1976) have classified possible mechanisms of plateau uplift as follows (Fig. 19.5):

- (a) addition of low-density material to the base of the crust from the mantle;
- (b) isochemical phase change with volume decrease in the lower crust or upper mantle;
- (c) intrusion of low-density plutons in the base of the crust;
- (d) tectonic thickening of the crust and/or upper mantle without subduction (as by plastic flow in the sub-crust);
- (e) hydration or other alteration involving volume increase in the lower crust or upper mantle;
- (f) heating of crust and upper mantle resulting in volume increase;
- (g) uplift due to thickening of lithosphere by underthrusting or by float-up of a previously subducted block.

Summary of mountain building

Mountains are made by uplift of continental areas. If the uplifted area remains undissected it may be a plateau if flat (like the Colorado Plateau) or a broad warp if somewhat bent. If the uplifted area is deeply dissected it will be a typical mountain chain with isolated peaks rather than a continuous high surface. In this way theories of plateau formation are really theories of mountain building.

Some uplifted areas are on the sites of old geosynclines, but many are not. Some mountain chains are parallel to the old geosynclines, but many are not. It is possible, as Hess pointed out long ago, that geosynclines have nothing to do with mountain building. Certainly the folds that are seen in the sedimentary rocks within mountain chains have nothing to do with the uplift of the mountains.

Some mountains are at the edges of the plates delineated in plate-tectonic theory, but many are not. Interaction at plate edges may give rise to mountains, but there is a great variety in the styles of mountains even within this situation. The Andes of Peru, for example, seem to have a very different geological and tectonic history from the Rockies and Coast Ranges of North America. Plate-tectonic theories of mountain formation relate in some way to the hypothesis of subduction, but the many mountain chains in positions where subduction is impossible, such as the Drakensberg of South Africa or the Eastern Highland of Australia, show that subduction is not necessary for mountain formation. It may be that subduction, like geosynclines, has nothing to do with mountain building.

Some mountains may relate to the changes in continental configuration in plan. Thus the opening of the Bay of Biscay may be related to compressional thickening under the Pyrenees; the opening of the Arabian Sea may relate to the Baluchistan orocline. But it is clear that many mountains are not in any such situation so this is merely an occasional accompaniment to mountain building.

Some mountains may relate to collision of continental blocks, like the Himalayan Mountains where India collided with Asia, or the Urals where Asia collided with Europe, but again there are far more mountain ranges where collision cannot be invoked.

All we know for certain is that mountains exist because they have been pushed up, generally after a period of planation at low altitude. Vertical tectonics describes the uplift. In certain instances we may be able to apply one of the numerous theories of mountain building and attempt to find a cause behind vertical tectonics. It is unlikely that we shall ever find one single cause of mountain building that accounts for vertical tectonics in all mountains.

20 Some regional examples

The details of tectonics and geomorphology reviewed so far do not add up to a generally accepted and unified whole. This is not an unsatisfactory state of affairs because theories of the earth, like the earth itself, are still evolving, and there is no reason to suppose that ideas of the present day are the last word. Indeed they cannot be because there is no general agreement at the present time: not everyone accepts plate tectonics, and even amongst believers there are great differences in the way particular features are interpreted. The following few examples are reviewed not to show the latest ideas on landform and tectonic evolution, but rather to show the range of interpretations that are possible for some of the major features of the surface of the earth.

The Alps

The Alpine orogenic cycle started its sedimentary phase after the Hercynian orogeny in what is now southern Europe, in geosynclines that were part of the Tethys Sea. The main folding phases of the Alpine system fall in the Early and Middle Tertiary, whereas terrestrial sediments (molasse) are common from the Oligocene or Miocene onwards. There are variations in time in the main folding, the Pyrenees and the Provence having been folded earlier than most of the other parts of the chain. Despite many variations all the mountain chains belong to the same orogeny, the Alpine.

In the Eastern Alps northward movement of nappes across the basin of molasse deposition ceased before the Upper Miocene. After the Miocene the development of the topography continued with differential vertical movements on steeply-dipping faults. Fault-bounded basins of marine or lacustrine deposition were formed in places, most importantly the fault-bounded Vienna basin which cuts across the Alps and buries their continuation to the Carpathians under several thousand metres of Miocene to Quaternary marine and later lacustrine sediments.

Prior to about 1880 the rocks of the Alps were thought to be more or less in the same place in which they had been originally deposited (autochthonous) although it was clear that they had been folded. By about 1900 it was clear that many of the rocks had travelled long distances from their site of deposition (allochthonous) and the term 'nappe' came to be applied to any large, more or less horizontal sheet of rock that had been displaced a considerable distance.

At this stage the classical theory of Alpine orogeny was evolved, based on nappes and the 'root zone' roughly along the Swiss–Italian border. The general concept, illustrated in Fig. 20.1, is of vice-like compression between the forelands of Europe and Africa, with squeezing out of the deposits of sediments in the middle, forming a root zone of vertical and metamorphosed rocks, together with the far-travelled nappes.

Fig. 20.1 Section across the European Alps according to the classical concept that the great recumbent folds and nappes were driven from the Tethys geosyncline by the vice-like approach of the African and European forelands, with the zone of roots as the central, most compressed part (after Holmes, 1965).

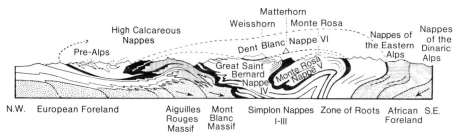

The root zone is nowhere connected geographically to the nappes, and probably has little direct geological connection. It is a zone, 400 km by 20 km or less, of vertical gneisses, highly metamorphosed and obscured by later plutons. It is a distinct structure separating the northern Alps from the southern. It is a major crustal cicatrice, but how and when it was formed and its relationship to nappes is not clear. The root zone is not to be confused with the root of the Alps postulated to explain negative Bouguer anomalies. The root zone is immediately to the south of the northern nappes, so if they are far travelled they have been followed closely by their own root zone – the root has travelled with them.

The Alps have also been explained on the undation hypothesis, and Van Bemmelen (1972) believes the Alps evolved through the following stages:

1. Subsidence of the Tethys–Adriatic geosyncline (pre-Senonian).
2. The Adriatic centre of subsidence was raised as a broad, domelike upwarp (mid-Cretaceous). The rise may have been compensated by a subsidence of the Pennine foreland to the north (Western Alps) and Tauern foredeep (Austria) in which mainly flysch sediments were deposited.
3. The thick sediments of stage 1 now slid (décollement) and accumulated in the Pennine–Tauern foredeep. These are the epidermal gravity slides, of which the Helvetian nappes are a fine example. Thick Permian evaporites at the base of the sedimentary pile aided the sliding.
4. Formation of the Austride nappes, overriding earlier ones, and involving the migmatic front beneath the Adriatic tumour. These are the mesodermal slides.
5. In places the base of the crust was injected into the overlying nappes.
6. A dome rose in the mid-Tertiary beneath the Tauern foredeep, compensated by the development of deeps to the north and south in which the molasse sediments were deposited.

7. From this dome there was further movement of the earlier-formed nappes – the Helvetides, Austrides and Pennines.

The Alpine summit levels form a very broad arch, uplifted between the Molasse basin in the northwest and the basin of the River Po in the southeast with superimposed minor undulations along the arch and across it. Each of the major Tertiary orogenic phases, and particularly that of the Miocene, was followed by an interval of several million years during which uplift failed to keep pace with lowering of the surface by denudation. Towards the close of the Pliocene the Alps had been reduced to a region of low relief, the complex underlying structures being truncated by the erosion surface. Powerful upwarping of the latter then made possible the carving of the peaks and valleys that make up the Alpine scenery of today.

In plate-tectonic terms the Alpine belt is a continent–continent collision belt formed by interaction of the African and Arabian plates with several European plates. Numerous plate-tectonic scenarios have been presented, including Dewy *et al.* (1973), and Windley (1977,

Fig. 20.2 Proposed plate tectonic evolution of the Alpine system (after Dewey, Pitman, Ryan and Bonnin, 1973).

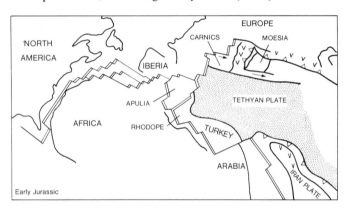

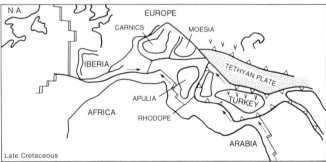

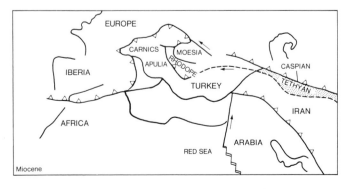

Ch. 18) presents a summary. The evolution is complicated even by plate-tectonics standards because the evolution took about 200 m.y. and involved a large number of plates and microplates (Fig. 20.2). For example, according to Dewey *et al.* (1973) there were six periods of basalt formation, seven of ophiolite formation, three of ophiolite subduction, eleven of deformation and seven of high T/P metamorphism taking place diachronously in different areas! A synopsis is as follows:

1. Fracturing of pre-Alpine continental Pangaea and extrusion of flood basalts in Triassic times.
2. Formation of new ocean crust between platelets.
3. Deposition of Triassic evaporites, carbonates and redbeds.
4. Collapse of carbonate shelves and deposition of deep water shales in the Jurassic.
5. Cretaceous development of island arcs with ophiolite mélanges, acid volcanism, blueschist metamorphism, deposition of flysch in trenches, thrusting and obduction.
6. Eocene–Oligocene continent–continent collision, with thrust sheets (nappes), further flysch deposition and opening of marginal basins in the western Mediterranean.
7. Later orogenic uplift, and deposition of Upper Tertiary molasse in foredeeps. Late Miocene (Messinian) evaporite deposits in Mediterranean and other basins.

There is a great deal of variation in plate-tectonic models, as indicated by interpretations of the Insubric Line, a suture dividing the southern Alps from the northern, Austro–Alpine nappes. Ernst (1973) considers the southern Alps overrode and metamorphosed the northern block; Oxburgh (1972) suggested the Austro–Alpine nappes were flaked off a northward descending southern plate, and Laubscher (1971a) regards the Insubric Line as a strike-slip fault with up to 300 km of dextral movement.

The Himalayas

At the simplest, the tectonic story is of a collision between India and Tibet, with the Himalayas between (Fig. 20.3). Many authors believe that India has underthrust Asia, and that their is a double thickness of crust beneath the Tibetan Plateau. If a subduction model is used, the subduction must have been flat at the start, though there may be a more steeply-dipping Benioff zone at present. The location of the Indus suture ophiolites rather suggests that thickening and thrusting were more confined (Fig. 20.4).

But the Himalayas have several complexities which confuse the simple picture.

In the first place they consist, to a large measure, of nappes, which moved from the northeast. The rocks are highly metamorphosed, but some authorities have detected tectonic dolomite between the major nappes, such as the Katmandu nappe and the Nawakot nappe. If the nappes are gravity structures they imply uplift of the Tibetan Plateau or other high ground as a geotumour from which nappes spread to a lowland further south.

A very striking feature of the Himalayas in plan is the perfection of the frontal arc. It is quite comparable with an island arc yet is firmly located on land, and furthermore the arc can be linked with the festoon of arcs that run from Indonesia, through the Himalayas, and on to the Mediterranean. It is not understood why simple underthrusting should

Fig. 20.3   (*a*) Tentative reconstruction of the Tethys in the Late Cretaceous. (*b*) Position of the Tethys suture between Eurasia and Gondwana plates.

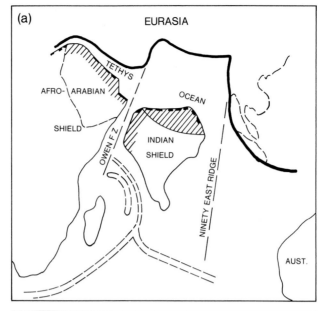

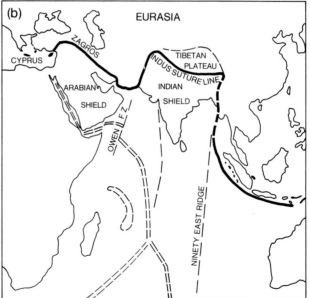

Fig. 20.4   Cross section of the Himalayas and Tibetan Plateau illustrating the hypothesis that there is a thickening of the crust in the region.

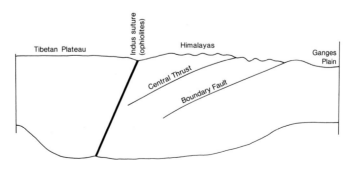

create an arc in plan, let alone why it should approximate to the size of an island arc and be continuous with the island-arc chain.

Major rivers from Tibet cross the Himalayas, so at an earlier stage the Tibetan Plateau must have been higher than the Himalayas. Later uplift along the Himalayas created higher ground, but the antecedent rivers kept pace with the uplift. The age of the uplift is not well known. An outside limit is provided by the Upper Cretaceous marine limestones that covers most of the Tibetan Plateau, which must therefore have been below sea level in the Upper Cretaceous. Another limit is placed by the molasse deposits in the Siwaliks (south of the Himalayas) which are at least 18 m.y. old, so mountains must have existed at that time. There is a good deal of evidence for considerable uplift in the Late Tertiary and Quaternary. According to some investigators only the latest glaciation is found in the Nan Shan and Tsaidam basin, suggesting ground was lower during earlier glacial periods.

Estimates of the uplift have been derived from comparison of the altitudes at which plant fossils of the Pliocene and of the interglacials are found now and at which the kind of vegetation they represent is now found on the southern flank. Thus *Cedrum deodar* and *Quercus semicarpifolia,* found in Pliocene gravels at 5 900 m on the north side where the mean annual temperature is $-9\,°C$, grow today in evergreen forest at 2 500 m on the south side with a temperature of 10 °C. Most of this difference is attributed to an uplift of about 3 000 m. A gross conclusion is that most of the substantial uplift since the Pliocene took place in the Late Pleistocene. This introduced a powerful new geographical factor in the pattern of climate. In the Early and Middle Pleistocene when the average elevation of the Himalayan range was about 4 400 m, the evidence from interglacial deposits shows that the north side was as warm as the south side at similar elevations. However, the uplift to bring the range to its present 6 000 m average elevation made the Himalayas a much more effective climatic barrier, preventing warm, moist air from entering the Tibetan Plateau.

Fig. 20.5 Simplified geography and tectonics of the Himalaya–Tibetan Plateau region. Convergence of the Indian Plate and Asia north of the Altyn Tagh Fault leads to a resultant eastward movement of the Tibetan Plateau plate (after Ni and York, 1978).

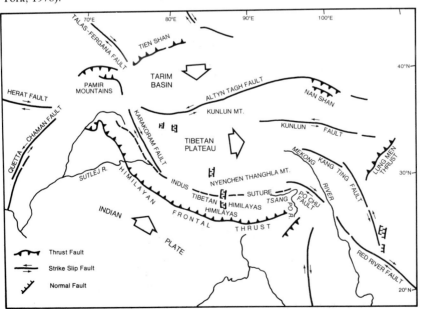

Exclusion of the Indian monsoon made the north side colder and drier and this restricted the growth of glaciers.

Active convergence of India and Eurasia does not manifest itself in Tibet by crustal shortening, and the dominant tectonic feature at present is east–west extension, with the formation of normal faults in a roughly north–south direction (Molnar and Tapponnier, 1978). The same situation prevailed in the Cenozoic (Ni and York, 1978).

According to Molnar and Tapponnier, Tibet is underlain by weak material (in contrast to the double-crust idea mentioned earlier) and the hydrostatic head that causes the high altitude of the plateau appears to be maintained by pressure applied by India to the rest of Eurasia. Tibet transmits this pressure to the regions to the north and east, and acts as 'the pressure gauge of Asia'.

Ni and York explain the extensional tectonics as resulting from the relative eastward motion of the Tibetan Plateau as it is wedged from between the converging Indian Plate and the stable Tarim Basin (Fig. 20.5). Beyond the Himalayas there is little sign of the compression that might be associated with the India–Asia collision.

The tectonics of northeast China is dominated by strike-slip and normal faulting, with right-lateral motion on north-northeast trending planes, left-lateral motion on west-northwest trending planes, and extension in approximately a northwest direction. Southeast China, in contrast, is relatively stable, Molnar and Tapponnier (1978) interpret these faults as a result of the India–Eurasia collision (Fig. 20.6). No simple plate boundaries can be recognized in northeast China.

Fig. 20.6 Distribution of tectonic styles in Asia.
Dark area = region of crustal thickening
Stipple = region of major strike-slip faulting
Diagonal shading = regions of normal faulting and crustal extension
Unshaded = regions of little deformation.

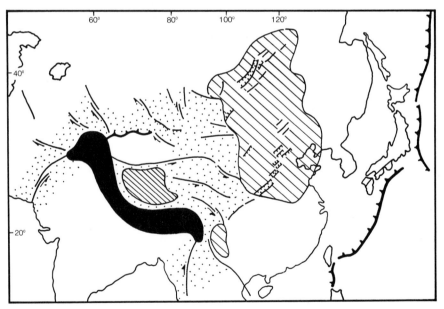

The Eastern Highlands of Australia

The Eastern Highlands of Australia consist of Palaeozoic rocks that have been planed down to a succession of erosion surfaces, and bowed up in a series of swells. The flat country of central Australia gradually rises to the Highlands, arches over and descends towards the eastern seaboard, but erosion from the sea has cut back considerably so scarps or ranges face the sea. Uplift was associated with faulting, minor for the most part, but occasionally defining distinct mountain ranges.

To the east the Tasman Sea is an extinct spreading site, and there is no indication of any trench, Benioff zone, or other indication of subduction, or indeed of any pressure from the sea. Further north the Coral Sea is an extinct sphenochasm where spreading has long since ceased, and again there is no evidence of subduction or pressure. Palaeomagnetic evidence and the evidence of hot spots (p. 109) indicate that Australia has moved in a direction about 10° east of north over the past 50 m.y. or thereabouts, which is roughly parallel to much of the axis of the Eastern Highlands and nowhere perpendicular to it, so the drift of the continental plate is not related to the uplift of the Highlands. Furthermore, the dating of the basalts that are related to the uplift indicates that the sinuous divide was already in existence on the drifting plate and at least some parts were already considerably uplifted, so the northern movement had nothing to do with the creation of the divide or the Eastern Highlands.

Neither the regional tectonic setting nor the details of geology and geomorphology would suggest that the Eastern Highlands are caused by compression but the combination of tension, vulcanicity and uplift, with occasional downwarped areas, marks the Eastern Highlands as a region of vertical tectonics (Ollier, 1978).

Australia appears to be surrounded by spreading sites on all sides, except the north. In this it is very like Africa, and the similarity may go further. Along the eastern side of Africa there is a chain of swells, commonly marked by rift valleys. Perhaps some of the higher parts of the Eastern Highlands are equivalent structurally to the swells of Africa, with the geocols equivalent to gaps between swells. The African swells commonly have a rise towards the rift valleys with their associated volcanism. The rift valley–swell landscape is attributed to vertical tectonics (Le Bas 1971).

Petrologically the rift valley volcanics differ from the eastern Australian basalts so we cannot pursue the analogy too far, but the map of swells in Africa is certainly reminiscent of the Eastern Highlands situation. The African rift valleys are tectonically continuous with the Red Sea, which is in turn continuous with sub-oceanic ridges: the rift valleys may be the initiation of spreading sites. If the Australian–Africa analogy has any validity, the Eastern Highlands may reflect a line of future splitting of the continent.

Despite the strong geological and geomorphic evidence for an extensional situation following doming, the situation is not completely understood. Measurements of residual stress in rock of southeast Australia show a considerable stress in an east–west direction in most places, which would be consistent with compression, though the source and mechanism of such compression remains unknown (Denham, Alexander and Worotnicki, 1979).

The Andes

As Gansser has pointed out (1973), plate-tectonic theories that use the Andes as a model adopt simplified assumptions that neglect the fact that only the recent morphogenic uplift made the apparently uniform Andes, masking a very complicated geological history. The Andes as a marginal chain are influenced by the shields to the east, which display resurgent tectonics by remobilization along old fracture zones, and by the Pacific plates in the west, which Gansser thinks more complicated than generally supposed.

The sub-Recent and Recent volcanic belts in the Andes are

approximately 250 km inland from the marginal oceanic trench, a figure that is remarkably constant except in middle Colombia. The rhyolitic-dacitic volcanism can have nothing to do with the melting of a downgoing ocean slab, but originates from continental material.

The mostly Mesozoic plutons constitute the largest mass of plutons on earth, covering about 465 000 km², about 15 per cent of the Andes surface. Their alignment parallel to the coast (Fig. 20.7) must surely be genetic, and Gansser thinks it relates to the 'birth' of the Pacific in Jurassic–Cretaceous time.

Fig. 20.7 The distribution of Mesozoic and Tertiary batholiths and young volcanics in the Andes (after Gansser, 1973).

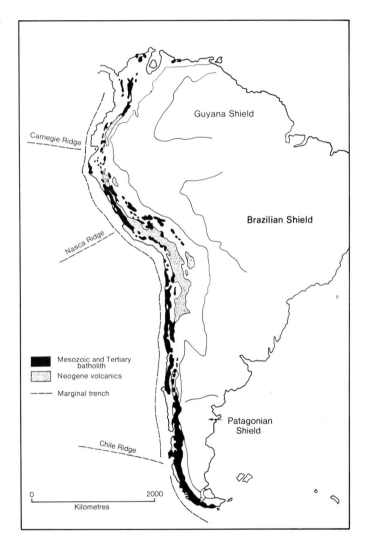

The plutons made their way upwards over 70 m.y. and it is remarkable that the orientation was maintained over all this time.

In several ways the Andes do not conform to the plate-tectonics models that they allegedly illustrate. In particular there is no evidence of compression along the ocean–continent interface. In the coastal belt block faulting is the most important tectonic process. Sediments in the Peru–Chile trenches are not compressed and show horizontal bedding. Fractures and graben in the Cocos plate suggest extension, and there is no indication that oceanic plates are being pushed from the widening oceanic ridges towards a subduction zone.

Even Cobbing, who believes the Andes represent a 'convergent regime' writes (1978): 'It is salutary to reflect that, in the area described [the Peruvian Andes], there is no indication from the sediments themselves that they were deposited at a convergent plate boundary. Only the volcanics of the eugeosyncline do this.'

According to Cobbing (1978) the Andean chain differs from Alpine chains: Andean chains are characterized by andesite volcanoes and tonalite batholiths. Alpine chains are characterized by ophiolite belts and sedimentary flysch. The difference probably reflects their tectonic setting, the Andes being a convergent regime between a continent and a large ocean, the Alpine chains have evolved under a regime involving the opening and closing of small seas.

A relatively narrow band of Coast Ranges, consisting mainly of Precambrian gneisses and Palaeozoic rocks, is separated from the Andes by the Para–Andean trough which is largely full of Tertiary sediments. The Andes are divided into the Eastern and Western Andes. The Western Andes is the most continuous, consisting of folded and faulted Palaeozoic and Mesozoic rocks, with intrusions and volcanics of Mesozoic and Cenozoic age. Several tectonic basins lie between the East and West Andes, including the Lake Titicaca basin. They are formed by normal faulting, and are sediment filled – not erosion surfaces. The eastern slope of the Andes drops abruptly to the Amazon lowlands, and several great tributaries of the Amazon have cut through the Andean front.

An east–west section of the Peruvian Andes reveals a series of blocks (Myers, 1975). From the coast these are the Paracas Block, the Paramonga Block, the Chavin Block, the Maranon Block, the East Peruvian Trough Block and the Brazilian shield (Fig. 20.8) The Maranon Block is a geoanticline or horst separating two belts of subsidence that were the West and East Peruvian troughs. The Paramonga Block is the site of the Coastal Batholith. Folding of relatively surficial sediments was related to essentially vertical movement of the major blocks, which were long, ribbon-like belts parallel to the coast. The folding appears to be of décollement type, with open folds and no large recumbent folds or nappes, and no ophiolites or mélanges.

In Bolivia the western Cordillera is separated from the eastern, here called the Cordillera Real, by the Quito–Cuenca graben. This depression is flanked on both sides by volcanoes – the 'Avenue of Volcanoes' of von Humboldt.

In Colombia the Andes fan out into three ranges, the Western, Central and Eastern, with tectonic lowland valleys between. Active volcanism is limited to a narrow strip of the Central Cordillera.

In the Andes crustal shortening during Mesozoic and Tertiary times was clearly trivial and could not account for the thickening of the crust (170 m thick), which must therefore be brought about by accretion from beneath. This may conceivably have been from a downgoing slab (Shackleton, discussion after Pitcher, 1977). Katz (1971) on the other hand presents evidence for considerable extension in the Andes. In this case the plate-tectonic model must be changed from one of compression to one of tension, as in the tensional models for back-arc basins, or for the extension of western USA (Elston, 1978).

Since Middle Miocene time the Andes of central and northern Peru

Fig. 20.8 (a) Simplified geologic and tectonic map of Peru.
1, Oceanic crust; 2, Paracas Geanticline; 3 and 4, West Peruvian Trough (3, Paramonga Block; 4, Chavin Block); 5, Maranon Geanticline; 6, East Peruvian Trough; 7, Brazilian shield. AB = line of section.
(b) Section AB of Fig. 20.8(a) across Peru 100 m.y. ago showing subduction of ocean floor and relative movement of blocks (after Myers, 1975).

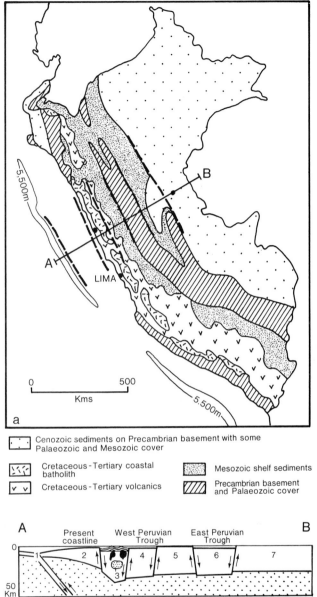

Cenozoic sediments on Precambrian basement with some Palaeozoic and Mesozoic cover

Cretaceous-Tertiary coastal batholith

Cretaceous-Tertiary volcanics

Mesozoic shelf sediments

Precambrian basement and Palaeozoic cover

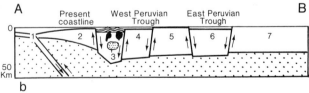

have undergone several distinct stages of uplift. A major erosion surface, the Puna surface, closely post-dates middle Miocene deformation. In some places the Puna was deeply dissected before deposition of volcanics 10 m.y. ago. Renewed uplift led to the canyon erosion stages, dated in northern Peru at 5–6 m.y. ago.

The succession of erosion surfaces preserved on the western flank of the Andes in Peru indicates sporadic uplift of a gently undulating landscape formed by erosion of the Coastal Batholith and its volcanic envelope (Myers, 1975). The planation surfaces increase in altitude with increasing age and distance from the western edge of the Paramonga Block. There is also an increase in the slope of the erosion surfaces with age, indicating uplift by arching of the Chavin Block and westward tilting of the Paramonga Block. The uplift and erosion took place between the emplacement of the last plutons of the Coastal Batholith, about 30 m.y. ago, and the eruption of an ignimbrite about 6 m.y. ago. In addition the Chavin Block has been split by faults that were intermittently active over the last 6 m.y.

Where extensive the high plains are known as the Altiplano, and they lie between the east and west Cordilleras. The western Cordillera in Bolivia is like the eroded edge of the Altiplano, complicated by a line of young volcanoes running along it. The main uplift in the Bolivian Andes appears to be Pliocene and Pleistocene.

Eastern North America

The Coastal Plains of the USA make a physiographic province which is tectonically significant as a wedge of sediment stretching from an inland margin where coast plain sediments lie unconformably on a variety of older rocks to the seaward and Gulf side where the sediments may be traced out under the continental shelves. The oldest rocks of the Coastal Plain sediments are of Cretaceous age. Along most of the margin these rest on Palaeozoic rocks, and the unconformity between them represents anything from 25 million to over a billion years.

The major effects of Palaeozoic folding were over in the Triassic, and no Jurassic sediments are found along the Atlantic seaboard, even at depth, though some shallow sea deposits with salt and gypsum represent the Jurassic in some Gulf Coast rocks.

Appalachians

Cooper (1968), quoted in Leet and Judson (1971, p. 467) found that sedimentation and deformation were going on at the same time within the Appalachian geosynclines. The structural features were produced while the sediments were being deposited and not late in the geosyncline's history. In other words, deformation was not confined to a climax at the close of the Palaeozoic. He concludes that the dominant forces involved in creating the folds and faults of the Appalachians were vertical.

It seems that all previous structures were planed off in the Jurassic. Deposition started again in the Cretaceous when the sea advanced extensively into North America.

The Appalachian region, which might once have had high relief, had been planed down to a subdued surface by Cretaceous times. But since the present-day mountains are not related to folding in the Appalachian 'orogeny' there is no need to postulate any direct association of folding and mountain building at all. It seems more probable, as Hall maintained in his early account of the Appalachian area, that folding took place during the filling of the Palaeozoic geosyncline and mountain uplift occurred much later. Any uplift that preceded the Cretaceous transgression was not necessarily on a mountain-building scale.

As sediment accumulated on the plain there was minor movement, warping, shifting coastlines, but no folding of significance. In general the seaward side sank while the land side rose, culminating in the Appalachian highlands or plateau. Accumulations of geosynclinal proportions developed from Newfoundland to Honduras, and the zone of maximum deposition has tended to shift seawards ever since the Cretaceous.

Some students of Appalachian geomorphology have suggested that the Cretaceous sea actually covered a large part of the region which is now the Appalachian Highlands, citing the apparent summit accordance.

Johnson (1931) postulated that the area was covered by a Cretaceous sea, and a series of consequent streams developed when it withdrew. The lack of any remnants of a Cretaceous cover makes this idea questionable. It seems equally if not more probable that the area was a Cretaceous peneplain bounding the Cretaceous transgression to the east.

Meyerhoff and Olmsted (1963) go back in time, and believe that the present drainage lines are direct descendents of Permian streams. They stress the supposed coincidence between present stream courses through ridges and structural sags and fault zones, a situation unlikely if the drainage lines are superimposed from a Cretaceous cover. Thornbury (1969, p. 230) suggests that the drainage may have been along present lines since Triassic times.

With more modern knowledge of the opening of the Atlantic this seems quite reasonable, for the streams would have been draining the Appalachian region to the line of rifts and lakes that were a Triassic precursor to the opening of the Atlantic.

Further south the plain around the Gulf of Mexico shows progressively younger sediments towards the coast. The sea transgressed in Lower Tertiary times, and since then the history has been one of marine regression. The Gulf Coast has an east–west axis, parallel to the edge of the continental shelf along the north side of the Gulf. In general, sediments change towards the axis from non-marine to marine. Deep subsidence has taken place between the coast and the continental shelf and Cenozoic sediments reach 14 000 m, but there is no associated upwarp of the land as in the Appalachians.

Western North America

The Colorado Plateau was at sea level at the end of the Cretaceous. It has since been uplifted about 2 000 m with a gentle tilt from south to north. The uplift was in two phases: first a small uplift in the Late Miocene, accompanied by much volcanic activity around the borders and intrusions in the interior, followed by a much bigger uplift in the Late Pliocene to Recent, with only minor volcanic activity.

The Sierra Nevada is a tilted fault-block mountain range, about 100 km by 600 km, high on the eastern fault scarp edge, sloping gently to the west. It consists largely of a batholith (with plutons of three different ages) intruded into Palaeozoic and Early Mesozoic rock. It was uplifted and eroded in the Late Mesozoic, and at a standstill in Eocene and Oligocene times when it was fairly low. The uplift and tilting occurred in the Late Miocene and Pliocene.

In the Late Cenozoic the Sierra Nevada was uplifted to twice the height of the Appalachian Range, although crustal thickness beneath the two ranges is the same. The seismically-observed thickened crust or root beneath the Sierra was most likely formed during Mesozoic times, yet the present range has been uplifted about 1 800 m within the past 10 m.y. (Christensen, 1966). Crough and Thompson (1977) suggest that uplift is related to the passage of a subducted slab under the ranges. The Mendocino triple junction is at present in line with Mt Lassen, the southernmost volcano in the Cascades, but 10 m.y. ago it was in line with Mt Whitney in the southern Sierra Nevada. The southern Sierra Nevada have risen the most, presumably because they have been rising for a longer time.

Axelrod (1962) used biogeographical evidence to determine the time of uplift of the Sierra Nevada. At present the Owens Valley, east of the Sierra Nevada, has a very different climate from the lowlands to the west of the range, a climate modified by the existence of the high mountains between. In the Early Pleistocene a similar vegetation, a pine-fir ecotone, was established right across the whole region, a situation that could only exist if the present climatic barriers were absent, so the major uplift, on this basis, is well into the Pleistocene.

The Rocky Mountains are distinguished by large thrusts, for which at least three mechanisms have been postulated. In southern Canada it is thought that the thrusts result from gravity sliding from a buoyant mass pushed up by an upwelling mass of hot mobile rocks in an orogenic core (Price and Mountjoy, 1971). Each local pulse of thrusting might be related to the emplacement of a tongue of gneiss in the orogenic core.

Western Wyoming is the type area for the pore-pressure hypothesis of Hubbert and Rubey (1959) described on p. 137 with gravity sliding again the visible effect. Smith (1976) has criticized the mechanism: 'Why fluids do not leak out at the rear of gravity slides fast enough to reduce the ratio to hydrostatic values before significant thrusting has occurred is unclear'. However, if gravity spreading rather than sliding is invoked (p. 132), this objection disappears.

Further south in Nevada and California the thrusts cut into Precambrian basement and the area did not form a high zone from which gravity slides could have spread. Burchfield and Davis (1972) attribute the structures to internal stress, possibly related to a subduction zone to the southwest.

The Basin and Range Province lies between the Sierra Nevada to the west and the Colorado Plateau and Rockies to the east.

During Mesozoic and Early Tertiary time the region was uplifted and eroded, providing vast amounts of sediments to the Rocky Mountains and Coast Range geosynclines. In Late Tertiary time and Quaternary times the crust was distended, with collapse into fault blocks, strike-slip movement, and volcanic activity. The crust broke into blocks, forming the basins and ranges, with the fault-block mountains being bounded by high-angle faults dipping from 45° to 70°, and the block surfaces are tilted. The faults are curved and flatten out at depth.

The Basin and Range Province has been the site of extensional orogeny for the past 40 m.y., according to Elston (1978). It is geologically complex, resulting from a three-part history: first an Andes-type volcanic arc, followed by a spreading ensialic back-arc basin, which was in turn succeeded by intraplate block faulting and rifting (Fig. 20.9). The total extension may have exceeded 100 per cent.

The first half of the Basin and Range evolution reached its climax in a great 'ignimbrite flareup' between 35 and 25 m.y. ago, when the province was inundated by 10 million km³ of silicic volcanic rocks. The second half was dominated by extension, faulting and rifting which gives rise to the present topography. The province overlies earlier fold belts and the boundaries between cratons, eugeosynclines and

Fig. 20.9 Proposed stages of the tectonic evolution and expansion of the Basin and Range Province over the past 40 m.y. The sections run from about the northern end of Baja California to north-central New Mexico. (a) Andean arc stage. Granodiorite forms at depth and andesite volcanoes erupt at the surface; (b) Late modified Andean arc stage. East Pacific Rise is near the trench and hot oceanic lithosphere is being subducted. Back-arc spreading makes room for plutons, and associated ash flow eruptions and calderas; (c) Main back arc extension. Subduction has ceased and plate is free to expand. Many plutons and volcanic eruptions (basaltic andesite); (d) Intraplate block faulting. Deep faults tap basaltic magma. Diapirs and magma bodies continue to rise (after Elston, 1978).

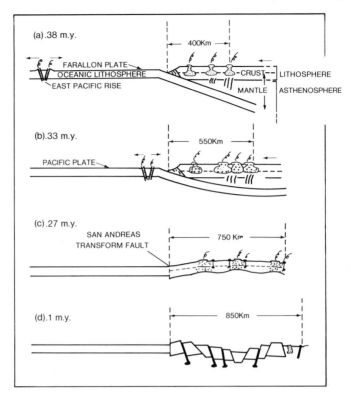

miogeosynclines, but seems to be quite unrelated to them. There was a quiet period between the Laramide orogeny (Late Cretaceous to Early Tertiary) and the Middle Tertiary events.

Elston's estimate of extension, over 100 per cent, is much greater than that of most other workers who have extension of about 5 or 10 per cent. The low estimates are based only on the geometry of Basin and Range faults, but Elston also considers the volume occupied by plutons and thinning of the lithosphere during extension. Plutons are thought to underlie at least one-third of the Basin and Range Province, and the crust has thinned from the average 40 km or more under the Sierra Nevada, Colorado Plateau and Great Plains, to 25–35 km under the Great Basin.

The extension of western United States is also brought out by Fig. 20.10 which shows the trend of extension, based on all normal faults up to 16 km long. The arrows are drawn at right angles to the faults and show the direction of extension but not the amount, which must be determined from other geological considerations (Gilluly, 1970).

The highest mountains in the Basin and Range Province are great volcanic plateaus, such as the San Juan Mountains and the Mogollon Plateau. They may have been raised to their present height (up to 4 500 m) by the buoyancy of underlying batholiths during regional extension.

In the Basin and Range Province of southwestern Arizona external drainage systems replace earlier internal drainage at some time between 10.5 and 6 m.y. ago, and since then progressive erosion has created the present landscape.

Fig. 20.10   Average trend of extension
of all normal fault segments up to 10 miles
(16 km) long, within each square degree
of the Basin and Range province (after
Gilluly, 1970).

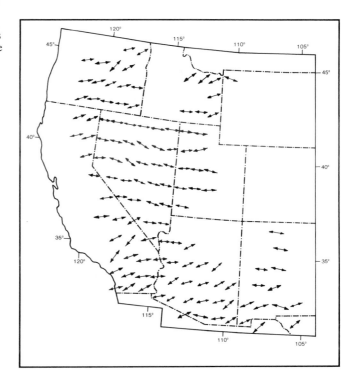

21 Tectonics and geomorphic theory

This review has shown that the relationship between geomorphology and tectonics has been little used to date although there is a lot of scope for interaction at this important interface between disciplines.

Geomorphology has been partly blinkered to tectonic relationships by ruling theories that consider that most landforms are very young. In his excellent and widely used textbook Thornbury (1969) could state as Fundamental Concept number 7, 'Little of the earth's topography is older than Tertiary and most of it no older than Pleistocene'. Ashley (1931) made the point even stronger when he wrote that 'most of the world's scenery, its mountains, valleys, shores, lakes, rivers, waterfalls, cliffs and canyons are post-Miocene, that nearly all details have been carved since the emergence of man, and that few if any land surfaces today have any close relation to pre-Miocene surfaces'. Although these references are rather old, the view they express is still commonly held amongst geomorphologists, and of course anyone holding such opinions is not likely to be concerned with major tectonic processes. Generally such views derive from Europe and North America where Quaternary glaciation gave a fresh start to much landscape formation, provided a basis for correlation, and created many spectacular landforms to study. Since most of the world's textbooks come from this region the Pleistocene bias had a world-wide influence, and although exceptional people like Lester King (1962) provided geomorphology books with a much greater time scale, they had surprisingly little effect on mainstream geomorpholoy in the northern hemisphere.

For the past three decades much of geomorphology has been concerned with process studies and with systems theory (see, for instance, Chorley and Kennedy, 1971) and even reached a pitch where historical studies of landscape were derided as old fashioned.

Chorley (1965) could write: 'The historical bias . . . has meant that landforms have been viewed in much the same manner as the light from a distant star, in which what is perceived is merely a reflection of happenings of past history . . . much of the former dominance claimed for denudation chronology in the field of geomorphology has now vanished'. Thus it was that when most branches of earth science were excited by the 'new geology' that accompanied ideas of sea-floor spreading and plate tectonics, geomorphology remained aloof. The revolution passed by.

It is now clear that an understanding of the new geology and of tectonics is essential to understand landforms, and not only first-order landforms. The lessons that King, Hills and others in the southern hemisphere tried to put across – that geomorphology could not be isolated from the rest of geological history and process – is now going back to Europe and North America, and there is an increasing concern with the older landscapes. In Europe, for instance, Büdel (1977) 'is convinced that the landscapes of today are pre-Pleistocene, little changed in the Holocene. The long post-Cretaceous period was warm and moist everywhere up to the ice age. The warm and moist areas (the tropical zone with summer rain) are zones of dominant planation. All landscapes are primarily of this type.' (Davis, 1978).

Nevertheless the change in outlook is only partial, and most textbooks still stick to the old theories of geomorphology, theories that should be reconsidered in the light of modern knowledge of tectonics, and the long time-scale of landscape evolution.

Application of geomorphology to tectonics

The tardiness with which geomorphologists have accepted tectonic ideas into their concepts is matched by the failure of most tectonicians to incorporate geomorphic data into their schemes, but the tectonicians have a better excuse.

The tectonicians, it seems, can cheerfully subduct a trench or throw up a mountain range without considering the effects on the landforms at the ground surface. Is the ground surface of an appropriate age? Have the rivers been diverted? Does the amount of uplift seem to be of the right order for the proposed mechanism? Are proposed strike-slip movements indicated also by displaced rivers? Occasionally a tectonic theory will use geomorphic evidence in a quite erroneous way. Brock, for instance, has the following idea for the formation of Lake Victoria: 'Lake Victoria is an excellent example of an equidimensional block which presumably has dropped. From its shores are radiating sutures . . .'. In fact Lake Victoria is an equidimensional lake formed by back-tilting in an area of dendritic drainage, the radiating sutures are ancient drainage lines with remarkably little structural control, and the geomorphic evidence is very clear, and had been deciphered many years before Brock's structural conclusions.

In any discussion of continental margins (such as the papers edited by Burk and Drake, 1974) there is usually a mass of data on geophysics, sedimentology, and offshore drilling, but what might be an equally valuable input – the evidence for the formation, history, structure and tectonics of the land area – is sadly missing. To understand the earth we shall need both sides of the story – continental and marine. To play a part in the problems of geology over the next few decades geomorphologists must forget their trivial catchments and see megaforests instead of trees.

The tectonician has a better excuse than the geomorphologist for not using the other's data or being cognizant with his ideas. Whatever theory a tectonician may prefer, it should be related to a certain amount of factual geological and geophysical data, and this data – seismic survey, gravity anomalies, heat-flow measurement or whatever – may be used by other people to test rival hypotheses.

In contrast, much geomorphology is concerned with a few aspects of a landscape that may support a particular interpretation, but the geomorphologist may not present a sufficient body of data to enable other people to draw their own conclusions.

For this gap to be filled, there is a need for a great deal more geomorphic mapping and descriptive factual geomorphology. This may be detailed, like the geomorphic maps of the French (see, for instance, Joly, 1963) or they may be more general. Terrain classification can be a quick way of producing a reconnaissance geomorphic map, and may be of value. Field mapping was used to solve tectonic-geomorphic problems by Doornkamp in southwest Uganda (1970), and the Japanese prepare detailed maps for the study of neotectonics, as well as for engineering purposes (see, for instance Kaizuka *et al.*, 1973).

There are still problems in deciphering geomorphic history in many parts of the world, mainly because of a lack of datable materials, but even relative dating puts some constraints on tectonic hypotheses, and eventually physical or stratigraphic methods will put firmer limits still.

Geomorphology has historically taken its tectonic basis from geologists with little question and little concern because it was generally thought that most geomorphology was too young compared with the major tectonic features of the world. Now we know that geomorphology is on the same time scale as continental drift, plate tectonics and biological evolution, and the science has an important part to play in the deciphering of earth history – it is not concerned merely with a little bit of sculpturing on the top of the geological column. Geomorphology can now put primary data into tectonic theories.

Paradigms of geomorphology

Some prevalent attitudes and some specific hypothesis and paradigms of geomorphology will now be reviewed, with an emphasis on their application in regions of long geomorphic history and tectonic activity.

Active process studies

It is clear that whatever processes may be acting on the ground surface at present, the gross features of many landscapes were formed long ago, under conditions very different from those of today. Process studies are interesting in their own right and tell us what is happening at present, which gives a valuable insight into possible rates of landscape evolution. They provide a basis for speculation, but they cannot be extrapolated to tell us what happened in the past, especially as we seem to be living in a time of unusually varied climate and high sea level.

Climatic geomorphology

When much of the landscape is of an age measured in geological periods the landscapes have inevitably experienced a variety of climates, but the mark these climates leave is not overwhelming and is often negligible. Certainly the present-day climates have little to do with major features of old landscapes. Throughout the Mesozoic the climate of the world was warm and temperate. Cooling and increasing variability set in at some time in the Upper Cenozoic. Some landforms are good climatic indicators, such as glacial landforms, desert dunes and coral reefs. Others, such as inselbergs and tafoni, have been associated with various climatic parameters but seem to have failed as real indicators, and common fluvial landscapes have so far not provided good data for climatic interpretations. There is no doubt that climate affects landform development, but it does not provide a good basis for a general theory of landscape evolution.

Dynamic equilibrium

Although it is commonly traced back to Gilbert (1877) this theory in its modern form was proposed by Hack (1960).

It is assumed that within a single erosional system all elements of topography are mutually adjusted so that they are downwasting at the same rate. The forms and processes are in a steady state of balance, and may be considered as time-independent. Equilibrium is achieved when all the slopes in a drainage basin are mutually adjusted to a common erosion rate. So long as uplift is maintained and the rivers do not reach base level the landscape can maintain the same form indefinitely, even though erosion is continuous.

It is necessary here to review the Davisian cycle to see just how the dynamic-equilibrium concept differs.

Davis assumed his landscape started development from an uplifted peneplain, and in the early stages considerable remnants of this plain were preserved. Davis called this a young landscape. As the rivers cut down the valleys may become wider until the valley sides intersect at sharp interfluves. This sort of landscape, with angular ridges and narrow valleys, was one of several landscape types that Davis called 'mature'. It has also been called 'ridge and ravine' topography, and by other terms such as 'angular and 'feral'. When the rivers have cut down to base level vertical erosion is no longer possible, but whatever processes act on the valley sides are still active, so the interfluves will be reduced, and eventually a new peneplain is formed near the new base level. Davis called this an 'old age' landscape.

The use of the anthropomorphic terms 'youth', 'maturity' and 'old age' leads to much confusion, and it is possible to describe landscapes without these stage names (Ollier, 1967). Landscapes combine valley profiles with interfluve profiles, and each of these can be angular, rounded or flat. Combining these gives the nine types shown in Fig. 21.1, which together with a simply flat landscape (plain or plateau) gives ten possible basic landscapes.

Fig. 21.1 Classification of landscape profiles based on interfluve profiles and valley profiles (after Ollier, 1967)

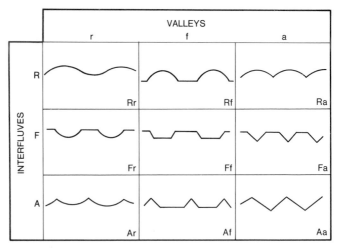

Now the dynamic-equilibrium theory only applies to the middle stage of landscape evolution with graded slopes. Hack denies that remnants of old erosion surfaces are present (although Bretz demonstrated one in Hack's own type area, see p. 155), so the first part of the cycle is denied. Nor can development of plains at a new base level be incorporated into dynamic equilibrium, and when base level is reached the landforms must change. This point is accepted by Hack himself, who writes that 'If the [rates of uplift and erosion] change, however, then the state of balance or equilibrium constant must change. The topography then undergoes an evolution from one form to another. Such an evolution might occur if diastrophic forces ceased to exert their influence, in which case the relief would gradually lower'.

To put the matter simply, dynamic equilibrium does not apply to landscapes with flat interfluves or valley floors. On the classification of Fig. 21.1 dynamic equilibrium can apply to only four of the ten possible landscape types (Rr, Ra, Ar and Aa – the last being the commonest in the real world), these four being what would be called 'mature' landscapes in Davisian cyclic terms. These four landscape types cover a significant part of the world and equilibrium considerations have a place there, but many of the world's landscapes cannot be fitted into the scheme and dynamic equilibrium cannot be a general theory of landscape evolution. The whole concept of dynamic equilibrium hinges on quite unwarranted assumptions about rates of uplift. 'Nevertheless as long as diastrophic forces operate gradually enough so that a balance can be maintained by erosive processes, then the topography will remain in a state of balance even though it may be evolving from one form to another. If, however, sudden diastrophic movements occur, relict landforms may be preserved in the topography until a new steady state is achieved.' Schumm and Lichty (1965) suggest that the difference between steady-state attitudes and cyclic attitudes is a matter of scale – the size of the landscape involved and the amount of time considered, and suggest that depending on one's viewpoint the landform is one stage in a cycle of erosion or a feature in dynamic equilibrium with the forces operative. These views are not mutually exclusive. It is just that the more specific we become the shorter is the time span with which we deal and the smaller is the space we can consider. They suggest that landform evolution can be considered during three time spans of different duration: cyclic, graded and steady. The cyclic time span encompasses a major period of geologic time, perhaps involving an erosion cycle. Over this long period the system will change because of significant removal of material by erosion, and possibly by tectonic movements. Schumm (1977) writes of cyclic time that 'when viewed from this perspective a fluvial system is undergoing continual change (dynamic equilibrium)', unfortunately compounding dynamic equilibrium with cyclic time, when the cycle concept is incompatible with dynamic equilibrium, at least as expressed by Hack. Graded time refers to a short period of cyclic time when any slight progressive change in landforms is masked by fluctuations about the average values. During a steady time span a static equilibrium may exist and the landforms are truly time-independent because they do not change.

These divisions may be helpful for the dilemma Hack finds when rates of uplift and erosion differ considerably, but they do not help the geologist or geomorphologist who is trying to discover actual relationships between landforms and time rather than philosophical ones. If he finds an Eocene valley, or a plateau that has remained undissected since Eocene times, he has an inherited landform and must interpret it, and the surrounding landscape in accordance with deduced events in landscape history, not as some variation on the equilibrium theme. On the long time span appropriate to tectonic geomorphology only cyclic time seems to be relevant.

Kennedy's models

Theoretical accounts of the relationships between tectonic uplift and landscape genesis have generally been of limited value, though there must be a theoretical background, either explicit or implicit if observations of landscape are to be used, somehow, to draw tectonic conclusions.

Penck (1924) believed that the form of valley sides reflected rates of uplift, with convex slopes indicating increasing rates of uplift and concave slopes a decreasing rate of uplift. Nobody believes this now. Davis assumed uplift occurred as a relatively quick event separated by long periods of erosion and peneplanation, and did not attempt detailed investigation of the relationship between uplift and erosion. Ahnert (1970) has calculated rates of uplift and erosion, with simple isostatic compensation, and drawn conclusions on the feasibility of attaining planation, and Bond (see Ch.16) has discussed the interrelationships of erosion, sedimentation and isostasy, relating the general uplift of continents but not discussing details of landforms.

Kennedy (1962) built a theoretical system on the three factors of tectonic movement (essentially uplift), erosion (by which he meant vertical erosion in river channels), and denudation (by which he meant the lowering of the total land surface between rivers). These factors can be combined in nine ways, as follows:

Case 1. (Rate of uplift > Rate of erosion > Rate of denudation.) Increasing relief developed at high altitude and without relationship to regional base-level ... consequent and reversed drainage.

Case 2. (Rate of uplift > Rate of erosion = Rate of denudation.) Static relief carried up and maintained at high level ... consequent and reversed drainage.

Case 3. (Rate of uplift > Rate of erosion < Rate of denudation.) Decreasing relief developed at high level, and leading eventually to the cutting of a high-level erosion-surface unrelated to regional base-level ... consequent and reversed drainage.

Case 4. (Rate of uplift = Rate of erosion > Rate of denudation.) Increasing relief, unrelated to regional base-level, but directly related to the levels of the pre-existing valley floors ... main drainage antecedent.

Case 5. (Rate of uplift = Rate of erosion = Rate of denudation.) Static relief, unrelated to regional base-level, but maintained at level of pre-existing valley floors ... main drainage antecedent.

Case 6. (Rate of uplift = Rate of erosion < Rate of denudation.) Decreasing relief resulting, eventually, in cutting of an (inclined) erosion-surface unrelated to regional base-level, but directly related to the level of pre-existing valley floors ... main drainage antecedent.

Case 7. (Rate of uplift < Rate of erosion > Rate of denudation.) Increasing relief adjusted, eventually, to regional base-level ... main drainage antecedent.

Case 8. (Rate of uplift < Rate of erosion = Rate of denudation.)

Static relief maintained indefinitely at base-level of erosion . . . main drainage antecedent.

Case 9. (Rate of uplift < Rate of erosion < Rate of denudation.) Decreasing relief with eventual cutting of a peneplain or erosion-surface, perfectly adjusted to regional base-level . . . main drainage antecedent.

The various combinations are shown diagrammatically in Fig. 21.2. From this analysis Kennedy draws some remarkable conclusions:

Fig. 21.2 Diagram to illustrate relative volumes of rock removed under different rate-ratios of uplift-erosion and erosion-denudation. The amount of uplift is the same in all cases. X, original form and final position of initial surface; Z, form and position of erosion surface developed; stippled areas = area (volume) of rock removed (after Kennedy, 1962).

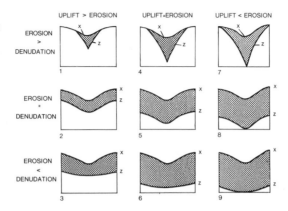

1. Peneplanation may take place during active tectonic uplift and does not necessitate static structural conditions.
2. Peneplains may not be related to regional base level, and may develop at high, intermediate or low levels.
3. An elevated peneplanated surface does not necessarily imply subsequent tectonic uplift.

Many objections may be made to this scheme, some of which have been pointed out by Crickmay (1972). The scheme involves major assumptions about the geomorphic processes of erosion and 'denudation'. It seems to assume a sort of 'dynamic equilibrium' so that all parts of a landscape are affected by denudation, and does not permit the observed fact of preserved ancient landscapes. The arguments might be appropriate to individual valleys to some extent, but ideas of general and local base level are so simplified that it seems improbable it could be applied to whole landscapes. Nevertheless Kennedy's scheme does provide some sort of model for relating tectonics to erosional geomorphology, and some sort of improved scheme along these lines, making more use of observational knowledge as well as theory, would be very welcome. We need a tectonic-landscape model which is not, in the words of Crickmay (1972) 'almost perfectly insulated from observable realities'.

The probabilistic theory of landscape evolution

This approach was developed by Leopold and Langbein (1962) who wrote:

the distribution of energy in a river system tends toward the most probable state. This principle . . . governs ultimately the paths of movement in the fluvial process and the spatial relations between different parts of the system at one time or stage. . . . The development of the landscape involves not only the total available energy, but its distribution as well, a factor that may appropriately be described as entropy . . .

As the theory is summarized by Christofoletti (1974):

The distribution of energy may be analysed as the probability of occurrence of a determinate distribution in relation to a set of all the possible alternative distributions. Each alternative represents a landscape. In this perspective, all the landscapes are integrant parts of the same stochastic process, and their individual differences are related to the variations of the energy intensity and to mass distribution. All the landscapes are instances of one and the same set; they are alternative states.

As I understand this, landscapes are analogous to poker hands (integrants) dealt by a random (stochastic) process, from a shuffled pack (variations in energy and mass distribution). It does not help me to understand landscapes – I seem to be stuck with the hand I have been dealt, and there are no instructions on how to play it.

The probabilistic theory has been elaborated in models by Scheidegger (1967) and Scheidegger and Langbein (1966), and has found favour with more theoretical geomorphologists such as Culling (1963) since the approach lends itself to mathematical analysis and computer simulation. The approach fails with long-term landscape evolution, and although it may have some success with process studies and those parts of landscapes that really are in dynamic equilibrium it does not deal with those many occasions when the system changes (as shown by field evidence) because of a change in external process or of internal (tectonic) processes.

Cyclical theories

Cyclical theories, whether Davisian or of any other kind such as the pedimentation cycle of King, have limitations in landscape explanation. A first problem arises from the fact that geomorphologists cannot agree on the existence, number or origin of erosion surfaces, and if they cannot agree on the field evidence there is little chance of agreement on the explanation. To some extent the situation arises from people working in different areas: it is much easier to recognize erosion surfaces in Africa or Australia than in much of North America or Europe. Another factor is the fact that for the last 40 years an anti-Davisian outlook pervading geomorphology has prevented or dissuaded people from examining field evidence in cyclic terms, just when field methods and dating methods were making a new look possible.

More fundamentally the existence of multiple planation surfaces indicates a succession of planation events, but as yet we are not sure that these correlate from place to place and to what extent they can be regarded as evidence of cycles rather than merely a succession of different events. In the Quaternary the world was affected by a succession of climatic changes, with corresponding changes in base level, glaciation, groundwater levels, run-off, dune mobility and erosion rates that have complicated many landscapes. There may be a vague cyclicity in these events, but not on the scale of Davisian cycles: some of them may be what Davis called 'climatic accidents.'

To find Davisian cycles we must look on a longer time-scale – to Mesozoic and Tertiary times. Since the Mesozoic there have been some world-wide changes of sea level (Ch. 16) and such changes should be conducive to rejuvenation and initiation of new cycles of erosion. Some sort of cyclical landscape formation took place during the Tertiary, but the situation is complicated by Tertiary tectonic movement, and by the fact that some landforms even pre-date these

Tertiary changes. There is little indication of cyclicity in pre-Tertiary geomorphology and geology. The Permo–Carboniferous glaciation was a unique event in the Gondwana landscape, as was the Cretaceous flood, and our ideas about pre-Permian geomorphology are highly speculative.

One of the major events that should initiate a new cycle of erosion is the formation of a new coastline by the splitting apart of continents. The opening of the Atlantic is now fairly well documented. It began as a rift valley in Triassic times and gradually widened, with the sea progressively entering, creating the salt deposits that border the Atlantic in the early stages, and eventually creating a continuous sea or ocean.

The splitting around Australia was progressive, and is recorded in the basins of deposition that border the continent: in Western Australia the Bonaparte Gulf and Canning Basins have lain at the continental margin facing an open ocean for 600 m.y., whereas the Perth Basin lay in the interior of Gondwanaland until it split apart in the Cretaceous. The Tasman Sea opened off eastern Australia about 80 m.y. ago, and the Eucla Basin originated with the separation of Australia and Antarctica in the Eocene. The disruption of the continent did indeed lead to the formation of new lowlands, like the Eastern Lowlands, separated from the upwarped plateaus of the Eastern Highlands by steep erosional escarpments. But the break-up was not all at the same time, we cannot distinguish distinct different peneplains associated with particular breaks, and the time scale involved seems to be greater than that envisaged by Davis.

As described in Chapter 11, weathering phenomena may be related to cyclic patterns. One must assume that weathering goes on all the time, but before the break-up of Gondwanaland or Pangaea there was one very perfect planation surface underlain by a great thickness of saprolite, weathered rock in place. After the break-up of the supercontinent and the formation of new continental margins, new surfaces were created. While erosion was mainly stripping the saprolite the formation of new planation surfaces was easy and rapid, and a major surface was formed which is essentially a stripping of old weathered material. The formation of subsequent planation surfaces was much more difficult because they had to cut across hard rock, so younger surfaces tend to be smaller and more irregular than older ones.

Uniformitarianism

The most important principle used to understand remote periods is the principle of uniformitarianism, which is, briefly, that the present is the key to the past. Although a powerful tool, it cannot be applied rigidly because it is quite clear that in some respects at any rate the present and the past are quite different states. For example, before about 250 m.y. ago there was no grassy cover to the open landscape, and processes of erosion were probably much faster than they are today. Other things that were different in the past include palaeolatitude, distance to the sea, climate, altitude (in many places), ocean currents and wind systems, vegetation cover and soil-farming processes. More fundamental variables may have been in day length, gravity, and the earth's radius.

Indeed wherever geomorphic histories are long there seems to be evidence that things were different in the past. Nevertheless uniformitarianism remains our chief guide to the past, but it begins to fail when we deal with large spans of time. If we want to make some use of uniformitarianism we have to relax present-day conditions – first little changes like changes of sea level or vegetation cover, later perhaps major changes such as a reducing atmosphere, a reversal of magnetism, or a reduction in the gravitational constant. The general feeling is that the best solution is the one that requires least relaxation, or the one in which the relaxation seems more plausible in some ill-defined and subjective way. Because people don't agree on what is the most plausible solution, there is still major disagreement on fundamental topics. Because some people feel that *their* subjective choice is better than that of their opponents, such controversies as those over continental drift and the expanding earth can become very personal, vehement and bitter. For geomorphologists especially, the lesson is that we can only discover what the past was like from consistent internal evidence, not from comparison with places that are thought to be somewhat similar today.

Evolutionary geomorphology

Dynamic equilibrium, climatic geomorphology and process studies have all been shown to have limited application to geomorphology wherever geomorphic history is measured in hundreds of millions of years. If we also reject cyclic ideas and even uniformitarianism, what have we left? The answer is evolutionary geomorphology.

By this phrase I do not mean that landforms evolve through a sequence of stages such as youth, maturity, and old age, but that the earth's landscapes as a whole are evolving through time. This concept is perhaps easiest to see in relation to the concept of an evolving earth, a concept brought out in some recent geology books such as Windley (1977).

The whole tectonic and geomorphic system was very different in the distant past. Wynne-Edwards (1977) believes that the Proterozoic was a time of ductile flow rather than brittle fracture and failure characteristic of Phanerozoic time.

The origin of Precambrian orogenic belts located within continents and bounded by older cratons can be examined by palaeomagnetic data (McElhinney and McWilliams, 1977, see Ch. 4). The consistency of the data strongly suggests that the cratons were not previously widely separated and then converged to form orogenic belts. Indeed the palaeomagnetic data precludes such plate-tectonic models with convergence of cratons to explain orogenesis. Precambrian orogeny must have resulted from a different mechanism, and it seems that the plate-tectonics regime of today must have originated at some time between the Upper Precambrian and the Mesozoic. Even the longer estimate does not give time for many splits and collisions, so the geomorphology associated with plate tectonics is fairly novel in the long history of the earth.

The early earth had a reducing atmosphere, and geomorphology would have been very different after the evolution of an oxygen-rich atmosphere. The early earth would have had a greater abundance of radioactive elements, especially those with relatively short half-lives, so generation of heat would have been greater, presumably affecting such things as tectonic and volcanic activity. The oceans evolved through time, and the amount of water falling on the land was probably very different in the past. The nature of landscape evolution would have been very different before the Devonian, when growth of a terrestrial cover of vegetation changed the geomorphic system.

Another important biological change was the spread of grassland in the Cretaceous.

Other great changes to the system may result from a supposed increase in the crustal thickness of cratonic shields. Chapman and Pollack (1977) predict a lithospheric thickness of over 300 km in shield areas, offering an explanation for what they claim is an observed retarded motion of plates with shields. If in the future the lithosphere continues to thicken the shields will become viscous anchors, plate motion will diminish and eventually cease, bringing to an end the plate-tectonic phase of earth evolution.

At present most tectonic activity seems to be related to the edges of the plates of plate tectonics. This includes the Mediterranean–Himalayan belt, but is mainly concerned with the 'active' continental margins around the Pacific. There is some evidence that the Pacific kind of activity was much less significant in the past, and early geosynclines and fold belts were mainly of Mediterranean type (Ziegler, 1977). Presumably many of the geomorphic features associated with active plate boundaries around the Pacific would also have been generally absent before the plate-tectonic regime of post-Jurassic times.

Carey (1976) has suggested a massive deviation from unifor-mitarianism in proposing that the earth has expanded, with creation of most of the ocean floors since the Jurassic. The period of expansion is roughly the same as that of landscape evolution, and if such expansion occurred it must surely have had a large and progressive effect on the evolution of geomorphic processes.

Even on a less controversial basis it is clear that several revolutions have occurred in the geomorphic system, and that landscape-forming processes today are not what they were in the Archaean, the Devonian, or the Cretaceous.

The prevailing tectonic view at present is that continents drift, occasionally collide to make larger land masses, which then split along new cracks from which new continents drift apart again. Island arcs and geosynclines may form at new continental edges, the rocks of which may ultimately be welded to the old continental fragment – a process called cratonization. Some theories suggest that continents are conserved by underthrusting at the edges of continents, others suggest that continents have become thicker as the earth evolved, which would have all sorts of geomorphic side-effects resulting from isostatic rise of the continents.

Whatever may be true in the long term it seems certain that the present era is one of continental drift which started in about Jurassic times, though some spreading sites are younger and some are currently splitting, like the Red Sea rift. Certainly Gondwanaland was a supercontinent before the Jurassic, and much of Laurasia – the northern continental assemblage – was also attached to this vast landmass. Geomorphology on a supercontinent would be very different from that of today, resulting from the existence of vast inland areas at great distance fom the sea, longer rivers, more inland deposition, and very different climatic patterns. The break-up of Gondwanaland would have important effects on each fragment, with rivers having shorter courses to the sea, rejuvenation and increased erosion on new continental edges, and associated tectonic features such as uplift at continental rims or the development of island arcs. The amalgam of fragments of land as in India or Papua New Guinea brings together areas with very different geomorphic histories.

Each fragment of Pangaea would have its own individual and distinct history, with many unique events such as formation of new continental edges, biological isolation and local evolution, changes in latitude and the development of new climatic patterns which would depend on size and shape of the fragment, its latitude and orientation, and the location of inherited or newly-formed highlands, and the effects of newly-formed seas.

Each Pangaea fragment and its history must be seen on the extended time scale that is appropriate for such major topics as continental drift, mountain building, and biological evolution. In this context some of the theories and fashions of geomorphology – process studies, dynamic equilibrium, and even cyclical theories – appear to have limited application and importance. What we see is evolutionary geomorphology, which is part of the story of an evolving earth.

Epilogue

The reader might well find this book unsatisfactory for a number of reasons. We do not finish with a lot of definite answers, but with a lot of questions; we do not find a systematic technique for future investigations, but only hopeful guidelines. A lot of the evidence is unsatisfactory, and much of it has been presented in a partial, biased and inconsistent way. What was fact in one part of the book is refuted elsewhere; what was an assumption in one place is presented as a fact somewhere else. Some features of landforms or tectonics are explained in different ways in different parts of the book.

But this is the state of the art. Facts, theories and speculations are confused to the extent that some geologists now regard their offering as 'geopoetry' but with all their reservations they are often hostile to the geopoetry of others. It is a stimulating time to be involved in tectonics and geomorphology and I only hope that the inadequacies of this book encourage geomorphologists to indulge in the entertaining and illuminating art of landscape study on the grand scale.

Bibliography

Adams, G. F. 1975. Planation Surfaces. *Benchmark Papers in Geology,* **Vol. 22**, Dowden, Hutchins and Ross.

Affleck, J. 1970. Definition of regional structures by magnetics, in H. Johnson and B. L. Smith (eds), 1970 *op. cit,* Ch. 1, pp 3–11.

Ahnert, F. 1970. Functional relationships between denudation, relief, and uplift in large mid-latitude drainage basins, *Am. J. Sci.,* **268**, 243–63.

Ala, M. A. 1974. Salt diapirism in southern Iran, *Bull. Am. Assoc. Petrol. Geol.,* **58**, 1758–70.

Amstutz, G. C. and **Bernard, A. J.** 1973. *Ores in Sediments,* Springer-Verlag, Berlin.

Anderson, R. N. and **Hobart, M. A.** 1976. The relation between heat flow, sediment thickness and age in the eastern Pacific, *J. Geophys. Res.,* **81**, 2968–89.

Armstrong, R. L., 1969. Control of sea level relative to the continents, *Nature,* **Vol. 221**, 1042–3.

Ashley, G. H. 1931. Our youthful scenery, *Bull. geol. Soc. Am.,* **42**, 537–46.

Atwater, T. 1970. Implications of plate tectonics for the Cenozoic tectonic evolution of western North America, *Bull. geol. Soc. Am.,* **81**, 3513–36.

Aubouin, J. 1965. *Geosynclines,* Elsevier, Amsterdam.

Audley-Charles, M. G., Curray, J. R. and **Evans, G.** 1977. Location of major deltas, *Geology,* **5**, 341–4.

Axelrod, D. I. 1962. Post-Pliocene uplift of the Sierra Nevada, California, *Bull. geol. Soc. Am.,* **73**, 183–98.

Badgley, P. C. 1965. *Structural and Tectonic Principles,* Harper and Row, New York.

Bailey, D. K. 1964. Crustal warping – a possible tectonic control of alkaline magmatism, *J. geophys. Res.,* **69**, 1103–11.

Bailey, D. K. 1977. Lithosphere control of continental rift magmatism, *Jl geol. Soc. Lond.,* **133**, 103–6.

Baker, B. H. and **Mitchell, J. G.** 1976. Volcanic stratigraphy and geochronology of the Kedong-Olorgesailie area and the evolution of the South Kenya rift valley, *Jl geol. Soc. Lond.,* **132**, 467–84.

Baker, B. H., Mohr, P. A. and **Williams, L. A. J.** 1972. Geology of the eastern rift system of Africa, *Geol. Soc. Am. Spec. Paper* 136.

Bamford, D. and **Prodehl, C.** 1977. Explosion seismology and the continental crust-mantle boundary, *Jl geol. Soc. Lond.,* **134**, 139–51.

Barazangi, M. and **Isacks, B. L.** 1976. Spatial distribution of earthquakes and subduction of the Nazca plate beneath South America, *Geology,* **4**, 686–92.

Barker, P. F. and **Griffiths, D. H.** 1972. The evolution of the Scotia Ridge and Scotia Sea, *Phil. Trans. Roy. Soc. Lond.,* **A 271**, 151–83.

Barnett, C. H. 1962. A suggested reconstruction of the land masses of the earth as a complete crust, *Nature,* **195**, 447–8.

Beavis, F. C. 1960. The Tawonga fault, northeast Victoria, *Proc. R. Soc. Vict.,* **72**, 95–100.

Beloussov, V. V. 1971. On possible forms of relationship between magmatism and tectogenesis, *Jl geol. Soc. Lond.,* **127**, 57–68.

Bender, P. L. and **Silverberg, E. C.** 1975. Present tectonic-plate motions from lunar ranging, *Tectonophysics,* **29**, 1–7.

Berry, M. J. and **Knopoff, L.** 1967. Structure of the upper mantle under the western Mediterranean basin, *J. Geophys. Res.,* **72**, 3613–26.

Blake, D. H. and **Ollier, C. D.** 1970. Geomorphological evidence of Quaternary tectonics in southwestern Papua, *Rev. Geomorph. dynamique.,* **19**, 28–32.

Blank, H. R. 1978. Fossil laterite on bedrock in Brooklyn, New York, *Geology,* **6**, 21–24.

Bloom, A. L. 1978. *Geomorphology,* Prentice-Hall, New Jersey.

Bond, G. 1976. Evidence for continental subsidence in North America during the Late Cretaceous global submergence, *Geology,* **4**, 557–60.

Bond, G. 1978. Speculations on real sea-level changes and vertical motions of continents at selected times in the Cretaceous and Tertiary Periods, *Geology,* **6**, 247–50.

Branch, C. D. 1966. Volcanic cauldrons, ring complexes and associated granites in the Georgetown Inlier, Queensland, *Bull. Bur. Miner. Resour. Geol. Geophys. Aust.*, 76.

Bretz, J. H. 1962. Dynamic equilibrium and the Ozark landforms, *Am. J. Sci.*, **260**, 427–38.

Brock, B. B. 1951. The Vredefort Ring, *Trans. Geol. Soc. S. Africa*, **53**, 131–44.

Brock, B. B. 1972. *A Global Approach to Geology*, Balkema, Capetown.

Broid, C. E., De Aubertin, F. and **Ravenne, C.** 1976. Structure and History of the Solomon–New Ireland Region, pp. 37–49 in *Geodynamics of the South West Pacific*, Orstom (ed).

Brosske, L. 1962. *Wachst die Erde mit Katastrophen?* Dusseldorf.

Brown, E. H. 1961. *The Relief and Drainage of Wales*, University of Wales Press, Cardiff.

Brown, L. D. and **Oliver, J. E.** 1976. Vertical crustal movement from levelling data and their relation to geologic structure in the eastern United States, *Rev. Geophys. and Space Phys.*, **14**, 13–35.

Büdel, J. 1965. The relief types of the South Indian sheet-wash zone on the eastern Deccan slope, near Madras, *Colloquium Geographicum* (Bonn), 8.

Büdel, J. 1977. *Klima-Geomorphologie*, Gebruder Borntraeger, Berlin.

Bullard, E., Everett, J. E. and **Smith, A. G.** 1965. The fit of the continents around the Atlantic, *Phil. Trans. R. Soc.*, A. **258**, 41–51.

Burchfield, B. C. and **Davis, G. A.** 1972. Structural framework and evolution of the southern part of the Cordille van ovogen, Western United States, *Am. J. Sci.*, **272**, 97–118.

Burk, C. A. and **Drake, C. L.** (eds). 1974. *The Geology of Continental Margins*, Springer-Verlag, Berlin.

Burke, K. and **Dewey, J. F.** 1973. Plume-generated triple junctions: key indicators in applying plate tectonics to old rocks, *J. Geol.*, **81**, 406–33.

Burke, K., Dewey, J. F. and **Kidd, W. S. F.** 1976. Precambrian palaeomagnetic results compatible with contemporary operation of the Wilson cycle, *Tectonophysics*, **33**, 287–99.

Butzer, K. W. 1976. *Geomorphology from the Earth*, Harper Row, New York.

Buwalda, J. P. 1936. Shutterridges, characteristic physiographic features of active faults (abs), *Proc. geol. Soc. Am.*, 1936, **307**.

Carey, S. W. 1954. The rheid concept in geotectonics, *J. geol. Soc. Aust.*, **1**, 67–117.

Carey, S. W. 1958. The tectonic approach to continental drift, *Continental Drift: a symposium*, University of Tasmania, Hobart.

Carey, S. W. 1963a. The asymmetry of the earth, *Aust. J. Sci.*, **25**, 369–84, 479–88.

Carey, S. W. (ed) 1963b. *Syntaphral Tectonics and Diagenesis: a symposium*, Geology Department, University of Tasmania, Hobart.

Carey, S. W. 1976. *The Expanding Earth*, Elsevier, Amsterdam.

Chapman, D. S. and **Pollack, H. N.** 1977. Regional geotherms and lithospheric thickness, *Geology*, **5**, 65–268.

Chappell, B. W. and **White, A. J. R.** 1974. Two contrasting granite types, *Pacific Geol.*, **8**, 173–4.

Chappell, J. 1974. Geology of coral terraces, Huon Peninsula, New Guinea: a study of Quaternary tectonic movements and sea-level changes, *Bull. geol. Soc. Am.*, **85**, 555–70.

Chevallier, J. M. and **Cailleux, A.** 1959. Essai de reconstitution géométrique des continents primitifs, *Z. Geomorph.*, **3**, 257–68.

Chorley, R. J. 1965. The application of quantitative methods to geomorphology, in R. J. Chorley and P. Haggett (eds) *Frontiers in Geographical Teaching*, Methuen, London, pp. 147–163.

Chorley, R. J. and **Kennedy, B. A.** 1971. *Physical Geography: A Systems Approach*, Prentice Hall, London.

Choukroune, P. and **Seguret, M.** 1973. Tectonics of the Pyrenees: role of compression and gravity, in K. A. De Jong and R. Scholten (eds) 1973. *Gravity and Tectonics*, Wiley, New York, pp. 144–56.

Christensen, M. N. 1966. Late Cenozoic crustal movements in the Sierra Nevada of California, *Bull. Geol. Soc. Am.*, **77**, 163–182.

Christofoletti, A. 1974. The explanatory finality of geomorphology, *Recent Researches in Geology*, **2**, 1–5. Department of Geology, University of Delhi, Delhi.

CLIMAP Project Members. 1976. Surface of the ice-age earth, *Science*, **191**, 1131–37.

Cloos, H. 1941. Bau und Tätigkeit von Tuffschloten: Untersuchungen an dem Schwabischen Vulkan, *Geol. Rundsch.*, **32**, 709–800.

Coats, R. R. 1962. Magma type and crustal structure in the Aleutian arc Crust of the Pacific Basin, *Geophys. Monogr.*, **6**, 92–109.

Cobbing, E. J. 1978. The Andean geosyncline in Peru, and its distinction from Alpine geosynclines, *Jl geol. Soc. Lond.*, **135**, 207–18.

Cogne, J. 1979. Principal stages in the creation and evolution of Armorican continental crust in the context of Western Europe, *Geol. Soc. Lond. Newsletter*, **8**, 13–4.

Coleman, R. G. 1971. Plate-tectonics emplacement of upper-mantle peridotites along continental edges, *J. geophys. Res.*, **76**, 1212–22.

Coleman, R. G. and **Irwin, W. P.** 1974. Ophiolites and ancient continental margins, in C. A. Burk and C.L. Drake (eds) 1974, *op. cit.*, pp. 221–310.

Compton, R. B. 1962. *Manual of Field Geology*, Wiley, New York.

Condie, K. C. 1976. *Plate Tectonics and Crustal Evolution*, Pergamon Press, New York.

Cook, E. F. 1966. Paleovolcanology, *Earth Sci. Rev.*, **1**, 155–74.

Cooper, B. N. 1968. Profile of the folded Appalachians of West Virginia, *Univ. Missouri Rolla J.*

Corbel, J. 1959. Vitesse de l'erosion, *Zeit. f. Geomorph.*, **3**, 1–28.

Cotton, C. A. 1944. *Volcanoes as Landscape Forms*, Whitcombe and Tombs, Christchurch.

Cox, A., Dalrymple, G. B. and **Doell, R. R.** 1967. Reversals of the earth's magnetic field, *Scientific American*, **216**, no. 2, 44–54.

Cox, D. P. and **Cox, H. R.** 1974. *Geology-Principles and Concepts,* Freeman, San Francisco.

Cox, K. G. 1972. The Karroo volcanic cycle, *Jl geol. Soc. Lond.*, **128**, 311–36.

Crawford, A. R. 1969. India, Ceylon and Pakistan: new age data and comparisons, *Nature*, **223**, 380–2.

Crawford, A. R. 1979. Gondwanaland and the Pakistan Region, in *Geodynamics of Pakistan*, A. Farah and K. A. De Jong (eds), Geological Survey of Pakistan, Quetta, pp. 3–100.

Creer, K. M. 1965. An expanding earth? *Nature,* **205**, 539–44.

Crickmay, C. H. 1933. The later stages of the cycle of erosion, *Geol. Mag.*, **70**, 337–47.

Crickmay, C. H. 1972. Discovering a meaning in scenery, *Geol. Mag.*, **109**, 171–7.

Crittenden, M. D. 1963. Effective viscosity of the earth derived from isostatic loading of Pleistocene Lake Bonneville, *J. geophys. Res.*, **68**, 5517–30.

Crough, S. T. and **Thompson, G. A.** 1977. Upper mantle origin of Sierra Nevada uplift, *Geology*, **5**, 396–9.

Culling, W. E. H. 1963. Soil creep and the development of hillside slopes, *J. Geol.*, **71**, 127–61.

Curray, J. R. and **Moore, D. G.** 1971. Growth of the Bengal deep-sea fan and denudation in the Himalayas, *Bull. Geol. Soc. Am.*, **82**, 563-72.

Curray, J. R. and **Moore, D. G.** 1974. Sedimentary and tectonic processes in the Bengal deep-sea fan and geosyncline, in C.A. Burk and C. L. Drake (eds) 1974, *op. cit.*, pp. 617–27.

Daly, R. A. 1905. The accordance of summit levels among the Alpine mountains: the fact and its significance, *J. Geol.*, **13**, 105–25.

Davies, H. L. 1978. Folded thrust fault and associated metamorphics in the Suckling-Dayman massif, Papua New Guinea, *Geological Survey of PNG Report*, 78/16.

Davis, D. J. 1978. Klima-Geomorphologie by J. Büdel (Review), *Geogr. J.*, **144**, 324–5.

Davis, G. H. 1975. Gravity-induced folding off a gneiss dome complex, Rincon Mountain, Arizona, *Bull. geol. Soc. Am.*, **86**, 979–90.

Davis, W. M. 1899. The geographical cycle, *Geogr. J.*, **14**, 481–504.

Dawson, J. B. 1977. Sub-cratonic crust and upper mantle models based on Xenolith suites in kimberlite and nephelinitic diatremes, *J. geol. Soc. Lond.*, **134**, 173–84.

Decker, R. W. and **Einarsson.** 1971. Rifting in Iceland, *Trans. Am. Geophysical Union*, **52**, 352.

Deffeyes, K. S. 1970. The axial valley: a steady-state feature of the terrain, in H. Johnson and B. L. Smith (eds), 1970, *op. cit.*, pp. 194–122.

Deiss, C. A. 1943. Structure of central part of Sawtooth Range, Montana, *Bull. geol. Soc. Am.*, **54**, 1123–67.

De Jong, K. A. and **Scholten, R.** (eds) 1973. *Gravity and Tectonics*, Wiley, New York.

Denham, D., Alexander, L. G. and **Worotnicki, G.** 1979. Stress field in the crust of southeast Australia, *Bur. Min. Res. Geol. Geophys. Rec.*, 1979/2, 25.

Denton, G. H. and **Armstrong, R. L.** 1969. Miocene-Pliocene glaciations in southern Alaska, *Am. J. Sci.*, **267**, 1121–42.

De Rezende, W. M. 1972. Post Paleozoic geotectonics of South America related to plate tectonics and continental drift, *Soc. Brasiliera de Geologia, Anais,* do XXVI. *Congresso Brasiliero de Geologia*, 209–10.

De Swart, A. M. and **Trendall, A. F.** 1970. The physiographic development of Uganda, *Overseas Geol. and Mineral Res.*, **10**, 241–88.

Dewey, J. F. and **Bird, J. M.** 1970. Mountain belts and the new global tectonics, *J. geophys. Res.*, **75**, 2625–47.

Dewey, J. F., Pitman, W. C., Ryan, W. B. F. and **Bonnin, J.** 1973. Plate tectonics and the evolution of the Alpine system, *Bull. geol. Soc. Am.*, **84**, 3137–80.

Dickinson, W. R. 1970. Second Penrose Conference: the new global tectonics, *Geotimes*, **15**(4), 18–22.

Dickinson, W. R. and **Seely, D. R.** 1979. Structure and stratigraphy of forearc regions, *Bull. Am. Assoc. Petrol. Geol.*, **63**, 2–31.

Dietz, R. S. 1961. Continent and ocean basin evolution by spreading of the sea floor, *Nature*, **190**, 854–7.

Dietz, R. S. 1963. Collapsing continental rises: an actualistic concept of geosynclines and mountain building, *J. Geol.*, **71**, 314–33.

Dietz, R. S. 1966. Passive continents, spreading sea floors and collapsing continental rises, *Am. J. Sci.*, **264**, 177–93.

Dietz, R. S. 1972. Geosynclines, mountains and continent building, *Scientific American*, **226**, 30–3.

Dietz, R. S. and **Holden, J. C.** 1970. The breakup of Pangaea, *Scientific American*, **223**, no. 4, 30–41.

Dietz, R. S. and **Sproll, W. P.** 1966. Equal areas of Gondwana and Laurasia (ancient supercontinents), *Nature*, **212**, 1196–8.

Dingle, R. V. 1977. The anatomy of a large submarine slump on a sheared continental margin (SE Africa), *Jl geol. Soc. Lond.*, **134**, 293–310.

Donovan, D. T., Horton, A. and **Ivimey-Cook, H. C.** 1979. The transgression of the Lower Lias over the northern flank of the London Platform, *Jl geol. Soc. Lond.*, **136**, 165–73.

Doornkamp, J. C. 1970. The geomorphology of the Mbarara area, Sheet SA-36-1, *Geol. Surv. and Min. Dept. Uganda*, 1970.

Dott, R. H. 1978. Tectonics and sedimentation a century later, *Earth Sci. Rev.*, **14**, 1–34.

Drake, C. L. 1976. *Geodynamics: Progress and Prospects,* American Geophysical Union, Washington.

Drake, D. E. 1976. Suspended sediment transport and mud deposition on continental shelves, in *Marine Sediment Transport and Environmental Management*, D. J. Stanky and D. J. P. Swift (eds), Wiley, New York, pp. 127–58.

Drake, E. T. 1976. Alfred Wegener's reconstruction of Pangea, *Geology*, **1**, 41–4.

Duncan, R. A. and **McDougall, I.** 1976. Linear volcanism in French Polynesia, *J. Volcan. and Geotherm. Res.*, **1**, 198–227.

Eardley, A. J. 1962. *Structural Geology of North America*, 2nd. ed., Harper and Row, New York.

Egyed, L. 1956. The change of the Earth's dimensions determined from palaeogeographical data, *Geofisica Pura et Applicata*, **33**, 42–8. Milan, Italy (in English).

Elliott, D. and **Johnson, M. R. W.** 1978. Discussion on structures found in thrust belts, *Jl geol. Soc. Lond.*, **135**, 259–60.

Elston, W. E. 1978. Rifting and volcanism in the New Mexico segment of the Basin and Range province, Southwestern USA, in *Petrology and Geochemistry of Continetal Rifts*, E. R. Neumann and I. B. Ramberg (eds) Reidel Publ. Co., Dordrecht, pp. 79–86.

Elston, W. E. 1979. The Rio Grande Rift in context of regional post-40 m.y. volcanic and tectonic events, *American Geophysical Union Special Publ.* (in press).

Elter, P. and **Trevisan, L.** 1973. Olistostromes in the tectonic evolution of the Northern Appenines, in *Gravity and Tectonics*, K. A. De Jong and R. Scholten (eds), Wiley, New York, pp. 175-88.

Engelen, G. B. 1963. Gravity tectonics in the northwestern Dolomites (N. Italy), *Geologica Ultraiectina*, no. 13. Rijksuniversiteit te Utrecht.

England, P. C. amd **Richardson, S. W.** 1977. The influence of erosion upon the mineral facies of rocks from different metamorphic environments, *Jl geol. Soc. Lond.*, **134**, 201–13.

Erickson, J. M. 1969. Geological rate units, *Compass*, **47**, 5–9.

Ernst, W. G. 1973. Interpretative synthesis of metamorphism in the Alps, *Bull. geol. Soc. Am.*, **84**, 2053–78.

Ernst, W. G. 1975. *Subduction Zone Metamorphism*. Benchmark Papers in Geology 19, Dowden, Hutchinson and Ross, Stroudsburg.

Eskola, P. E. 1949. The problem of mantled gneiss domes, *Q. Jl. geol. Soc. Lond.*, **104**, 461–76.

Evans, R. 1978. Origin and significance of evaporities in basins around Atlantic Margin, *Bull. Am. Assoc. Petrol. Geol.*, **62**, 233–234.

Evamy, D. D., **Haremboure, J.**, **Kamerling, P.**, **Knaap, W. A.**, **Moly, F. A.** and **Rowlands, P. H.** 1979. Hydrocarbon habitat of Tertiary Niger Delta, *Bull. Am. Assoc. Petrol. Geol.*, **62**, 1–39.

Eyles, R. J. and **Ho, R.** 1970. Soil creep on a humid tropical slope, *J. trop. Geogr.*, **31**, 40–2.

Fairbridge, R. W. and **Finkl, C. W.** 1978. Geomorphic analysis of the rifted cratonic margins of Western Australia, *Z. Geomorph*, **22**, 369–89.

Farhoudi, G. and **Karig, D. E.** 1977. Makran of Iran and Pakistan as an active arc system, *Geology*, **5**, 664–8.

Farrar, E. and **Noble, D. C.** 1976. Timing of Late Tertiary deformation in the Andes of Peru, *Bull. geol. Soc. Am.*, **87**, 1247–50.

Farrell, W. E. and **Clark, J. A.** 1976. On postglacial sea level, *Geophys. J.*, **46**, 647–67.

Fergusson, J. 1978. Kimberlites in southeastern Australia, *B.M.R J. Aust. Geol. Geophys.*, **4**, 13–16.

Findlay, A. L. 1974. The structure of foothills south of the Kubor Range, Papua New Guinea, *Aust. Petrol. Exp. Assoc. J.*, **14**, 14–20.

Finkl, C. W. and **Fairbridge, R. W.** 1979. Paleogeographic evolution of a rifted cratonic margin: S.W. Australia, *Palaeogeog. Palaeoclim. Palaeocol.*, **26**, 221–52.

Finlayson, D. M., **Muirhead, K. J.**, **Webb, J. P.**, **Gibson, G.**, **Furumoto, A. S.**, **Cooke, R. J. S.** and **Russell, A. J.** 1976. Seismic investigation of the Papuan Ultramafic Belt, *Geophys. J. R. astr. Soc.*, **44**, 45–60.

Fischer, A. G. 1969. Geological time-distance rates: the Bubnoff unit, *Bull. Geol. Soc. Am.*, **80**, 549–52.

Forney, G. G. 1975. Permo-Triassic sea-level change, *J. Geol.*, **83**, 773–9.

Fournier, F. 1960. *Climat et érosion: la relation entre l'érosion du sol par l'eau et les précipitations atmospheriques*, Presses Universitaires de France, Paris.

Fuller, R. E. and **Waters, A. C.** 1929. The nature and origin of the horst and graben structure of southern Oregon, *J. Geol.*, **37**, 204–39.

Fyfe, W. S. 1970. Some thoughts on granitic magmas, in *Mechanisms of Igneous Intrusion*, G. Newall and N. Rast (eds), *Geol. J. Spec. Issue*, 2, pp. 201–16.

Gansser, A. 1973. Facts and theories on the Andes, *Jl geol. Soc. Lond.*, **129**, 93–131.

Gardner, J. V. 1970. Submarine geology of the western Coral Sea, *Bull. geol. Soc. Am.*, **81**, 2599–614.

Garrels, R. M. and **Mackenzie, F. T.** 1971. *Evolution of Sedimentary Rocks*, Norton, New York.

Gass, I. G. 1977. The evolution of the Pan African crystalline basement in NE Africa and Arabia, *Jl geol. Soc. Lond.*, **134**, 129–38.

Gass, I. G. and **Masson-Smith, D.** 1963. The geology and gravity anomalies of the Troodos Massif, Cyprus, *Phil. Trans. Roy. Soc. London.* A, **255**, 417–67.

Gastil, G. 1960. Distribution of mineral dates in time and space, *A. J. Sci.*, **258**, 1–35.

Gautier, A. 1965. Relative dating of peneplains and sediments in the Lake Albert Rift area, *A. J. Sci.*, **263**, 537–47.

Gilbert, G. K. 1877. Report on the Geology of the Henry Mountains in US Geographical and Geological Survey of the Rocky Mountain Region (Powell).

Gilbert, G. K. 1890. Lake Bonneville, *US Geol. Survey Mon.*, 1.

Gilluly, J. 1964. Atlantic sediments, erosion rates, and the evolution of the continental shelf: some speculations, *Bull. geol. Soc. Am.*, **75**, 483–92.

Gilluly, J. 1970. Crustal deformation in the western United States, in H. Johnson and B. L. Smith (eds), 1970 *op. cit.*, pp. 47–73.

Gilluly, J. 1973. Steady plate motion and episodic orogeny and magmatism, *Bull. Geol. Soc. Am.*, **84**, 499–514.

Gilluly, J., **Reed, J. C.** and **Cady, W. M.** 1970. Sedimentary volumes and their significance. *Bull. geol. Soc. Am.*, **81**, 353–76.

Glangeaud, L. 1957. Essai de classification géodynamique des chaines et des phénomenes orogeniques, *Rev. Géog. Phys. Géol. Dynam.*, **1**, 214.

Goetze, G. 1978. The mechanism of creep in olivine, *Phil. Trans. R. Soc. Lond.* **ser A, 288**, 99–119.

Goodwin, A. M. 1974. The most ancient continental margins, in C. A. Burk and C. L. Drake (eds), 1974 *op cit.*, pp. 767–80.

Grabert, H. 1971. Die Prae-Andine Drainage des Amazonas Stromsystems, *Muenster Forsch. Geol. Palaeontol,* **20**, 51–60.

Grasty, R. L. 1967. Orogeny, a cause of world-wide regression of the seas, *Nature*, **216**, 779–80.

Green, D. H. and **Ringwood, A. E.** 1968. Genesis of the calc-alkaline igneous rock suite, *Contr. Mineral and Petr.*, **18**, 105–62.

Green, J. (in press). *Encyclopedia of Volcanoes and Volcanology*, Dowden, Hutchinson and Ross, Stroudsburg.

Green, J. and **Short, N. M.** 1971. *Volcanic Landforms and Surface Features*, Springer-Verlag, Berlin.

Griggs, D. 1939. A theory of mountain building, *Am. J. Sci.*, **237**, 611–50.

Hack, J. T. 1960. Interpretation of erosional topography in humid temperate regions, *Am. J. Sci.*, **258**, 80–97.

Hager, B. R. 1978. Oceanic plate motions driven by lithospheric thickening and subducted slabs, *Nature*, **276**, 156–8.

Haggett, P., Chorley, R. J. and **Stoddart, D. R.** 1965. Scale standards in geographical research: a new measure of areal magnitude, *Nature*, **205**, 844–7.

Haile, N. S., McElhinney and **McDougall I.** 1977. Palaeomagnetic data and radiometric ages from the Cretaceous of West Kalimantan (Borneo), and their significance in interpreting regional structure, *Jl geol. Soc. Lond.*, **133**, 133–44.

Hallam, A. 1963. Major epeirogenic and eustatic changes since the Cretaceous, and their possible relationship to crustal structure, *Am. J. Sci.*, **261**, 397–423.

Haman, P. J. 1975. A lineament analysis of the United States, *West Can. Res. Publ. Geol. Relat. Sci.*, Ser. 4, no. 1, 27.

Hamilton, W. 1977. Subduction in the Indonesian region, in M. Talwani and W. C. Pitman (eds), 1977 *op. cit.*, pp. 11–31.

Harper, C. T. 1973. *Geochronology,* Dowden, Hutchinson and Ross, Stroudsburg.

Harrison, J. V. and **Falcon, N. L.** 1934. Collapse structures, *Geol. Mag.*, **71**, 529–39.

Harrison, J. V. and **Falcon, N. L.** 1936. Gravity collapse structures and mountain ranges, as exemplified in south-western Persia. *Q. Jl. geol. Soc. Lond.*, **92**, 91–102.

Hatherton, T. 1974. Active continental margins and island arcs, in C. A. Burk and C. L. Drake (eds), 1974 *op. cit.*, pp. 93–103.

Hays, J. D. and **Pitman, W. C.** 1973. Lithospheric plate motion, sea level changes and climatic and ecological consequences, *Nature*, **246**, 18.

Heezen, B. C. and **Drake, C. L.** 1963. Gravity tectonics, turbidity currents and geosynclinal accumulations in the continental margin of eastern North America, in *Syntaphral Tectonics and Diagenesis,* University of Tasmania, Hobart, D1–D10.

Heezen, B. C. and **Wilson, L.** 1968. Submarine geomorphology, in *Encyclopedia of Geomorphology*, R. W. Fairbridge (ed), Reinhold, New York, pp. 1079–97.

Heirtzler, J. R. 1968. Seafloor spreading, *Scientific American,* **219**, no. 6, 60–70.

Hess, H. H. 1962. History of the ocean basins, in *Petrological Studies: A volume to honour A. F. Buddington*, A. E. J. Engel, H. L. James and B. F. Leonard (eds), *Geol. Soc. Am.*

Hilgenberg, O. C. 1933. *Vom wachsenden*, Erdball, Berlin.

Hills, E. S. 1953. *Outlines of Structural Geology*, Methuen, London.

Hills, E. S. 1956. A contribution to the morphotectonics of Australia, *J. geol. Soc. Aust.*, **3**, 1–15.

Hills, E. S. 1975. *The Physiography of Victoria*, Whitcombe and Tombs, Melbourne.

Hine, R., Williams, I. S., Chappell, B. and **White, A. J. R.** 1978. Contrasts between I- and S-type granitoids of the Kosciusko Batholith, *J. geol. Soc. Aust*, **25**, 219–34.

Hobbs, B. E., Means, W. D. and **Williams, P. F.** 1976. *An Outline of Structural Geology*, Wiley, New York.

Hodgson, J. H. 1962. Movements of the earth's crust as indicated by earthquakes in S. K. Runcorn (ed), *Continental Drift*, Academic Press, New York.

Holeman, J. N. 1968. The sediment yield of major rivers of the world, *Water Resources Research*, **4**, 737–41.

Hollingworth, S. E. 1938. The recognition and correlation of high level erosion surfaces in Britain: a statistical study, *Q. Jl. geol. Soc. Lond.*, **94**, 55–84.

Hollingworth, S. E., Taylor, J. H. and **Kellaway, G. A.** 1944. Large-scale superficial structures in the Northampton Ironstone Field, *Q. Jl. geol. Soc. Lond.*, **100**, 1–44.

Holmes, A. 1965. *Principles of Physical Geology*, Nelson, London.

Hoyle, F. and **Narlikar, J. V.** 1971. On the nature of mass, *Nature*, **233**, 41–4.

Hsu, K. J. 1972. When the Mediterranean dried up, *Scientific American*, **227**, 26–36.

Hubbert, M. K. and **Rubey, W. W.** 1959. Role of fluid pressure in mechanics of overthrust faulting: I. Mechanics of fluid-filled porous solids and its application to overthrust faulting, *Bull. geol. Soc. Am.*, **70**, 115–66.

Hudleston, P. J. 1977. Similar folds, recumbent folds and gravity tectonics in ice and rocks, *J. Geol.*, **85**, 113–22.

Illies, J. H. 1972. The Rhine graben rift system – Plate tectonics and transform faulting, *Geophys. Surveys*, **1**, 27–60.

Inman, D. L. and **Nordstrom, C. E.** 1971. On the tectonic and morphologic classification of coasts, *J. Geol.*, **79**, 1–21.

Irving, E. 1964. *Palaeomagnetism*, Wiley, New York.

Isaacson, P. E. 1975. Evidence for a western extra-continental land source during the Devonian Period in the central Andes, *Bull. geol. Soc. Am.*, **86**, 39–46.

Isachsen, Y. W. 1975. Possible evidence for contemporary doming of the Adirondack Mountains, New York, and suggested implications for regional tectonics and seismicity, *Tectonophysics*, **29**, 169–81.

Isacks, B. L. and **Barazangi, M.** 1977. Geometry of Benioff zones: lateral segmentation and downwards bending of the subducted lithosphere, in M. Talwani and W. C. Pitman (eds), 1977 *op. cit.*, pp. 94–114.

Jeffreys, H. 1931. On the mechanics of mountains, *Geol. Mag.*, **68**, 433–42.

Johnson, B. D., Powell, C. McA and **Veevers, J. J.** 1976. Spreading history of the eastern Indian Ocean and India's northward flight from Antarctica, *Bull. geol. Soc. Am.*, **87**, 1560–66.

Johnson, D. W. 1905. The Tertiary history of the Tennessee River, *J. Geol.*, **13**, 194–231.

Johnson, D. W. 1919. *Shore Processes and Shoreline Development*, Wiley, New York.

Johnson, D. W. 1931. A theory of Appalachian geomorphic evolution, *J. Geol.*, **39**, 497–508.

Johnson, H. and **Smith, B. L.** 1970. *The Magatectonics of Continents and Oceans*, Rutgers University Press, New Brunswick.

Johnson, M. R. W. and **Stewart, F. H.** 1963. *The British Caledonides*, Oliver and Boyd, Edinburgh.

Joly, F. 1963. Projet de legende pour une carte morpholique a petit échelle des regions arides et semi-ardes du monde, *I.G.U. Commission on the Arid Zones*.

Jordan, P. 1971. *The Expanding Earth*, Pergamon, Oxford.

Joyce, E. B. 1975. Quaternary volcanism and tectonics in southeastern Australia, in *Quarternary Studies*, R. P. Suggate and M. M. Cresswell (eds), Royal Society of New Zealand, Wellington, pp. 169–76.

Judson, S. and **Ritter, D. F.** 1964. Rates of regional denudation in the United States, *J. geophys. Res.*, **69**, 3395–401.

Kahle, C. F. (ed.). 1974. Plate tectonics – Assessments and reassessment, *Am. Assoc. Petrol. Geol.*, Memoir 23.

Kahle, C. F. and **Pawlowicz, E. F.** 1977. Penrose Conference Report: Geodynamics of continental interiors, *Geology*, **5**, 431–2.

Kahle, H. G. and **Werner, D.** 1975. Gravity and temperature anomalies in the wake of drifting continents, *Tectonophysics*, **29**, 487–504.

Kaizuka, S., Matsuda, T., Nogami, M. and **Yonekura, N.** 1973. Quaternary tectonic and recent seismic crustal movements in the Arauco Peninsula and its environs, Central Chile, *Geogr. Repts. Tokyo Metropolitan University*, no. 8.

Karapetian, K. I. 1964. Some regularities in areal volcanism, *Bull. volcan.*, **27**, 381–3.

Karig, D. E. 1971. Origin and development of marginal basins in the western Pacific, *J. geophys. Res.*, **76**, 2542–61.

Katsui, Y. 1971. List of the World's Active Volcanoes, with Map. *Spec. Issue of Bull. Volc. Eruption.* 160 p.

Katz, H. R. 1971. Continental margin in Chile – is tectonic style compressional or extensional? *Bull. Am. Assoc. Petrol. Geol.*, **55**, 1753–8.

Kay, M. 1947. Geosynclinal nomenclature and the craton, *Bull. Am. Assoc. Petrol. Geol.*, **31**, 1289–93.

Kay, M. 1951. North American geosynclines, *Geol. Soc. Am.*, Mem 48.

Kennedy, W. Q. 1946. The Great Glen Fault, *Q. Jl Geol. Soc. Lond.*, **102**, 41–76.

Kennedy, W. Q. 1962. Some theoretical factors in geomorphological analysis, *Geol. Mag.*, **99**, 304–12.

Kennett, J. P., Houtz, R. E., Andrews, P. B., Edwards, A. R., Gostin, V. A., Hajos, M. A., Jenkins, D. G., Margolis, S. V., Ovenshine, A. T. and **Perch-Nielson, K.** 1974. Development of the Circum-Antarctic current, *Science*, 186, 14–47.

Kent, P. E. 1977. The Mesozoic development of aseismic continental margins, *Jl geol. Soc. Lond.*, **134**, 1–18.

King, B. C. 1949. The Napak area of southern Karamoja, Uganda, *Mem. geol. Surv. Uganda*, 5.

King, B. C. 1976. The Baikal Rift, *Jl geol. Soc. Lond.*, **132**, 348–9.

King, B. C., Le Bas, M. J. and **Sutherland, D. S.** 1972. The history of the alkaline volcanoes and extrusive complexes of eastern Uganda and western Kenya, *Jl geol. Soc. Lond.*, **128**, 173–205.

King, L. C. 1953. Canons of landscape evolution, *Bull. Geol. Soc. Am.*, **64,** 721–51.

King, L. C. 1962. *The Morphology of the Earth*, Oliver and Boyd, Edinburgh.

King, L. C. 1976. Planation remnants upon high lands, *Z. Geomorph.*, **20**, 133–48.

King, P. B. 1965. Tectonics of Quaternary time in middle North America, in *The Quaternary of the United States*, H. E. Wright and D. G. Frey (eds). pp. 831–70.

King, P. B. 1969. The tectonics of North America – a discussion to accompany the tectonic map of North America, Scale 1 : 5,000,000, *US Geol. Surv. Prof. Paper 628*.

Kingma, J. T. 1958. Possible origin of piercement structures, local unconformities, and secondary basins in the eastern geocncline, New Zealand, *N.Z.J. Geol. Geophys.*, **1**, 269–74.

Kirkby, M. J. 1967. Measurement and theory of soil creep, *J. Geol.*, **75**, 359–78.

Knopoff, L. 1974. Geophysical constraints on geodynamic models – summary, in C. F. Kahle (ed). 1974 *op. cit.*, pp. 409–10.

Kobayashi, K. and **Isezaki, N.** 1976. Magnetic anomalies in the Sea of Japan and the Shikoku Basin: possible tectonic implications, in G. H. Sutton, M. H. Manghnani and R. Moberley (eds). 1976 *op. cit.*, pp. 235–51.

Korn, H. and **Martin, H.** 1959. Gravity tectonics in the Naukluft Mountains of South-West Africa, *Bull. geol. Soc. Am.*, **70**, 1047–78.

Korsch, R. J. 1977. A framework for the Palaeozoic geology of the southern part of the New England geosyncline, *J. geol. Soc. Aust.*, **25**, 339–55.

Kossinna, E. 1933. Die Erdoberfläche, in *Handbuch der Geophysik*, Vol. 2, B. Gutenberg (ed), pp. 869–954.

Krebs, W. 1975. Formation of southwest Pacific island arc-trench and mountain systems: plate or global vertical tectonics, *Bull. Am. Assoc. Petrol. Geol.*, **59**, 1639–66.

Krynine, P. D. 1941 (abstract) *Bull. Geol. Soc. Am.* **52** (1) Differentiation of sediments during the life history of a landmass, p. 1915; (2) Paleogeographic and tectonic significance of sedimentary quartzites, pp. 1915–16; (3) Paleogeographic and tectonic significance of graywackes, p.1916; (4) Paleogeographic and tectonic significance of arkoses, pp. 1918–19.

Kulm, L. D. and **Fowler, G. A.** 1974. Oregon continental margin structure and stratigraphy; a test of the imbricate thrust model. pp. 261–63 in Burk, C. A. and Drake, C. L. (eds). The geology of continental margins. Springer-Verlag, Berlin.

Kuno, H. 1959. Origin of Genozoic petrographic provinces of Japan and surrounding areas, *Bull. volcan.*, **20**, 37–76.

Kuno, H. 1966. Lateral variation of basalt magma across continental margins and island arcs, *Bull. volcan.*, **29**, 195–222.

La Breque, J. L., Kent, D. V. and **Cande, S. C.** 1977. Revised magnetic polarity time scale for Late Cretaceous and Cenozoic time, *Geology*, **5**, 330–35.

Langbein, W. B. and **Schumm, S. A.** 1958. Yield of sediment in relation to mean annual precipitation, *Trans. Am. geophys. Union*, **39**, 1076–84.

Larson, R. L. and **Hilde, T. W. C.** 1975. A revised time scale of magnetic reversals for the Early Cretaceous and Late Jurassic, *J. geophys. Res.*, **80**, 2586–94.

Laubscher, H. P. 1971a. The large-scale kinematics of the western Alps and the northern Apennines and its palinspastic implications, *Am. J. Sci.*, **271**, 193–226.

Laubscher, H. P. 1971b. Das Alpen-dinariden Problem und die Palinspastik der südlichen Tethys, *Geol. Rdsch.*, **60**, 813–33.

Laubscher, H. 1973. Jura Mountains, in *Gravity Tectonics*, K. A. De Jong and R. Scholten (eds). 1973 *op. cit.*, pp. 217–27.

Le Bas, M. J. 1971. Per-alkaline volcanism, crustal swelling, and rifting, *Nature: Physical Science*, **230**, 85–7.

Lee, T. C. 1979. Erosion, uplift, exponential heat source distribution and transient heat flux, *J. geophys. Res.*, **84**, 585–90.

Lees, G. M. 1952. Foreland folding, *Q. Jl geol. Soc. Lond.*, **108**, 1–34.

Lees, G. M. and **Falcon, N. L.** 1952. The geographical history of the Mesopotamian plains, *Geogr. J.*, **118**, 24–39.

Leet, L. D. and **Judson, S.** 1971. *Physical Geology*, Prentice Hall, New Jersey.

Le Masurier, W. E. 1972. Volcanic record of Antarctic glacial history: Implications with regard to Cenozoic sea levels, in *Polar Geomorphology*, R. J. Price and D. E. Sugden (eds), Inst. Brit. Geographers Spec. Pub., no. 4, 59–74.

Lensen, G. 1968. Analysis of progressive fault displacement during downcutting at the Branch River Terrace, South Islands, New Zealand, *Bull. geol. Soc. Am.*, **79**, 545–56.

Leopold, L. B. and **Langbein, W. B.** 1962. The concept of entropy in landscape evolution, *US Geol. Surv. Prof. Paper* 500-A, 20.

Lucchitta, I. 1972. Early history of the Colorado River in the Basin and Range Province, *Bull. geol. Soc. Am.*, **83**, 1933–48.

Lilienthal, Th. Ch. 1756. *Die Gute Sache der Göttlichen offenbarung.* Hartung, Könisberg.

Luyendyk, B. P. and **Rennick, W.** 1977. Tectonic history of aseismic ridges in the eastern Indian Ocean, *Bull. geol. Soc. Am.*, **88**, 1347–56.

McClay, K. R. 1977. Pressure solution and Coble creep in rocks, a review, *Jl geol. Soc. Lond.*, **134**, 71–5.

McConnell, R. B. 1968. Planation surfaces in Guyana, *Geogr. J.*, **134**, 506–20.

McElhinney, M. W. and **Brock, A.** 1975. A new palaeomagnetic result from East Africa and estimates of the Mesozoic palaeoradius, *Earth Plan. Sci. Rev. Let.*, **27**, 321–8.

McElhinney, M. W. and **McWilliams, M. O.** 1977. Precambrian geodynamics – a palaeomagnetic view, *Tectonophysics*, **40**, 137–59.

Macgregor, A. M. 1951. Some milestones in the Precambrian of Southern Rhodesia, *Trans. Geol. Soc. S. Africa*, **54**, xxvii–lxxiv.

Macintyre, I. G., Burke, R. B. and **Stuckenrath, R.** 1977. Thickest recorded Holocene reef section, Isla Perez core hole, Alacran Reef, Mexico, *Geology*, **5**, 749–54.

McKee, E. D. and **McKee, E. H.** 1972. Pliocene uplift of the Grand Canyon Region – time of drainage adjustment, *Bull. geol. Soc. Am.*, **83**, 1923–32.

McKee, E. D. , Wilson, R. F., Breen, W. J. and **Breed, C. S.** 1967. *Evolution of the Colorado River in Arizona*, Museum of North Arizona, Flagstaff.

McKerrow, W. S. 1979. Ordovician and Silurian changes in sea level, *Jl geol. Soc. Lond.*, **136**, 137–45.

McMillian, N. J. 1973. Shelves of Labrador Sea and Baffin Bay, Canada, in The future petroleum provinces of Canada – their geology and potential, R. G. McCrossan (ed), *Canadian Soc. Petrol. Geol., Mem.* 1, pp. 473–517.

Mammerickx, J. 1978. Re-evaluation of some geophysical observations on the Caroline Basins, *Bull. geol. Soc. Am.*, **89**, 192–6.

Manser, W. 1973. Geological setting and geology of the islands, in *New Guinea Barrier Reefs*, W. Manser (ed.), Univ. Papua New Guinea, Geol. Dept. Occ. Paper No. 1, Ch. 3.

Marinatos, S. N. 1960. Helice, a submerged town of classical Greece, *Archaeology*, **13**, 186–93.

Marlow, M. S., Scholl, D. W. and **Cooper, A. K.** 1977. St George Basin Bering Sea Shelf: A collapsed Mesozoic margin, in M. Talwani and W. C. Pitman (eds.), *op. cit.*, pp. 211–20.

Martin, H. 1975. Structural and palaeogeographical evidence for an Upper Palaeozoic sea between southern Africa and South America, in *Gondwana*

Geology, K. S. W. Campbell (ed), ANU Press, Canberra, pp. 37–59.

Mason, B. 1966. *Principles of Geochemistry*, Wiley, New York.

Maxwell, J. C. 1970. The Mediterranean, ophiolites and continental drift, in H. Johnson and B. L. Smith (eds) 1970, *op. cit.,* 167–93.

Maxwell, J. C. 1974. The new global tectonics – an assessment, in C. F. Kahle (ed) 1974. *op. cit.,* pp. 24–42.

Meade, B. K. 1975. Geodetic surveys for monitoring crustal movements in the United States, *Tectonophysics*, **29**, 103–12.

Menard, H. W. 1961. Some rates of regional erosion, *J. Geol.,* **69**, 154–61.

Menard, H. W. 1964. *Marine Geology of the Pacific.* McGraw-Hill, New York.

Menard, H. W. 1974. *Geology, Resources and Society,* Freeman, San Francisco.

Meservey, R. 1969. Topological inconsistency of continental drift on the present-sized earth, *Science*, **166**, 609–11.

Meyerhoff, A. A. and **Meyerhoff, H. A.** 1972. The new global tectonics: major inconsistencies, *Bull. Am. Assoc. Petrol. Geol.,* **56**, 269–336.

Meyerhoff, H. A. and **Olmsted, E. W.** 1963. The origins of Appalachian drainage, *Am. J. Sci.,* **332**, 21–41.

Miyashiro, A. 1972. Metamorphism and related magmatism in plate tectonics, *Am. J. Sci.,* **272**, 629–56.

Miyashiro, A. 1973. Paired and unpaired metamorphic belts, *Tectonophysics,* **17**, 241–54.

Molnar, P. and **Burke, K.** 1977. Erik Norin Penrose Conference on Tibet, *Geology*, **5**, 461–3.

Molnar, P. and **Tapponnier, P.** 1978. Active tectonics of Tibet, *J. geophys. Res.,* **83**, 5361–75.

Molnar, P. and **Wang-Ping, Chen.** 1978. Evidence of large Cainozoic crustal shortening of Asia, *Nature*, **78**, 218–20.

Montadert, L., Winnock, E., Deltiel, J. R. and **Grau, G.** 1974. Continental margins of Galicia-Portugal and Bay of Biscay, in C. A. Burk and C. L. Drake (eds), 1974 *op. cit.,* pp. 323–42.

Morisawa, M. 1973. Plate tectonics and geomorphology, *Recent Researches in Geology*, Vol. 1, V. K. Verma (ed), Hindustan Publ. Co., Delhi.

Mortimer, C. 1973. The Cenozoic history of the southern Atacama Desert, Chile, *Jl geol. Soc. Lond.,* **129**, 505–26.

Moseley, F. 1972. A tectonic history of northwest England, *Jl geol. Soc. Lond.,* **128**, 561–98.

Mulcahy, M. J. 1966. Peneplains and pediments in Australia, *Aust. J. Sci.,* **28**, 290–1.

Myers, J. S. 1975. Vertical crustal movement of the Andes in Peru, *Nature*, **254**, 672–4.

Nagumo, S. and **Kasahara, J.** 1976. Ocean bottom seismographic study of the western margin of the Pacific, in G. H. Sutton, M. H. Manghnani and R. Moberly (eds), 1976 *op. cit.,* pp. 155–67.

Newell, G. and **Rast, N.** 1970. *Mechanisms of Igneous Intrusion*, Gallery Press, Liverpool.

Ni, J. and **York, J. E.** 1978. Late Cenozoic tectonics of the Tibetan Plateau, *J. geophys. Res.,* **83**, 5377–84.

Nikolaev, N. I. and **Schultz, S. S.** 1959. Printsipy i metody sostavleniya karty noveishei tektoniki SSR. *Otd. geol.-geogr. nauk ANSSR. Materialy 2nd Geomorf. soveshchaniya.* Moscow, 20 pp.

Noble, D. C. and **McKee, E. H.** 1977. Spatial distribution of earthquakes and subduction of the Nazca plate beneath South America: Comment, *Geology*, **5**, 576–78.

O'Driscoll, E. S. 1977. The double helix in geotectonics (abstract), *Carey Appreciation symposium*, University of Tasmania, Hobart.

Okada, H. 1974. Migration of ancient arc-trench systems, in Ancient and Modern Geosynclinal Sedimentation, R. H. Dott and R. H. Shaver (eds), *Soc. Econ. Pal. Mineral. Sp. Publ.,* **19**, 311–20.

Ollier, C. D. 1959. A two-cycle theory of tropical pedology, *J. Soil Sci.,* **10**, 137–48.

Ollier, C. D. 1960. The inselbergs of Uganda, *Z. Geomorph.,* **4**, 43–52.

Ollier, C. D. 1967. Landscape description without stage names, *Aust. geogr. Studies*, **5**, 73–80.

Ollier, C. D. 1969. *Volcanoes*, ANU Press, Canberra.

Ollier, C. D. 1973. *Earth History in Maps and Diagrams,* Longman, Australia.

Ollier, C. D. 1974. Phreatic eruptions and maars, in *Physical Volcanology*, L. Civetta, P. Casparini, G. Luongo and A. Rappala (eds), Elsevier, Amsterdam, pp. 289–311.

Ollier, C. D. 1977a. Applications of Weathering Studies, in *Applied Geomorphology*, J. R. Hails (ed), Elsevier, Amsterdam, pp. 9–50.

Ollier, C. D. 1977b. Terrain classification: methods, applications and principles, in *Applied Geomorphology*, J. R. Hails (ed), Elsevier, Amsterdam, pp. 277–316.

Ollier, C. D. 1977c. Early landform evolution, in *Australia: a geography*, D. Jeans (ed), pp. 85–98.

Ollier, C. D. 1978. Tectonics and geomorphology of the eastern highlands, in *Landform Evolution in Australasia*, J. L. Davies and M. A. J. Williams (eds), ANU Press, Canberra, pp. 5–47.

Ollier, C. D. 1980. Evolutionary geomorphology of Australia and Papua New Guinea, *Trans. Inst. Brit. Geogr.* (in press).

Ollier, C. D. and **Brown, M. J. F.** 1971. Erosion of a young volcano in New Guinea, *Z. Geomorph.,* **15**, 12–28.

Ollier, C. D. and **Pain, C. F.** 1980. Actively rising gneiss domes in Papua New Guinea, *J. geol. Soc. Aust.* (in press).

Orstom, 1977. *Geodynamics in South-West Pacific*, International Symposium, Noumea, 1976. Editions Technip, 27 Rue Ginoux, 75737 Paris.

Osmaston, M. F. 1973. Limited lithosphere separation as a main cause of continental basins, continental growth and epeirogeny, in D. N. Tarling and S. K. Runcorn (eds). 1973. *op. cit.,* pp. 649–74.

Osmaston, M. F. 1977. Discussion, *Jl geol. Soc. Lond.,* **133**, 360.

Owen, H. G. 1976. Continental displacement and expansion of the earth during the Mesozoic and Cenozoic, *Phil. Trans. Roy. Soc. Lond.*, **A. 281**, 223–91.

Oxburgh, E. R. 1968. *The geology of the Eastern Alps*, The Geologists' Association.

Oxburgh, E. R. 1972. Flake tectonics and continental collision, *Nature*, **239**, 202–4.

Oxburgh, E. R. and **Turcotte, D. L.** 1970. The thermal structure of island arcs, *Bull. geol. Soc. Am.*, **81**, 1665–88.

Penck, W. 1924. *Morphological Analysis of Landforms*, English translation by H. Czech and K. C. Boswell, London, 1953.

Phillips, W. E. A., Stillman, C. J. and **Murphy, T.** 1976. A Caledonian plate tectonic model, *Jl geol. Soc. Lond.*, **132**, 579–609.

Pitcher, W. S. 1970. Ghost stratigraphy in intrusive granites: a review, in N. Rast and G. Newell (eds), *op. cit*, pp. 123–40.

Pitcher, W. S. and **Bussell, M. A.** 1977. Structural control of batholithic emplacement in Peru: a review, *Jl geol. Soc. Lond.*, **133**, 249–56.

Pitman, W. C. and **Hayes, D. E.** 1968. Sea-floor spreading in the Gulf of Alaska, *J geophys. Res.*, **73**, 6571–80.

Potter, P. E. 1978. Significance and origin of big rivers, *J. Geol.*, **86**, 13–33.

Potter, D. B. and **McGill, G. E.** 1978. Valley anticlines of the Needles District, Canyonlands National Park, Utah, *Bull. geol. Soc. Am.*, **89**, 952–60.

Press, F. 1973. The gravitational instability of the lithosphere, in De Jong and Scholten (eds) 1973 *op. cit.*, pp. 7–16.

Price, R. A. and **Mountjoy, E. W.** 1971. The Cordilleran foreland thrust and folded belt in the Southern Canadian Rockies (abst.) *Geol. Soc. Amer. Abst.* **3**, 404–5.

Quennell, A. M. 1959. Tectonics of the Dead Sea, *Int. Geol. Cong.*, **20** (Mexico), 385–405.

Ramberg, H. 1967. *Gravity, Deformation and the Earth's Crust*, Academic Press, London.

Ramsay, J. 1963. Structure and metamorphism of the Moine and Lewisian rocks of the north-west Caledonides, in *British Caledonides*, M. R. W. Johnson and F. H. Stewart (eds). Oliver and Boyd, Edinburgh, pp. 147–75.

Ramsbottom, W. H. C. 1979. Rates of transgression and regression in the Carboniferous of N. W. Europe, *Jl geol. Soc. Lond.*, **136**, 147–53.

Rast, N. 1970. The initiation, ascent and emplacement of magmas, in N. Rast and G. Newell (eds), *op. cit.*, pp. 339–62.

Raymond, C. F. 1978. Mechanics of glacier movement, in *Rockslides and Avalanches, 1 Natural Phenomena*, B. Voight (ed)., Elsevier, Amsterdam, pp. 793–833.

Read, H. H. and **Watson, J.** 1966. *Beginning Geology*, Macmillan, London.

Reeves, C. V. 1978. A failed Gondwana spreading axis in southern Africa, *Nature*, **273**, 222–3.

Rickard, M. J. 1969. Relief of curvature on expansion – a possible mechanism for geosynclinal formation and orogenesis, *Tectonophysics*, **8**, 129–44.

Ringwood, A. E. 1974. The petrological evolution of island arc systems, *Jl geol. Soc. Lond.*, **130**, 183–204.

Rittmann, A. 1962. *Vocanoes and their Activity*, trans. E. A. Vincent, Wiley, New York.

Roeder, D. 1977. Philippine arc system – collision or flipped subduction zones? *Geology*, **5**, 203–6.

Rubey, W. W. 1951. Geological history of seawater: an attempt to state the problem, *Bull. geol. Soc. Am.*, **62**, 1111–48.

Rubke, N. A. 1970. Continental Drift before 1900. *Nature*, **227**, 349–350.

Rutten, L. M. R. 1949. Frequency and periodicity of orogenic movements, *Bull. geol. Soc. Am.*, **60**, 1755–70.

Rutten, M. G. 1969. *The Geology of Western Europe*, Elsevier, Amsterdam.

Ruxton, B. P. and **McDougall, I.** 1967. Denudation rates in northeast Papua from potassium-argon dating of lavas, *Am. J. Sci.*, **265**, 545–61.

Salisbury, R. D. 1919. *Physiography*, Henry Holt and Co., New York.

Sass, J. H. 1971. The earth's heat and internal temperatures, in *Understanding the Earth*, I. G. Cass, P. J. Smith and R. C. L. Wilson (eds), Artmis Press, Sussex. pp. 81–70.

Schaer, J. P., Reimer, G. M. and **Wagner, G. A.** 1975. Actual and ancient uplift rate in the Gotthard region, Swiss Alps: A comparison between precise levelling and fission-track apatite age, *Tectonophysics*, **29**, 293–300.

Scheidegger, A. E. 1967. A complete thermodynamic analogy for landscape evolution, *Bull. Int. Scientific Hydrology*, **12**, 57–62.

Scheidegger, A. E. and **Langbein, W. B.** 1966. Probability concepts in geomorphology, *U.S. Geol. Surv. Prof. Paper* 500-C, 14.

Scholl, D. W. and **Marlow, M. S.** 1974. Global tectonics and the sediments of modern and ancient trenches: some different interpretations, in C. F. Kahle (ed) 1974 *op. cit.*, pp. 255–72.

Scholl, D. W., Marlow, M. S. and **Cooper, A. K.** 1977. Sediment subduction and offscraping at Pacific margins, in M. Talwani and W. C. Pitman (eds), 1977 *op. cit.*, pp. 199–210.

Scholl, D. W., Von Heune, R. and **Ridlon, J. B.** 1968. Spreading of the ocean floor: undeformed sediments in the Peru-Chile trench, *Science*, **159**, 869–71.

Scholz, C. H. 1977. Transform fault systems of California and New Zealand: similarities in their tectonic and seismic styles, *Jl geol. Soc. Lond.*, **133**, 215–29.

Schubert, C. and **Henneberg, H. G.** 1975. Geological and geodetic investigations on the movement along the Bocono Fault, Venezuelan Andes, *Tectonophysics*, **29**, 199–207.

Schumm, S. A. 1963. The disparity between present rates of denudation and orogeny, *U.S. Geol. Surv. Prof. Paper* 454-H.

Schumm, S. A. 1977. *The Fluvial System,* Wiley, New York.

Schumm, S. A. and **Lichty, R. W.** 1965. Time, space and causality in geomorphology, *Am. J. Sci.,* **263**, 110–19.

Seed, H. B. 1968. Landslides during earthquakes due to soil liquefaction, *J. Soil Mech. Foundations Div., Proc. Am. Soc. Civ. Eng.,* **94**, 1055–123.

Sellwood, B. W. and **Jenkyns, H. C.** 1975. Basins and swells and the evolution of an epeiric sea (Pliensbachian-Bajocian of Great Britain). *Jl Geol. Soc. Lond.,* **131**, 373–88.

Shackleton, discussion after Pitcher and Bussell, 1977.

Shand, S. J. 1938. *Earth Lore,* Dutton, New York.

Shepard, F. P. 1973. *Submarine Geology,* Harper and Row, New York.

Sheppard, S. M. F. 1977. The Cornubian batholith, SW England: D/H and $^{18}O/^{16}O$ studies of kaolinite and other alteration minerals, *Jl geol. Soc. Lond.,* **133**, 573–91.

Shields, O. 1977. A Gondwanaland reconstruction for the Indian Ocean, *J. Geol.,* **85**, 236–42.

Shields, O. 1979. Evidence for initial opening of the Pacific Ocean in the Jurassic, *Palaeogeog. Palaeoclim. Palaeoecol.,* **26**, 181–220.

Shirinian, K. G. 1968. Endogenetic conditions of areal volcanism (on the example of Armenia), *Bull. Volcan.,* **32**, 283–95.

Sissons, J. B. 1954. The erosion surfaces and drainage system of South-West Yorkshire, *Proc. Yorks. Geol. Soc.,* **29**, 305–42.

Slaymaker, O. and **McPherson, H. J.** 1977. An overview of geomorphic processes in the Canadian Cordillera, *Z. Geomorph,* **21**, 169–86.

Sleep, N. and **Toksoz, M. N.** 1971. Evolution of marginal basins, *Nature,* **233**, 548–50.

Smith, A. G. 1976. Plate tectonics and orogeny – a review, *Tectonophysics,* **33**, 215–85.

Smith, P. J. 1978. Volcano spacing is related to crustal thickness, *Open Earth,* **1**, 36–7.

Smith, P. J. 1979. Mantle convection: shallow or deep? *Nature,* **278**, 305–6.

Snider, A. 1858. *La création et ses mystires dévoilées.* Frank & Dentu, Paris.

Somov, V. I. and **Kuznetsova, V. G.** 1975. Results of geodetic and geophysical investigations of recent crustal movements in the Soviet part of the eastern Carpathians, *Tectonophysics,* **29**, 377–82.

Sparks, B. W. 1972. *Geomorphology,* 2nd edn, Longman, London.

Spencer, E. W. 1965. *Geology: A Survey of Earth Science,* Crowell, New York.

Spjeldnaes, N. 1973. Comment in D. N. Tarling and S.K. Runcorn (eds), 1973 *op. cit.,* p. 864.

Steiner, J. 1977. An expanding Earth on the basis of sea-floor spreading and subduction rates, *Geology,* **5**, 313–8.

Stephansson, O. 1977. Granite diapirism in Archaean rocks, *Jl geol. Soc. Lond.,* **133**, 357–61.

Stevens, G. R. 1974. *Rugged Landscape: the Geology of Central New Zealand,* Reed, Wellington.

Stewart, I. C. F. 1976. Mantle plume separation and the expanding earth, *Geophys. J. R. Astron. Soc.,* **46**, 849–80.

Stille, H. 1936. The present tectonic state of the earth, *Bull. Am. Assoc. Petrol. Geol.,* **20**, 849–80.

Stille, H. 1955. Recent deformations of the earth's crust in the light of those of earlier epochs. In Crust of the earth – a symposium, A. Poldervaart (ed.), *Geol. Soc. Am. Spec. Pap,* **62**, 171–91.

Stoddart, D. R. 1969. World erosion and sedimentation, in *Water, Earth and Man,* R. J. Chorley (ed.), Methuen, London, pp. 43–64.

Stoneley, R. 1974. Evolution of the continental margins bounding a former southern Tethys, in C. A. Burk and C. L. Drake (eds). 1974 *op. cit.,* pp. 889–903.

Strakhov, N. M. 1948. *Outlines of Historical Geology,* Moscow, USSR Govt. Publ.

Sullivan, W. 1974. *Continents in Motion,* Macmillan, London.

Sutton, G. H., Manghnani, M. H. and **Moberly, R.** 1976. The geophysics of the Pacific Ocean basin and its margin, *American Geophysical Union Monograph,* 19, Washington.

Suzuki, T. 1968. Settlement of volcanic cones, *Bull. Volcan. Soc. Japan,* **13**, 95–108.

Suzuki, T. 1977. Volcano types and their global population percentages, *Bull. Volcan. Soc. Japan,* **22**, 27–40.

Swann, D. H. 1963. Classification of Genevievian and Chesterian (Late Mississippian) rocks of Illinois, *Ill, Geol. Surv. Rept. Inv.,* 216.

Sweeting, M. M. 1966. The weathering of limestones, in *Essays in Geomorphology,* G. H. Dury (ed), Heinemann, London.

Talwani, M. and **Pitman, W. C.** (eds). 1977. *Island Arcs, Deep Sea Trenches and Back-arc Basins,* American Geophysical Union, Washington.

Tarling, D. N. and **Runcorn, S. K.** (eds), 1973. *Implications of Continental Drift to the Earth Sciences,* Academic Press, London.

Tarney, J. and **Windley, B. F.** 1977. Chemistry, thermal gradients and evolution of the lower continental crust, *Jl geol. Soc. Lond.,* **134**, 153–72.

Taylor, F. B. 1910. Bearing of the Tertiary Mountain Belt on the Origin of the Earth's Plan, *Bull. geol. Soc. Am.,* **21**, 179–226.

Taylor, R. L. S. and **Smalley, I. J.** 1969. Why Britain tilts, *Science J.,* 55–9.

Taylor, T. G. 1911. Physiography of eastern Australia, *Bull. Bur. Met. Aust.,* 8.

Te Punga, M. T. 1957. Live anticlines in western Wellington, *N.Z.J. Sci. Tech* **B.38**, 433–46.

Termier, H. and **Termier, G.** 1952. *Histoire géologique de la biosphere,* Masson, Paris.

Termier, H. and **Termier, G.** 1969. Global palaeogeography and earth expansions, in *Application of Modern Physics to Earth and Planetary Interiors,* K. Runcorn (ed.), Wiley, New York, pp. 87–101.

Thomas, W. A. 1977. Appalachian-Ouachita salients and recesses from re-entrants and promontories in the continental margin, *Am. J. Sci.,* **277**, 1233–78.

Thompson, R. 1977. Stratigraphic consequences of palaeomagnetic studies of Pleistocene and Recent sediments, *Jl geol. Soc. Lond.*, **133**, 51–9.

Thornbury, W. D. 1969. *Principles of Geomorphology*, 2nd edn Wiley, New York.

Toksöz, M. N. and **Bird, P.** 1977. Formation and evolution of marginal basins and continental plateaus, M. Talwani and W. C. Pitman (eds), 1977 *op. cit.*, pp. 379–93.

Trümpy, R. 1960. Palaeotectonic evolution of the central and western Alps, *Bull.geol. Soc. A,.*, **71**, 843–908.

Trusheim, F. 1960. Mechanism of salt migration in northern Germany, *Bull. Am. Assoc. Petrol. Geol.*, **44**, 1519–40.

Uffen, R. J. and **Jessop, A. M.** 1963. The stress relief hypothesis of magma formation, *Bull. volcan.*, **26**, 57–66.

Uyeda, S. 1977. Some basic problems in the trench-arc-back arc system, in M. Talwani and W. C. Pitman (eds) 1977 *op. cit.*, pp. 1–14.

Vail, P. R., Mitchum, R. M. and **Thompson, S.** 1977. Seismic stratigraphy and global changes of sea level, in *Seismic Stratigraphy – Applications to Hydrocarbon Exploration*, C. E. Peyton (ed) Am. Assoc. Petrol. Geol., Mem. 26.

Vail, P. R. and **Todd, R. G.** 1977. Interpreting stratigraphy and eustatic cycles from seismic reflection patterns (abstract), in *Orthodoxy and Creativity at the Frontiers of Science*, University of Tasmania, Hobart.

Valentin, H. 1970. Principles and problems of a handbook on regional coastal geomorphology of the world. Paper read at the *Symposium of the IGU Commission on Coastal Geomorphology, Moscow*, 10.

Valentine, J. W. and **Moores, E. M.** 1970. Plate tectonic regulation of faunal diversity and sea level: a model, *Nature*, **228**, 657–9.

Van Bemmelen, R. W. 1930. The volcano-tectonic origin of Lake Toba (North Sumatra), *Proc. Pacif. Sci. Congr.*, **2**, 115–24.

Van Bemmelen, R. W. 1954. *Mountain Building*, Martinus Jijhoff, The Hague.

Van Bemmelen, R. W. 1972. *Geodynamic Models*, Elsevier, Amsterdam.

Van Bemmelen, R. W. 1976. Plate tectonics and the undation model: a comparison, *Tectonophysics*, **32**, 145–82.

Van de Graafe, W. J. E., Crowe, R. W. A., Bunting, J. A. and **Jackson, M. J.** 1977. Relict Early Cainozoic drainages in arid Western Australia, *Z. Geomorph*, **21**, 379–400.

Van Houten, F. B. 1976. Late Variscan nonmarine deposits, northwestern Africa: implications for pre-drift north Atlantic reconstructions, *Am. J. Sci.*, **276**, 671–93.

Van Houten, F. B. and **Brown, R. H.** 1977. Latest Paleozoic–Early Mesozoic Paleography, Northwestern Africa, *J. Geol.*, **85**, 143–56.

Veevers, J. J. 1974. Western continental margin of Australia, in C. L. Burk and C. A. Drake (eds), 1974 *op. cit.*, pp. 606–16.

Veevers, J. J. and **Cotterill, D.** 1978. Western margin of Australia: Evolution of a rifted arch system, *Bull. geol. Soc. Am.*, **89**, 337–55.

Vening Meinesz, F. A. 1947. Shear patterns in the earth's crust, *Trans. Am. geophys. Union*, **28**, 1–61.

Vine, F. J. and **Matthews, D. H.** 1963. Magnetic anomalies over oceanic ridges, *Nature*, **199**, 947–9.

Vita-Finzi, C. 1979. Rates of Holocene folding in the coastal Sagros near Bandar Abbas, Iran, *Nature*, **278**, 632–4.

Voight, B. (ed.) 1976. *Mechanics of Thrust Faults and Décollement*, Benchmark Papers in Geology, 32, Dowden, Hutchinson and Ross, Stroudsburg.

Voight, B. 1978. *Rockslides and Avalanches, 1 Natural Phenomena*, Elsevier, Amsterdam.

Voight, B. and **Cady, W. M.** 1978. Transported rocks of the Taconide Zone, eastern North America, in B. Voight (ed.) 1976 *op. cit.*, pp. 505–61.

Von Engeln, O. D. 1942. *Geomorphology*, Macmillan, New York.

Wager, L. R. 1937. The Arun River drainage pattern and the rise of the Himalaya, *Geogr. J.*, **89**, 239–50.

Walcott, R. I. 1970a. An isostatic model for basement uplift, *Can. J. Earth. Sci.*, **7**, 931–7.

Walcott, R. I. 1970b. Flexural rigidity, thickness and viscocity of the lithosphere, *J. geophys. Res.*, **75**, 3941–54.

Walcott, R. I. 1972. Gravity, flexure and the growth of sedimentary basins at a continental edge, *Bull. geol. Soc. Am.*, **83**, 1845–8.

Walsh, P. T., Boulter, M. C., Ijtaba, M. and **Urbani, D. M.** 1972. The preservation of the Neogene Brassington Formation of the southern Pennines and its bearing on the evolution of upland Britain, *Jl geol. Soc. Lond.*, **128**, 519–59.

Wanless, H. J., Tubb, D. G. and **Winer, J.** 1963. Mapping sedimentary environments of Pennsylvanian cycles, *Bull. geol. Soc. Am.*, **74**, 437–86.

Watanabe, T. 1977. Heat flow in back-arc basins of the Western Pacific, in M. Talwani and W. C. Pitman (eds) 1977 *op. cit.* pp. 137–61.

Waterschoot van der Gracht, W. A. J. 1931. Permo-Carboniferous orogeny in the south-central United States, *Bull. Am. Assoc. Petrol. Geol.*, **15**, 991–1057.

Watts, A. B. and **Talwani, M.** 1975. Gravity effects of downgoing lithospheric slabs beneath island arcs, *Bull. geol. Soc. Am.*, **86**, 1–4.

Weeks, L. G. 1948. Palaeogeography of South America, *Bull. Geol. Soc. Am.*, **69**, 249–82.

Wegener, A. 1929. *The Origin of Continents and Oceans*, trans. from 4th revd German edn, by J. G. Skerl, Methuen, London.

Wellman, H. W. 1955. New Zealand Quaternary tectonics, *Geol. Rdsch.*, **43**, 248–57.

Wellman, P. and **McDougall, I.** 1974. Cainozoic igneous activity in eastern Australia, *Tectonophysics*, **23**, 49–65.

Wells, J. W. 1963. Coral growth and geochronometry, *Nature*, **197**, 948–50.

White, S. H. and **Knipe, R. J.** 1978. Transformation –

and reaction-enhanced ductility in rocks, *Jl geol. Soc. Lond.*, **135**, 513–6.

Whitford, D. J., Compston, W. and Nicholls, I. A. 1977. Geochemistry of late Cenozoic lavas from eastern Indonesia: Role of subducted sediments in petrogenesis, *Geology*, **5**, 571–5.

Williams, G. E. and Goode, A. D. T. 1978. Possible western outlet for an ancient Murray River in South Australia, *Search*, **9**, 443–7.

Williams, H. and McBirney, A. R. 1964. Petrological and structural contrast of the Quaternary volcanoes of Guatemala, *Bull. volcan.*, **27**, 61.

Williams, H. R. and Williams, R. A. 1977. Kimberlites and plate-tectonics in West Africa, *Nature*, **270**, 507-8.

Williams, M. A. J. 1973. The efficacy of creep and slopewash in tropical and temperate Australia, *Aust. geogr. Stud.*, **11**, 62–78.

Wilson, L. 1973. Variations in mean annual sediment yield as a function of mean annual precipitation, *Am. J. Sci.*, **273**, 335–49.

Windley, B. F. 1977. *The Evolving Continents*, Wiley, New York.

Winkler, C. D. and Howard, J. D. 1977. Correlation of tectonically deformed shorelines on the southern Atlantic coastal plain, *Geology*, **5**, 123–7.

Wise, D. U. 1963. An outrageous hypothesis for the tectonic pattern of the North American cordillera, *Bull. geol. Soc. Am.*, **74**, 357–62.

Wise, D. U. 1974. Continental margins, freeboard and volumes of continents and oceans through time, in C. A. Burk and C. L. Drake (eds) 1974 *op. cit.*, pp. 45–58.

Woodward, L. A. 1977. Rate of crustal extension across the Rio Grande Rift, *Geology*, **15**, 269-72.

Wooldridge, S. W. and Linton, D. L. 1955. *Structure, Surface and Drainage in South-east England*, 2nd edn. Philip, London.

Worzel, J. L. 1976. Gravity investigations of the subduction zone, in G. H. Sutton, M. H. Manghnani and R. Moberley (eds) 1976 *op. cit.*, pp. 1–16.

Wright, C. A. 1977. Distribution of Cainozoic Foraminiferida in the Scott Reef no. 1 Well, Western Australia, *J. geol. Soc. Aust.*, **24**, 269–77.

Wright, J. B. 1971. Volcanism and the earth's crust, in *Understanding the Earth*, I. G. Gass, P. J. Smith and J. B. Wilson (eds). Artemis Press, Sussex, pp. 301–13.

Wyllie, P. J. 1974. Plate tectonics, sea floor spreading and continental drift: an introduction, in C. F. Kahle (ed.) 1974 *op. cit.*, pp. 5–15.

Wyllie, P. J., Huang, W. L., Stern, C. R. and Maaloe, S. 1976. Granitic magmas: possible and impossible sources, water contents and crystallization sequences, *Canadian J. Earth Sci.*, **13**, 1007–19.

Wynne-Edwards, H. R. 1976. Proterozoic ensialic orogenesis: the millipede model of ductile plate tectonics, *Am. J. Sci.*, **276**, 927–53.

Wynne-Edwards, H. R. 1977. Metallogenic implications of the millipede model – ductile ensialic orogenesis in the Proterozoic (abstract), *Orthodoxy and Creativity at the Frontiers of Earth Science*, University of Tasmania, Hobart.

Yoder, H. S. 1952. Change of melting point of diopside with pressure, *J. Geol.*, **60**, 364–74.

Young, A. 1960. Soil movement by denudational processes on slopes, *Nature*, **188**, 120–2.

Young, A. 1969. Present rate of land erosion, *Nature*, **224**, 851–2.

Young, A. 1974. The rate of slope retreat, in *Progress in Geomorphology*, Inst. Brit. Geogr. Spec. Publ., E. H. Brown and R. S. Waters (eds) pp. 65–78.

Young, R. A. and McKee, E. H. 1978. Early and Middle Cenozoic drainage and erosion in west-central Arizona, *Bull. geol. Soc. Am.*, **89**, 1745–50.

Zaruba, Q. and Mencl, V. 1969. *Landslides and their Control*, Elsevier, New York.

Ziegler, W. L. 1977. Summing up on seismically active margins, *Jl geol. Soc. Lond.*, **134**, 84–6.

Index